AF598614

Host-Parasite Interactions

EXPERIMENTAL BIOLOGY REVIEWS

Series Advisors:

D.W. Lawlor
AFRC Institute of Arable Crops Research, Rothamsted Experimental Station, Harpenden, Hertfordshire AL5 2JQ, UK

M. Thorndyke
School of Biological Sciences, Royal Holloway, University of London, Egham, Surrey TW20 0EX, UK

Environmental Stress and Gene Regulation
Sex Determination in Plants
Plant Carbohydrate Biochemistry
Programmed Cell Death in Animals and Plants
Biomechanics in Animal Behaviour
Cell and Molecular Biology of Wood Formation
Molecular Mechanisms of Metabolic Arrest
Environment and Animal Development: genes, life histories and plasticity
Brain Stem Cells – SEB Symposium Series Vol. 53
Endocrine Interactions of Insect Parasites and Pathogens
Vertebrate Biomechanics and Evolution
Osmoregulation and Drinking in Vertebrates
Host-Parasite Interactions

Forthcoming titles include
The Nuclear Envelope

Host-Parasite Interactions

Edited by

G. F. WIEGERTJES
Cell Biology and Immunology Group, Wageningen Institute of Animal Sciences, Wageningen University, PO Box 338, 6700 AH, Wageningen, The Netherlands

and

G. FLIK
Department of Animal Physiology, Faculty of Science, University of Nijmegen, Toernooiveld 1,6525 ED Nijmegen, The Netherlands

First published 2004

A CIP catalogue record for this book is available from the British Library.

ISBN 1 85996 298 X

Garland Science/BIOS Scientific Publishers
4 Park Square, Milton Park, Abingdon, Oxon, OX14 4RN, UK and
29 West 35th Street, New York, NY 10001-2299, USA
World Wide Web home page: www.bios.co.uk

Garland Science/BIOS Scientific Publishers is a member of the Taylor & Francis Group.

Distributed in the USA by
Fulfilment Center
Taylor & Francis
10650 Toebben Drive
Independence, KY 41051, USA
Toll Free Tel.: +1 800 634 7064; E-mail: taylorandfrancis@thomsonlearning.com

Distributed in Canada by
Taylor & Francis
74 Rolark Drive
Scarborough, Ontario M1R 4G2, Canada
Toll Free Tel: +1877 226 2237; E-mail: tal_fran@istar.ca

Distributed in the rest of the world by
Thomson Publishing Services
Cheriton House
North Way
Andover, Hampshire SP10 5BE, UK
Tel: +44(0)1264 332424; E-mail: salesorder.tandf@thomsonpublishingservices.co.uk

Library of Congress Cataloging-in-Publication Data

Host-parasite interactions / edited by G. Wiegertjes and G. Flik.
p. cm.
Includes bibliographical references.
Includes four chapters presented at an intersociety meeting of the American Physiological Society held in San Diego, Calif., Aug. 24–28, 2002. With invited papers.
ISBN 1-85996-298-X
1. Host-parasite relationships. I. Wiegertjes, G. (Geert) II. Flik, G. (Gert) III. American Physiological Society (1887–)

QL757.H668 2004
577.8'57--dc22

2004005857

Production Editor: Catherine Jones
Typeset by Saxon Graphics Ltd, Derby, UK
Printed by Cromwell Press, Trowbridge

Contents

Contributors

Alarcon-Chaidez, F.J., Center for Microbial Pathogenesis, School of Medicine, University of Connecticut Health Center, 263 Farmington Avenue, Farmington, Connecticut 06030, USA

Belosevic, M., CW-405 Biological Sciences, University of Alberta, Edmonton, AB T6G 2E9, Canada

Bresciani, J., The Royal Veterinary and Agricultural University, Department of Ecology, Section of Zoology, Thorvaldsensvej 40, DK-1871 Frederiksberg, Denmark

Buchmann, K., The Royal Veterinary and Agricultural University, Department of Veterinary Pathobiology, Laboratory of Fish Diseases, Grønnegårdsvej 15, DK-1870 Frederiksberg, Denmark

Fast, M.D., Institute for Marine Biosciences, National Research Council, 1411 Oxford Street, Halifax, Nova Scotia B3H 2Z1 & Department of Biology, Dalhousie University, 1355 Oxford Street, Halifax, Nova Scotia B3H 4J1, Canada

Flik, G., University of Nijmegen, Department Animal Physiology, Toernooiveld 1, 6525 ED Nijmegen, The Netherlands

Hoole, D., School of Life Sciences, Huxley Building, Keele University, Keele, Staffordshire, ST5 5BG, United Kingdom

Joerink, M., Wageningen University, Cell Biology & Immunology group, Marijkeweg 40, 6709 PG, P.O. Box 338, 6700 AH Wageningen, The Netherlands

Johnson, S.C., Institute for Marine Biosciences, National Research Council, 1411 Oxford Street, Halifax, Nova Scotia B3H 2Z1, Canada

Lindenstrøm, T., The Royal Veterinary and Agricultural University, Department of Veterinary Pathobiology, Laboratory of Fish Diseases, Grønnegårdsvej 15, DK-1870 Frederiksberg, Denmark

Nielsen, M.E., The Royal Veterinary and Agricultural University, Department of Veterinary Pathobiology, Laboratory of Fish Diseases, Grønnegårdsvej 15, DK-1870 Frederiksberg, Denmark

Saeij, J.P.J., Department of Microbiology and Immunology, Fairchild Science Building, Room D305, Stanford University School of Medicine, Stanford CA 94305-5124, USA

Sigh, J., The Royal Veterinary and Agricultural University, Department of Veterinary Pathobiology, Laboratory of Fish Diseases, Grønnegårdsvej 15, DK-1870 Frederiksberg, Denmark

Stafford, J.L., CW-405 Biological Sciences, University of Alberta, Edmonton, AB T6G 2E9, Canada

Steinhagen, D., School of Veterinary Medicine, Fish Disease Research Unit, Bünteweg 17, P.O. Box 711180 Hannover 30545, Germany

Vermeulen, A.N., Head Parasitology R&D, Intervet International BV, PO Box 31, 5830 AA Boxmeer, The Netherlands

Walker, P.D., University of Nijmegen, Department of Animal Ecophysiology, Toernooiveld 1, 6525 ED Nijmegen, The Netherlands

Wendelaar Bonga, S.E., University of Nijmegen, Department of Animal Ecophysiology, Toernooiveld 1, 6525 ED Nijmegen, The Netherlands

Wiegertjes, G.F., Wageningen University, Cell Biology & Immunology group, Marijkeweg 40, 6709 PG, P.O. Box 338, 6700 AH Wageningen, The Netherlands

Wikel, S.K., Center for Microbial Pathogenesis, School of Medicine, University of Connecticut Health Center, 263 Farmington Avenue, Farmington, Connecticut 06030, USA

Williams, G.T., School of Life Sciences, Keele University, Keele, Staffordshire, ST5 5BG, United Kingdom

Woo, P.T.K., Axelrod Institute of Ichthyology and Department of Zoology, College of Biological Science, University of Guelph, Guelph Ontario N1G 2W1, Canada

Abbreviations

ADP	adenosine diphosphate
AMP	adenosine monophosphate
APAF1	apoptotic protease activating factor 1
APC	antigen-presenting cell
ATP	adenosine triphosphate
CBH	cutaneous basophil hypersensitivity
CCM	cell-conditioned medium
CD	cell differentiation
CL	chemoluminescence assay
CRMA	cytokine response modifier A
dd	degree days
DIC	disseminated intravascular coagulation
dpi	days post infection
DTH	delayed-type hypersensitivity
EGC	eosinophilic granular cell
EHNV	epizootic haematopoietic necrosis virus
EST	expressed sequence tag
GPI	glycosylphosphatidylinositol
HBP	histamine-binding protein
HPLC	high-pressure liquid chromatography
HSP	heat shock protein
IAP	inhibitor of apoptosis proteins
IEL	intra-epithelial leukocyte
ICE	interleukin-converting enzyme
IDI	invertebrate developmental inhibitor
IL	interleukin
iNOS	inducible nitric oxide synthase
IPNV	infectious pancreatic necrosis virus
IRE	iron-responsive element
IRP	iron-regulatory protein
IVDKM	*in vitro*-derived kidney macrophage
LMP	latent membrane protein
LP	lamina propria
LPS	lipopolysaccharide
lsr	large subunit of ribosome
MAC	membrane attack complex
MAF	macrophage-activating factor
MHC	major histocompatibility complex
mMCP	murine mast cell protease
NFκB	nuclear factor kappa B
NK	natural killer
NO	nitric oxide
NRAMP	natural resistance-associated macrophage protein

NVI	necrotic volume increase
OUC	ornithine–urea cycle
PAF	platelet-activating factor
PAMP	pathogen-associated molecular pattern
PCD	programmed cell death
PG	prostaglandin
PKC	protein kinase C
PMA	phorbol 12-myristate 13-acetate
PS	phosphatidylserine
PTK	protein tyrosine kinase
RNI	reactive nitrogen intermediate
ROI	reactive oxygen intermediate
ROS	reactive oxygen species
SOD	superoxide dismutase
SRBC	sheep red blood cells
ssr	small subunit of ribosome
SVCV	spring viraemia in carp virus
TGF	transforming growth factor
TLR	Toll-like receptor
TNF-α	tumour necrosis factor alpha
VEGF	vascular endothelial cell growth factor
VSOR	volume-sensitive outwardly rectifying

Preface

This book gives an up-to-date overview of host–parasite interactions, using a comparative approach and covering vertebrates from fish to mammals. Four of the chapters in this book, or at least parts of them (Chapters 6, 7, 9 and 10) were originally presented at an intersociety meeting of the American Physiological Society, '*The Power of Comparative Physiology: Evolution, Integration and Application*', held in San Diego, California, August 24–28, 2002. It was decided then to invite partners of a European Union Research Training Network called 'Parity', working on host–parasite interactions in fish, to contribute with chapters and publish the joined papers in this book format. The first few chapters deal with more fundamental questions regarding host–parasite interactions and, as such, in Chapter 1, Steinhagen, and in Chapter 2 Hoole and Williams address fundamental aspects of apoptosis and necrosis. Although, for fish structural changes following cell death are well documented, studies on the physiological responses preceding cell death remain scarce. However, even if these changes would be better documented, the roles of apoptosis in the interactions between parasites and hosts are multifaceted and highly dependent on individual associations between the two organisms involved. Still, with the information obtained from studies in mammals and invertebrates, the next decade promises to be both exciting and productive with respect to our knowledge of the relationship between apoptosis in non-mammalian animals and infection.

Chapter 3 by Nielsen, Lindenstrøm, Sigh and Buchmann considers the role of thionine-positive cells in relation to parasites. Thionine-positive cells comprise such apparently diverse types as mast cells, goblet cells, basophils, eosinophils and cells from amyloid cartilage and mucous tissues. Thionin-positive cells are unique in the sense that they contain granules in their cytosol that are metachromatically stained with the cationic dye thionine (cf. toluidine blue, methylene blue, gallocyanin and pinacyanole). These cells derive from $CD34^+$ haematopoietic progenitors and play an important role in the innate response to parasites. Anti-parasitic activity of the cells is discussed in the context of cellular pathways including the NF-κβ activation via Toll-like receptors, phagocytosis and reciprocal signals between hosts and parasites. A review is presented on the vertebrate evolution of thionine-positive cells, with an emphasis on anti-parasitic mechanisms in fish.

In their chapter (Chapter 4), Joerink, Saeij, Stafford, Belosevic and Wiegertjes look at protozoan infections in fish. Kinetoplastid parasites such as *Trypanosoma* and *Leishmania* species are well studied in mammalian models but little is as yet known about the immune response against comparable parasite species in fish. Carp are host to at least two kinetoplastid species (*T. borelli* and *T. danilewskyi*) and leeches act as vectors. The roles of NRAMP (natural resistance-associated macrophage protein), transferrin and the omnipresent, but still only partly understood, messenger NO is discussed as well as the role of cytokine profiles in macrophage polarization and function. In-depth analysis of the immune response to infection(s) by protozoan kinetoplastid parasites in fish may shed light on the evolution of host–parasite relationships.

In Chapter 5, Woo also addresses kinetoplastid haemoflagellates, i.e. two groups that cause disease in fish or in cattle, with an emphasis on the former. *Cryptobia* species affect freshwater and seawater fish species, transmission occurs by leeches. *Trypanosoma* species are a nuisance for cattle, with blood-feeding flies their best-known vector. Susceptibility, tolerance and resistance to *Cryptobia* are addressed, as well as trypanosomiasis-related anaemia. Several aspects of immunosuppression are dealt with as well as the interaction of the parasites with the endocrine system of the host.

In Chapter 6, Walker, Wendelaar Bonga and Flik review the biology of the freshwater louse genus *Argulus* and its interactions with its hosts. They describe this obligate ectoparasite from an ecological point of view with a focus on the host–parasite interaction. The life cycle, morphology of developmental stages, locomotion, attachment strategies, feeding and sensory systems are addressed as well as geographic distribution and seasonality. The host–parasite interaction section focuses on host choice and specificity, with an emphasis on fish as host and their stress and immune responses to this parasite. Suggestions for parasite control and prevention are given.

In Chapter 7, Johnson and Fast describe the new developments in our understanding of host–parasite interactions between a seawater louse: the salmon louse *Lepeophtheirus salmonis* and its hosts. The salmon louse is a major problem in intensive marine salmon aquaculture, the costs of losses due to this parasite exceed over $40 million annually. The ultimate goal is development of a vaccine to this parasite, but classical (identification of protective antigens) approaches are unsuccessful so far. Biochemical, genomic, proteomic and immunological approaches are now used to unravel the interaction between this parasite and its host. Salivary components of the parasite are tested for their effects on the immune response of the host. New techniques such as the screening of subtractive cDNA libraries are predicted to provide the required resolution.

In Chapter 8, Buchmann, Lindenstrøm and Bresciani focus on fish and monogeneans. Helminths are parasites with a particular appetite for fish. There are an estimated 30 000 species of these parasites, including nematodes, acanthocephalans, trematodes, cestodes and monogeneans. They are almost exclusively ectoparasites, although some are found in cloaca, body cavity and heart. The chapter gives an overview of diversity, anatomy, gut and feeding, nervous system, osmoregulation, life cycle and reproduction of these parasites. Both the parasite and the host produce molecules for mutual communication, albeit that this communication is only poorly understood. However, the immune system of the host seems fully activated when infected by these parasites: both cellular (innate) and humoral (adaptive) responses with their particular machinery and signalling pathways are seen. At the end of the chapter measures to control these parasites are presented.

Alarcon-Chaidez and Wikel address, in Chapter 9, the topic of the tick–host relationship. Immunobiology, genomics and proteomics are the keywords for this chapter. They discuss life cycles and ecology of ticks, present the latest insights on tick-borne pathogens and host-acquired immunity to tick infestation. What is it in tick saliva that makes these parasites so nasty, how do they modulate host immunity? Ticks are vectors of diseases (Lyme), and a further in-depth analysis of host cytokine response to infection by this parasite is indicated to facilitate vaccination against this nuisance. Proteomic and genomic analysis will certainly bring more details on the parasite, possibly allowing control in the not too distant future.

In Chapter 10, Vermeulen describes avian coccidiosis. Intensive chicken rearing is a major meat industry yielding 35 billion broilers per year worldwide. *Eimeria* species inhabit the intestinal lining of the chick, enteritis, disruption of intestinal villi and death being the classical symptoms of this parasite infection. Even when the parasite is not life-threatening, it causes poor food absorption and water resorption resulting in impeded growth. Chickens become readily immune to the parasite but the mechanism is still poorly understood. Newly discovered antigens that stimulated T-cell subpopulations and have protective capacity are targets for the development of new vaccines to improve the relationship between host and parasite.

Special thanks go to all contributors to this volume for their constructive interaction with the editors. Significant financial support was received from the SEB, London, and the Osmoregulation group of the SEB, which we gratefully acknowledge. Daisy Maurits, Nijmegen, The Netherlands, is acknowledged for excellent secretarial support. Kay McNamee, Wageningen, The Netherlands is acknowledged for excellent editing support. Special thanks go to my co-editor and co-ordinator of the EU-RTN Parity programme, Geert Wiegertjes. Without his invaluable help and persistent drive the realization of this book would have been at serious risk.

Gert Flik
Geert F. Wiegertjes

1

Structural and physiological aspects of cell death

Dieter Steinhagen

1 Introduction

Cells are active participants in their environment and thus constantly adjusting structure and function to extracellular demands or stressors resulting from organismal needs or from outside. Beside structural or functional adjustments, developmental processes or physiological adaptations can cause death of cells. It is of significant importance during development of organisms, and it occurs in the context of stressors such as toxicology, oxygen deprivation or infection, as a result of irreversible injury. A concept of cell death was developed in the 19th century together with cell theory; and during the 20th century, cell death was characterized morphologically (reviewed by Majno and Joris, 1995).

Associated with cell death, alterations of subcellular organelles, as well as morphological changes appear in cells or tissues subjected to injury. Subcellular alterations include loss of cellular differentiation (cell junctions, pseudopodia), swelling and disruption of mitochondria, formation of membrane blebs, and deviation in cytoplasm density.

In histology, dead cells show eosinophilia and have a more glassy appearance, the basophilia of nuclear chromatin may fade (karyolysis), or nuclear shrinkage and increased basophilia occurs (pyknosis). Pyknotic nuclei may fragment, a pattern, which is called karyorrhexis. In fishes, these post-mortem changes were noted, for instances in hepatocytes (Braunbeck *et al.*, 1992), splenocytes (Spazier *et al.*, 1992), or branchial pillar cells (Speare *et al.*, 1999) in response to infection or exposure to various chemical compounds.

Once masses of necrotic cells occur, further distinctive morphological patterns are recognized, depending on whether enzymatic digestion of the cell or protein denaturation predominates. In processes that are dominated by protein denaturation, the basic outline of the dead cell is preserved, the cells appear acidophilic, coagulated. This 'coagulative necrosis' is characteristic of hypoxic cell death in various tissues. This occurs, when blood supply to a tissue area is lost (ischaemia), for instance in the context of fish tubercles or renal infection with amoebae (Lom and Dykova, 1992, Timur *et al.*, 1977).

Host–Parasite Interactions, edited by G. F. Wiegertjes & G. Flik.

When enzymatic digestion of the dead cells predominates, colliquative or liquefactive necrosis is the result. This often is seen in context of bacterial or fungal infections in fish, for instance in infections with *Vibrio anguillarum* (Ransom *et al*., 1984). These morphological changes are results of processes, which follow cell death and have to be regarded as post-mortem changes, but do not refer to physiological processes which resulted in cell death (for review see Majno and Joris, 1995; Sastry and Rao, 2000). A synthesis about cell physiological and biochemical processes resulting in cell death, however, was achieved only very recently. In this chapter major physiological events that determine the route to and how they lead to cell death are summarized. In addition, implications of cell death for local immune responses are discussed. Structural changes, which follow cell death are well documented in fish under pathogen infection. Biochemical or physiological processes resulting in cell death and its significance for pathological conditions in the context of fish diseases still needs considerable attention.

2 Cell death: necrosis or apoptosis?

In many occasions, cell death is associated with an orderly disassembly of cell structures, orchestrated by the action of biochemical pathways and associated with several distinct biochemical, molecular and structural characteristics. This process is called 'apoptosis' (Kerr *et al*., 1972). Morphologically apoptosis is associated with shrinkage of cell volume, chromatin condensation, DNA fragmentation, and membrane blebbing. Cells undergoing 'shrinkage necrosis', as this process was labelled by Kerr, kept a physiologically intact cell membrane. As this process is mediated by proteins with no other known functions, apoptosis was regarded as active. It was postulated, that cells have the ability to self-destruct by the activation of an intrinsic, genetically encoded cellular suicide program, which becomes active when the cell is no longer needed or seriously damaged. This idea is now accepted and the term programmed cell death (PCD) has been coined (McConkey and Orrenius, 1994) for a cell death, that is mediated by a built-in suicide program. It is activated by a wide variety of stimuli (reviewed by Sastry and Rao, 2000) and involves a cascade of cysteine proteases (cleaving after particular aspartate residues) and proteins of the Bcl family (reviewed by Fiers *et al*., 1999; McConkey, 1998). Although observed under some pathological conditions, apoptosis is now recognized as a normal feature in development and ageing and thus is regarded as the physiological form of cell death (Nicotera, 2002; Sastry and Rao, 2000).

In contrast to this, accidental cell injury is considered to induce cell death by a passive, loss-of-function phenomenon, for which the term necrosis is retained (Leist and Nicotera, 1997). It is considered to result from primary energy depletion (Leist and Nicotera, 1997), which leads to failure of ion pumps and other ATP-driven processes, loss of ion homeostasis, membrane blebbing, and finally cell lysis occurs (Okada *et al*., 2001). Proteins known to be involved in necrosis also participate in other physiological functions such as volume regulation or energy metabolism. Therefore, necrosis is regarded as accidental and non-specific (Leist and Nicotera, 1997).

There is increasing evidence that it is not always possible to distinguish between apoptosis and necrosis. In neuronal cells for instance, oxidative stress induced by a treatment with glutamate initiates events which lead to a form of cell death distinct from either necrosis or apoptosis (Tan *et al.*, 1998). The inhibition of oxidative phosphorylation induces processes, which abruptly increase the permeability of mitochondrial inner membrane to solutes of high molecular mass, induced necrotic killing of cells, when intracellular ATP levels fell profoundly. When ATP levels were partially maintained, apoptosis followed mitochondrial permeability transition, and in many cells, features of both apoptosis and necrosis occurred together after toxic stresses or death signals (Lemasters *et al.*, 1999). This supports the assumption that classical apoptosis and necrosis represent the extreme ends of a wide range of possible forms of death with morphological and biochemical different parameters (Nicotera *et al.*, 1998). In addition, the primary cause of cell death, such as exposure to reactive oxygen species, nitric oxide, or binding of tumour necrosis factor alpha, not always determines whether apoptosis or necrosis results from the insult. This has been shown in cell death induced, for instance, by oxidative stress (Bonfoco *et al.*, 1995; Dypbukt *et al.*, 1994; Kim *et al.*, 2000), or by activation of specific death receptors (Denecker *et al.*, 2001; Vercammen *et al.*, 1998). Important for the final outcome was the severity of the insult, and among others, the type of the cell and its metabolic status rather than the nature of the insult. Thus apoptosis or necrosis are considered as 'different execution of the same death' (Nicotera *et al.*, 1999). Can some events be identified, which determine that cell death is executed as necrosis?

3 Intracellular ion and energy levels

Necrotic cells death is paralleled by cell swelling (Kroemer *et al.*, 1998), termed necrotic volume increase (NVI). Living cells need to regulate their volume, because the presence of membrane-impermeable polyvalent anion macromolecules within the cytosol causes colloid-osmotic pressure which leads to a cationic leak influx. The volume regulation is coupled to Na^+ pump-mediated mechanisms (Okada *et al.*, 2001). Accidental cell damage or injury induces Na^+ uptake and ATP release due to membrane leakage as well as dissipation of ATP by constrained overworking of the Na^+ pump. The resulting ATP depletion then leads to a reduction of Na^+ pumping and cell swelling. Cell swelling by this mechanism can be induced, for instance by hypoxia or ischaemia (Hoffmann and Simonsen, 1989; Lipton, 1999). In addition, reactive oxygen species (ROS) including superoxide anion, hydroxyl radical or hydrogen peroxide may activate non-selective cation channels (Barros *et al.*, 2001b; Herson and Ashford, 1997) and by this induce Na^+ overload. The significance of Na^+ overload for the induction of necrosis was shown in experiments on hepatocytes exposed to menadione or KCN (Carini *et al.*, 1999). Both substances induced a drop of intracellular ATP concentration. In the presence of sodium in the bathing medium, the cell volume increased and the cells suffered necrosis, while cells in sodium-depleted medium did not swell or lose viability. Sodium was not toxic itself, as cells exposed to ouabian wherein intracellular Na^+ can reach extracellular levels did not lose viability (Orlov *et al.*, 1999). Thus, a combination of Na^+ overload and ATP depletion might be required for the induction of cell necrosis.

Intracellular ATP-deficient conditions, induced by the disruption of mitochondrial respiration, provoked neuronal cell swelling (Patel *et al.*, 1998). Under low intracellular ATP conditions, this cell swelling must persist, because volume-regulating, volume-sensitive outwardly rectifying (VSOR) anion channels depend on the binding of non-hydrolytic intracellular ATP (Okada, 1997; Okada *et al.*, 2001). In neuronal cells under hypoxia or mitochondrial inhibition, a marked inhibition of Cl^- handling VSOR channels was observed (Patel *et al.*, 1998). It is assumed that ATP depletion or metabolic inhibition induces a persistent cell swelling and finally cell rupture (Barros *et al.*, 2001b; Okada *et al.*, 2001).

In principle, any changes which lead to ATP depletion may be detrimental for the cell and may direct the execution of cell death towards necrosis. Pre-emptying human T-cells of ATP switches demise by apoptotic triggers such as staurosporine or CD95 stimulation from apoptosis to necrosis (Nicotera *et al.*, 1998). In these experiments, the death program was driven towards necrosis when mitochondrial or glycolytic generation of ATP was blocked and re-directed towards apoptosis with repletion of the cytosolic ATP pool with glucose supplementation (Nicotera *et al.*, 1998).

In free-radical-induced necrosis, the mechanism of intracellular energy depletion is not always obvious. It can be explained by metabolic exhaustion by the activity of ion pumps in addition to an inhibition of gycolysis or mitochondrial functions (Leist *et al.*, 1999; Redegeld *et al.*, 1990). Furthermore, excessive DNA damage causes poly (ADP-ribose) polymerase 1 (PARP-1) hyperactivation (reviewed by Chiarugi, 2002), an abundant nuclear enzyme involved in DNA-repair, DNA stability and transcriptional regulation. PARP-1 is activated by DNA strand breaks, and overactivation of PARP after cellular insults can lead to cell death caused by depletion of the enzyme's substrate β-nicotinamide adenine dinucleotide (NAD^+) and ATP. Caspases, in particular caspase-3 and -7, cleave PARP-1 and thus prevent the recruitment of the enzyme to sites of DNA damage (Cohen, 1997), a hallmark of apoptosis. Fibroblasts from PARP-deficient ($PARP^{-/-}$) mice were protected from necrotic cell death and ATP depletion, which was induced in fibroblasts from PARP ($^{+/+}$) mice upon hydrogen peroxide exposition (Ha and Snyder, 1999). In fibroblasts expressing caspase-resistant PARP, enhanced necrosis coupled with depletion of NAD^+ and ATP was induced by treatment with tumour necrosis factor alpha. The PARP inhibitor 3-aminobenzamide prevented the NAD^+ drop and concomitantly inhibited necrosis and elevated apoptosis (Herceg and Wang, 1999). Activation or cleavage of PARP is suggested to function as a molecular switch between apoptotic and necrotic modes of death receptor-induced cell death (Los *et al.*, 2002).

A current concept of cell death induced by free radicals or activation of death receptors assumes that ion homeostasis of cells is challenged by an activation of non-selective cation H channels, and cells experience a sodium influx (see *Figure 1*). In addition, the cell experiences a K^+ efflux. Rising cytosolic concentrations of sodium are met by, increased Na^+/K^+ pumping consuming extra energy, and ATP depletion together with other events such as calpain activation (Wang, 2000) can induce a swelling of the cell until it bursts (Barros *et al.*, 2001a). If the ATP depletion is less severe, Na^+/K^+ pumping maintains the cytosolic level of sodium low, the efflux of potassium would prevail and together with mitochondrial changes such as cytochrome c release (Büki *et al.*, 2000) the cell will be directed towards apoptosis. This concept of physiological events leading to cell death explains why frequently a combination of apoptosis and necrosis is found in the same tissue. Crucial for the execution of death appear to be the severity of the insult and the degree of ATP depletion associated with the insult (Leist *et al.*, 1997).

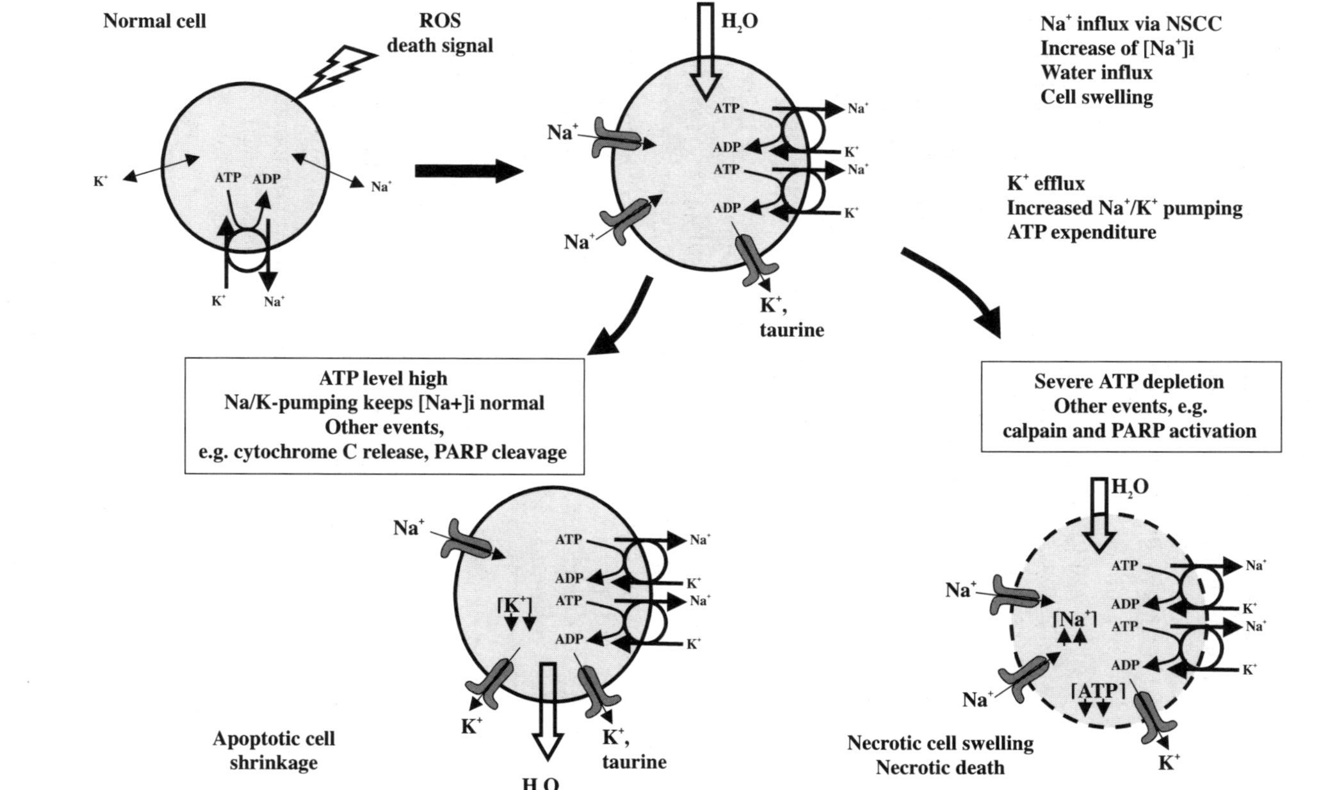

Figure 1 *Schematic summary of physiological mechanisms associated with necrotic versus apoptotic cell death. The physiological status of cells with a pump-leak balance is shown in the upper left. In response to stressors, the ionic homeostasis of cells is challenged, which is met by energy consuming ion pumping mechanisms. The severity of the insult appears to have an influence on energy demand for ion pumping and DNA repair. When energy depletion is severe, cell demise frequently is directed towards necrosis and towards apoptosis, if the depletion is less severe.*

4 Immunological implications

Whether cells die by apoptosis or by necrosis can have profound implications for local immune responses such as tissue inflammation. Apoptosis *in vivo* is followed almost inevitably by the rapid uptake of dead cells into adjacent phagocytic cells. Binding to, or uptake of, apoptotic cells by macrophages inhibited the secretion of inflammatory mediators such as interleukin (IL)-1β, IL-8, IL-10, and tumour necrosis factor alpha (TNF-α) as well as leukotriene C4 and thromboxane B2. In contrast, production of transforming growth factor (TGF)-β1, prostaglandin E2, and platelet-activating factor (PAF) was increased (Fadok *et al.*, 1998, 2001).

Macrophages exposed to necrotized tumour cells (Reiter *et al.*, 1999) or lysed neutrophils stimulated the induction of inflammatory signals such as IL-8, or IL-10, and TNF-α (Fadok *et al.*, 1998; Gough *et al.*, 2001). These pro-inflammatory effects were markedly reduced when serine protease inhibitors were added, or when separated neutrophil membranes only were used (Fadok *et al.*, 2001). Experiments with human monomyelocytes, which do not express phospatidylserine (PS) on the outer leaflet of the plasma membrane during apoptosis showed that suppression of inflammatory mediators and promotion of TGF-β1 secretion appeared to require PS on apoptotic cells (Huynh *et al.*, 2002). The presence of TGF-β1, prostaglandin E2, or PAF appeared to be involved in the suppression of pro-inflammatory cytokine production because exogenous supplementation of these substances inhibited lipopolysaccharide-stimulated cytokine production (Fadok *et al.*, 1998).

Necrosis is pro-inflammatory, while apoptosis is anti-inflammatory. The anti-inflammatory property of apoptotic cells may be instrumental to avoid tissue disruption by the high number of cell deaths, which occur in multicellular organisms during development, or in normal life. In human life, for instance, 100 000 cell deaths occur every second (Vaux and Korsmeyer, 1999). In other words, multicellular organisms with a high cell turn over and high numbers of cell deaths occurring in normal life reduce the risk of inflammation by the inflammation-resolving properties of apoptosis.

In the context of infection or tumour defence, however, inflammation plays an adaptive role. In tumour defence, for instance, inflammation-resolving properties of apoptosis have an unwanted side in the sense of reduced anti-tumoural activity of macrophages exposed to apoptotic cells (Reiter *et al.*, 1999). The ability of *Entamoeba histolytica*, a gut-dwelling and tissue-invading parasite of man, to kill and phagocytose host cells correlates with parasite virulence. The parasite induces apoptotic killing of host cells, and the subsequent phagocytosis of these cells may limit inflammation and enable amoebas to evade the host immune response (Huston *et al.*, 2003). In such contexts, necrosis would be adaptive, or even more physiological.

5 Detection of necrosis

Detection of necrotic cell death mainly relies on assays which include morphology, plasma membrane integrity, DNA content, and cell surface expression of phosphatidylserine. Morphological features of necrosis include cell swelling and development of translucent cytoplasm, as seen in human monocytes and in cells of a monocytic cell line by light microscopy and flow cytometry (see *Table 1*; Warny and Kelly, 1999). In contrast to this, apoptotic cells show shrinkage and fragmentation

(Sastry and Rao, 2000). In cell cycle analysis, necrotic cells are characterized by random DNA fragmentation compared to patterned DNA cleavage in apoptotic cells (Evans *et al.*, 2000). In addition, the plasma membrane of necrotic cells loses integrity, which can be determined by uptake of indicating dyes, or by leakage of intracellular enzymes. Uptake of chemicals frequently is detected by flow cytometry using propidium iodine (Warny and Kelly, 1999), or by light microscopy using trypane blue (Bonfoco *et al.*, 1995). Enzyme leakage frequently is traced *in vitro* by measuring lactate dehydrogenase activity in the culture medium (Bonfoco *et al.*, 1995; Kim *et al.*, 2000).

Table 1 *Some features of cell undergoing necrotic versus apoptotic cell death*

Necrosis	Apoptosis
Cell swelling Development of translucent cytoplasms Chromatin clumping Organellar swelling	Cell shrinkage Fragmentation, membrane blebbing Chromatin condensation
Loss of membrane integrity Leakage of intracellular enzymes	Intact cell membrane Cell surface expression of phosphatidyl serine
Random DNA fragmentation	Interchromosomal patterned DNA cleavage

6 Necrotic cell death in the context of fish diseases

In fish, morphological features of necrosis frequently are associated with infections with pathogens of viral, bacterial or parasitic origin. In viral diseases, affected organs are, for instance, nervous and retinal tissues in nodavirus infections of marine finfish (Munday and Nakai, 1997), liver and renal tubules in halibut (*Hippoglossus hippoglossus*) infected with an aquareovirus (Cusack *et al.*, 2001), haematopoietic tissues in the kidney in rainbow trout (*Oncorhynchus mykiss*) infected with the epizootic haematopoietic necrosis virus (EHNV; Reddacliff and Whittington, 1996), or pancreatic tissues in rainbow trout infected with the infectious pancreatic necrosis virus (IPNV; Hong *et al.*, 1998). In IPNV, cells from a fish cell line initially displayed structural and biochemical indications of apoptotic demise (Hong *et al.*, 1998), but at a later stage of the process leading to cell death, pores were formed in the outer cell membrane (Hong *et al.*, 1999). By this, the cell membrane of infected cells lost their integrity, and cellular compounds were leaking into the environment. Membrane leakage was considered to be a core event in necrotic cell death, mainly involved in the promotion of inflammation, which was found to be associated with this form of cell death.

Accumulations of necrotic cells in various tissues of fish were associated with inflammation. This was mainly noted in histology as an infiltration of granulocytes, macrophages and lymphocytes into tissue areas where cell necrosis had appeared. In was seen for instance in the liver of medaka *Oryzias latipes* (Braunbeck *et al.*, 1992; Hoft *et al.*, 2003), or spleen of eels (Spazier *et al.*, 1992), as well as in carp infected with the blood fluke *Sanguinicola.* Here, eggs and emigrating miracidia caused necrosis of vascular cells and induced inflammatory infiltration of epithelial tissue, which encapsulated and subsequently eliminated parasite stages by degradation (Kirk and Lewis, 1998). In rainbow trout infected with the blood flagellate *Cryptobia salmositica*, a generalized inflammation with lesions in connective tissue and in the reticulo-endothelial system was associated with tissue necrosis in liver and kidney. The severity of this response was related to parasitaemia, and most likely induced by occlusion of blood vessels due to parasites and mononuclear cells (Bahmanrokh and Woo, 2001). In carp infected with *Trypanoplasma borreli*, a parasite closely related to *Cryptobia salmositica*, renal tubule cells experienced mitochondrial damage and subsequent necrosis (Rudat *et al.*, 2000). In infected carp, an extensive proliferation of the lymphoid renal tissue and a congestion of blood vessels with parasites and inflammatory cells occurred (Bunnajirakul *et al.*, 2000). This might have caused changes in renal microcirculation and thus induced hypoxic conditions in renal tubule cells. In addition, this parasite induced a strong activation of phagocytic cells, which resulted in high secretion of nitric oxide and high production of ROS (Saeij *et al.*, 2000, 2002). Production of nitric oxide and ROS in excess could induce oxidative stress to cells and tissues, as suspected by Saeij *et al.* (2003), who found that peripheral blood derived leucocytes were highly susceptible to NO-induced cell death. In addition, rainbow trout phagocytes were suspected to be sensitive to high production of ROS in association with high intracellular Ca^{++} levels (Betoulle *et al.*, 2000). Chemical depletion of glutathione-S-transferase in cells of medaka significantly elevated the induction of necrotic cell death by methyl-nitro-nitrosoguanidine (Kwak *et al.*, 2000).

Taken together, a link between necrotic cell death and inflammation is well established in fish. In local responses, like in infections with *Sanguinicola inermis* (Kirk and Lewis, 1998) or with gut-dwelling coccidia (Jendrysek *et al.*, 1994 Landsberg and Paperna, 1987) inflammation induced by necrotic cells helps eliminating the parasite. Responses which are associated with ischaemic conditions in vital organs, such as liver or kidney, appear to promote oxidative stress to cells and tissues, which in consequence may result in progressive cell death by necrotic pathways. The significance of inflammation for the induction of necrotic cell death, and the significance of necrosis for pathological conditions, however, has hardly been addressed in the context of fish diseases and still needs considerable attention.

7 Summary and perspective

In the context of infectious diseases of fish, cell death is described as part of pathology on the basis of structural alterations in cells and tissues. Morphological changes, which can be observed by microscopy, however, are results of processes that follow cell death and thus have to be regarded as post-mortem changes. In mammalian cells, stressors such as oxygen deprivation (hypoxia), or oxidative damage due to the production of ROS or NO most frequently were found to induce cell injury and subsequently cell death. Oxidative damage has the capacity to trigger both main pathways of cell death, necrosis and apoptosis. While high levels of oxidative damage can cause necrosis, lesser degrees of oxidative stress induce cell death via the apoptotic cascade. Necrotic cell demise was found to be associated with energy depletion, cell swelling, membrane rupture and release of intracellular enzymes. Immunological consequences of necrotic cell death were a pronounced induction of pro-inflammatory effects such as the release of IL-8, IL-10 or TNF-α by macrophages or neutrophils exposed to necrotized cells.

High levels of reactive oxygen species were also released from inflammatory cells such as neutrophils and macrophages in fish infected with various infections agents and induced oxidative stress to cells or tissues (Saeij *et al.*, 2003). Physiological responses of fish cells confronted with reactive oxygen species are not studied in detail. It remains unclear whether cells die or can repair the damage resulting from the insult. Can a high-level release of nitric oxide induce necrotic cell death, which then could be responsible for functional deficiencies, or does it help to clear the pathogen from the tissue? Studies on the induction of cell death and on immunological implication of pathogen-induced demise of cells are completely lacking for fish, but would be highly needed for an understanding of pathogen–host interactions on the physiological level. Here, a wide field is open for future work.

Acknowledgements

This work was supported in part by the European Community's Improving Human Potential Programme under contract (HPRN-CT-2001-00214), (PARITY).

References

Bahmanrokh, M. and Woo, P.T.K. (2001) Relations between histopathology and parasitaemias in *Oncorhynchus mykiss* infected with *Cryptobia salmositica*, a pathogenic haemoflagellate. *Dis. Aquat. Org.* **46:** 41–45.

Barros, L.F., Hermonsilla, T. and Castro, J. (2001a) Necrotic volume increase and the early physiology of necrosis. *Comp. Biochem. Physiol. A* **130:** 401–409.

Barros, L.F., Stutzin, A., Calixto, A., Catalan, M., Castro, J., Hetz, C. and Hermonsilla, T. (2001b) Non-selective cation channels as effectors of free radical-induced rat liver cell necrosis. *Hepatology* **33:** 114–122.

Betoulle, S., Duchiron, C. and Deschaux, P. (2000) Lindane increases in vitro respiratory burst and intracellular calcium levels in rainbow trout (*Oncorhychus mykiss*) head kidney phagocytes. *Aquatic Toxicol.* **48:** 211–221.

Bonfoco, E., Krainc, D., Ankarcrona, M., Nicotera, P. and Lipton, S.A. (1995) Apoptosis and necrosis: two distinct events induced, respectively, by mild and intense insults with N-methly-D-aspartate or nitric oxide/superoxide in cortical cell cultures. *Proc. Natl Acad. Sci.* **92:** 7162–7166.

Braunbeck, T.A., The, S.J., Lester, S.M. and Hinton, D.E. (1992) Ultrastructural alterations in liver of medaka (*Oryzias latipes*) exposed to diethylnitrosamine. *Toxicol. Pathol.* **20:** 179–196.

Büki, A., Okonkwo, D.O., Wang, K.K. and Povlishok, J.T. (2000) Cytochrome c release and caspase activation in traumatic axonal injury. *J. Neurosci.* **20:** 2825–2834.

Bunnajirakul, S., Steinhagen, D., Hetzel, U., Körting, W. and Drommer, W. (2000) A study of sequential histopathology of *Trypanoplasma borreli* (Protozoa: Kinetoplastida) in susceptible common carp *Cyprinus carpio*. *Dis. Aquat. Org.* **39:** 221–229.

Carini, R., Autelli, R., Bellomo, G. and Albano, E. (1999) Alterations of cell volume regulation in the development of hepatocyte necrosis. *Exp. Cell Res.* **248:** 280–293.

Chirarugi, A. (2002) Poly(ADP-ribose) polymerase: killer or conspirator? The 'suicide hypothesis' revisited. *Trends Pharmocol. Sci.* **23:** 122–129.

Cohen, G.M. (1997) Caspases: the executioners of apoptosis. *Biochem. J.* **326:** 1–16.

Cusack, R.R., Groman, D.B., MacKinnon, A.M., Kibenge, F.S., Wadowska, D. and Brown, N. (2001) Pathology associated with an aquareovirus in captive juvenile Atlantic halibut *Hippoglossus hippoglossus* and an experimental treatment strategy for a concurrent bacterial infection. *Dis. Aquat. Org.* **44:** 7–16.

Denecker, G., Vercammen, D., Steemans, M., ***et al.*** (2001) Death receptor-induced apoptotic and necrotic cell death: differential role of caspases and mitochondria. *Cell Death Differ.* **8:** 828–840.

Dypbukt, J.M., Ankarcrona, M., Burkitt, M., Sjöholm, A., Ström, K., Orrenius, S. and Nicotera, P. (1994) Different prooxidant levels stimulate growth, trigger apoptosis, or produce necrosis of insulin-secreting RINm5F cells. The role of intracellular polyamines. *J. Biol. Chem.* **269:** 30553–30560.

Evans, D.L., Bishop, R. and Jaso-Friedmann, L. (2000) Methods for cell cycle analysis and detection of apoptosis of teleost cells. *Methods Cell Sci.* **22:** 225–231.

Fadok, V.A.; Bratton, D.L., Konowal, A., Freed, P.W., Westcott, J.Y. and Henson, P.M. (1998) Macrophages that have ingested apoptotic cells in vitro inhibit proinflammatory cytokine production through autocrine/paracrine mechanisms involving TGF-β, PGE_2, and PAF. *J Clin Invest* **101**: 890-898

Fadok, V.A., Bratton, D.L., Guthrie, L. and Henson, P.M. (2001) Differential effects of apoptotic versus lysed cells on macrophage production of cytokines: role for proteases. *J Immunol* **166**: 6847-9854

Fiers, W., Beyaert, R., Declercq, W. and Vandenabeele, P. (1999) More than one way to die: apoptosis and reactive oxygen damage. *Oncogene* **18:** 7719–7730.

Gough, M.J., Melcher, A.A., Ahmed, A., Crittenden, M.R., Riddle, D.S., Linardakis, E., Ruchath, A.N., Emiliusen, L.M. and Vile, R.G. (2001) Macrophages orchestrate the immune response to tumour cell death. *Cancer Research* **61**: 7240-7247

Ha, H.C. and Snyder, S.H. (1999) Poly(ADP-ribose) polymerase is a mediator of necrotic cell death by ATP depletion. *Proc. Natl. Acad. Sci.* **96**: 13978-13982

Hamm, J.T., Wilson, B.W. and Hinton, D.E. (1998) Organophosphate-induced acetylcholinesterase inhibition and embryonic retinal cell necrosis in vivo in the teleost (*Orzyias latipes*). *Neurotoxicology* **19:** 853–869.

Herceg, Z. and Wang, Z.Q. (1999) Failure of poly(ADP-ribose) polymerase cleavage by caspase leads to induction of necrosis and enhanced apoptosis. *Mol Cell Biol* **19**: 5124-5133

Herson, P.S. and Ashford, M.L.J. (1997) Activation of a novel non-selective cation channel by alloxan and H_2O_2 in rat insulin-secreting cell line CRI-G1. *J. Physiol.* **501:** 59–66.

Hofmann, E.K. and Simonsen, L.O. (1989) Membrane mechanisms in volume and pH regulation in vertebrate cells. *Physiol. Rev.* **69:** 315–382.

Hoft, P.T., van Dongen, W., Esmans, E.L., Blust, R. and de Coen, W.M. (2003) Evaluation of the toxicological effects of perfluorooctane sulfunic acid in the common carp (*Cyprinus carpio*). *Aquat. Toxicol.* **62:** 349–359.

Hong, J.R., Lin, T.L., Hsu, Y.L. and Wu, J.L. (1998) Apoptosis precedes necrosis of fish cell line with infectious pancreas necrosis virus infection. *Virology* **250:** 76–84.

Hong, J.R., Lin, T.L., Hsu, Y.L. and Wu, J.L. (1999) Dynamics of nontypical apoptotic morphological changes visualized by green fluorescent protein in living cells with infectious pancreatic virus infection. *J. Virol.* **73:** 5056–5063.

Huston, C.D., Boettner, D.R., Miller-Sims, V. and Petri, W.A. Jr. (2003) Apoptotic killing and phagocytosis of host cells by the parasite *Entamaeba histolytica. Infect. Immun.* **71:** 964–972.

Huynh, M.L., Fadok, V. and Henson, P.M. (2002) Phospatidylserine-dependent ingestion of apoptotic cells promotes TGF beta secretion and the resolution of inflammation. *J Clin Invest* **109**: 41-50

Jendrysek, S., Steinhagen, D., Drommer, W. and Körting, W. (1994) Carp coccidiosis: intestinal histo- and cytopathology under *Goussia carpelli* infection. *Dis. Aquat. Org.* **20:** 171–182.

Kerr, J.F.R., Wyllie, A.H. and Currie, A.R. (1972) Apoptosis: a basic biological phenomenon with wide ranging implications in tissue kinetics. *Br. J. Cancer* **26:** 239–257.

Kim, Y.M., Chung, H.A.T., Simmons, R.L. and Billar, T.R. (2000) Cellular non-heme iron content is a determinant of nitric oxide-mediated apoptosis, necrosis, and caspase inhibition. *J. Biol. Chem.* **275:** 10954–10961.

Kirk, R.S. and Lewis, J.W. (1998) Histopathology of *Sanguinicola inermis* infection in carp, *Cyprinus carpio*. *J. Helminthol.* **72:** 33–38.

Kroemer, G., Dellaporta, B. and Resche-Rigon, M. (1998) The mitochondrial death/life regulator in apoptosis and necrosis. *Ann. Rev. Physiol.* **60:** 619–642.

Kwak, H.I., Lee, M.H. and Cho, M.H. (2000) Interrelationship of apoptosis, mutation, and cell proliferation in N-methyl-N'-nitro-N-nitrosoguanidine (MNNG)- induced medaka carcinogenesis model. *Aquatic Toxicol.* **50:** 317–329.

Kwak, H.I., Bae, M.O., Lee, M.H., *et al.* (2001) Effects of nonylphenol, bisphenol A, and their mixture on the viviparous swordfish (*Xiphophorus helleri*). *Environ. Toxicol. Chemist* **20:** 787–795.

Landsberg, J.H. and Paperna, I. (1987) Intestinal infections by *Eimeria* (s.l.) *vanasi* n. sp. (Eimeriidae, Apicomplexa, Protozoa) in chichlid fish. *Ann. Parasitol. Hum. Comp.* **62:** 283–293.

Leist, M., Single, B., Castoldi, A.F., Kühnle, S. and Nicotera, P. (1997) Intracellular adenosine triphosphate (ATP) concentration: a switch in the decision between apoptosis and necrosis. *J. Exp. Med.* **185:** 1481–1486.

Leist, M., Single, B., Naumann, H., Fava, E., Simon, B., Kuhnle, S. and Nicotera, P. (1999) Inhibition of mitochondrial ATP generation by nitric oxide switches apoptosis to necrosis. *Exp. Cell Res.* **249:** 396–403.

Leist, M. and Nicotera, P. (1997) The shape of cell death. *Biochem. Biophy. Res. Commun.* **236:** 1–9.

Lemasters, J.J., Qian, T., Bradham, C.A., Brenner, D.A., Cascio, W.E., Trost, L.C., Nishimura, Y., Nieminen, A.L. and Herman, B. (1999) Mitochondrial dysfunction in the pathogenesis of necrotic and apoptotic cell death. *J. Bioenerg. Biomembr.* **31:** 305–319.

Lipton, P. (1999) Ischemic cell death in brain neurons. *Physiol. Rev.* **79:** 1431–1568.

Lom, J. and Dykova, I. (1992) Protozoan parasites of fish. *Develop. Aqua. Fish. Sci.* **26:** 315.

Los, M., Mozoluk, M., Ferrari, D., Stepczynska, A., Stroh, C., Renz, A., Herceg, Z., Wang, Z.Q. and Schulze-Osthoff, K. (2002) Activation and caspase-mediated inhibition of PARP: a molecular switch between fibroblast necrosis and apoptosis in death receptor signalling. *Mol. Biol. Cell.* **13:** 978–988.

Majno, G. and Joris, I. (1995) Apoptosis, oncosis, and necrosis. An overview of cell death. *Am. J. Pathol.* **146:** 3–15.

McConkey, D.J. (1998) Biochemical determinants of apoptosis and necrosis. *Toxicol. Lett.* **99:** 157–160.

McConkey, D.J. and Orrenius, S. (1994) Calcium and cyclosporin A in the regulation of apoptosis. *Curr. Top. Microbiol. Immunol.* **200:** 95–105.

Munday, B.L. and Nakai, T. (1997) Nodavirus as pathogens in larval and juvenile marine finfish. *World J. Microbiol. Biotechnol.* **13**: 375–381.

Nicotera, P. (2002) Apoptosis and age related disorders: role of caspase dependent and caspase-independent pathways. *Toxicol. Lett.* **127:** 189–195.

Nicotera, P., Leist, M. and Ferrando-May, E. (1998) Intracelluar ATP, a switch in the decision between apoptosis and necrosis. *Toxicol. Lett.* **102–103:** 139–142.

Nicotera, P., Leist, M. and Ferrando-May, E. (1999) Apoptosis and necrosis: different execution of the same death. *Biochem. Soc. Symp.* **66:** 69–73.

Okada, Y. (1997) Volume expansion-sensitising outward rectifier Cl- channel: fresh start to the molecular identity and volume sensor. *Am. J. Physiol.* **273**: C755–789.

Okada, Y., Maeno, E., Shimizu, T., Dezaki, K., Wang, J. and Morishima, S. (2001) Receptor-mediated control of regulatory volume decrease (RVD) and apoptotic volume decrease (AVD). *J. Physiol.* **532:** 3–16.

Orlov, S.N., Thorin-Trescases, N., Kotelevtsev, S.V., Tremblay, J. and Hamet, P. (1999) Inversion of the intracellular Na^+/K^+ ratio blocks apoptosis in vascular smooth muscle at a site upstream of caspase-3. *J Biol Chem* **274**: 16545-16552

Patel, A.J., Lauritzen, I., Lazdunski, M., and Honore, E. (1998) Disruption of mitochondrial respiration inhibits volume-regulated anion channels and provokes neuronal cell swelling. *J Neurosci* **18**: 3117-3123

Ransom, D.P., Lannan, C.N., Rohovec, J.S. and Fryer, J.L. (1984) Comparison of histopathology caused by *Vibrio anguillarum* and *Vibrio ordalii* in three species of Pacific salmon. *J. Fish Dis.* **7**: 107–115.

Reddacliff, L.A. and Whittington, R.J. (1996) Pathology of epizootic haematopoietic necrosis virus (EHNV) infection in rainbow trout (*Oncorhynchus mykiss* Walbaum) and redfin perch (*Perca fluviatilis* L.). *J. Comp. Pathol.* **115:** 103–115.

Redegeld, F.A., Moison, R.M., Barentsen, H.M., Koster, A.S. and Noordhoek, J. (1990) Interaction with cellular ATP generating pathways mediates menadione-induced cytotoxicity in isolated rat hepatocytes. *Arch. Biochem. Biophys.* **280:** 130–136.

Reiter, I., Krammer, B. and Schwamberger, G. (1999) Cutting edge: differential effect of apoptotic versus necrotic tumour cells on macrophage antitumour activities. *J. Immunol.* **163:** 1730–1732.

Rudat, S., Steinhagen, D., Hetzel, U., Drommer, W. and Körting, W. (2000) Cytopathological observations on renal epithelium cells in common carp *Cyprinus carpio* under *Trypanoplasma borreli* (Protozoa: Kinetoplastida) infection. *Dis. Aquat. Org.* **40:** 203–209.

Saeij, J.P.J., Stet, R.J.M., Groeneveld, A., Verburg-van Kemenade, B.M.L., van Muiswinkel, W.B. and Wiegertjes, G.F. (2000) Molecular and functional characterisation of a fish inducible-type nitric oxide synthase. *Immunogenetics* **51:** 339–346.

Saeij, J.P.J., van Muiswinkel, W.B., Groeneveld, A. and Wiegertjes, G.F. (2002) Immune modulation by fish kinetoplastid parasites: a role for nitric oxide. *Parasitology* **124:** 77–86.

Saeij, J.P.J., van Muiswinkel, W.B., van de Ment, M., Amaral, C. and Wiegertjes, G.F. (2003) Different capacities of carp leukocytes to encounter nitric oxide mediated stress: a role for the intracellular reduced glutathione pool. *Dev. Comp. Immunol.* **27:** 555–568.

Sastry, P.S. and Rao, K.S. (2000) Apoptosis and the nervous system. *J. Neurochem.* **74:** 1–20.

Spazier, E., Storch, V. and Braunbeck, T. (1992) Cytopathology of spleen in eel *Anguilla anguilla* exposed to a chemical spill in the Rhine River. *Dis. Aquat. Org.* **14:** 1–22.

Speare, D.J., Carajal, V. and Horney, B.S. (1999) Growth suppression and branchitis in trout exposed to hydrogen peroxide. *J. Comp. Pathol.* **120:** 391–402.

Tan, S., Wood, M. and Maher, P. (1998) Oxidative stress induces a form of programmed cell death with characteristics of both apoptosis and necrosis in neuronal cells. *J. Neurochem.* **71:** 95–105.

Timur, G., Roberts, R.J. and McQueen, A. (1977) The experimental pathogenesis of focal tuberculosis in plaice (*Pleuronectes platessa* L.). *J. Comp. Pathol.* **87:** 89–96.

Vercammen, D., Brouckaert, G., Denecker, G., Van den Craen, M., Declercq, W., Fiers, W. and Vandenabeele, P. (1998) Dual signalling of the Fas receptor: initiation of both apoptotic and necrotic cell death pathways. *J. Exp. Med.* **188:** 919–930.

Vaux, D.L. and Korsmeyer, S.J. (1999) Cell death in development. *Cell* **96:** 245–254.

Warny, M. and Kelly, C.P. (1999) Monocytic cell necrosis is mediated by potassium depletion and caspase-like proteases. *Am. J. Physiol.* **276** (Cell. Physiol. **45**): C717–C724.

Wang, K.K. (2000) Calpain and caspase: can you tell the difference? *Trends Neurosci.* **23:** 20–26.

2

The role of apoptosis in non-mammalian host–parasite relationships

Dave Hoole and Gwyn T. Williams

1 Introduction

The study of cell death goes back many years especially if one includes the analysis of its occurrence in embryonic development (reviewed by Saunders, 1966). Much of the initial work was naturally restricted to cytological observations and, although it was clear that cell death was an important component of developmental processes, it was assumed by many to be of marginal importance in cell biology as a whole.

It was probably in neurobiology that the functional importance of cell death was first widely appreciated. The production of excess neurons during embryonic development is followed by the death of many of them (e.g. Barde, 1989), and the importance of this was that it allowed selection of useful cells and elimination of unwanted cells through physiological cell death by a process which left the remaining cells completely healthy.

The common thread running through many of the studies on cell death, i.e., primarily the common morphological features (*Figure 1*) often present, was pointed out by Wyllie, Kerr and Currie (reviewed by Wyllie *et al.*, 1980). They introduced the term apoptosis (Kerr *et al.*, 1972) as a unifying concept of cell death operating in opposition to mitosis to maintain stable cell population sizes. This represented a crucial development in this area, and the beginning of the process of recognition of cell death control as a central process in cell biology. This was an important factor in the conceptual advance that gradually incorporated the control of cell populations by regulation of apoptosis into mainstream thinking in cell biology (e.g. Williams *et al.*, 1992). It has since become impossible to produce an adequate analysis of any metazoan cellular system that ignored the possible role of apoptosis in its control. Analysis of cellular proliferation and differentiation alone was no longer acceptable – it now appeared to be rather like calculating the depth of water in the bath by extrapolation of the rate of flow of water though the taps and forgetting about whether the plug is in or not!

Host–Parasite Interactions, edited by G. F. Wiegertjes & G. Flik.

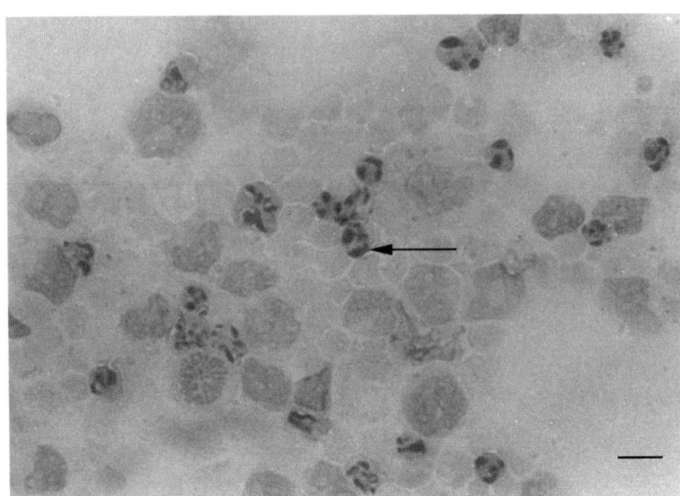

Figure 1 *Mouse haematopoietic cell line induced apoptosis by withdrawal of interleukin 3 (IL3). Note condensed chromatin in apoptotic cells. Stained with Giemsa. Scale bar 10 μm.*

Since cell biology provides the ground rules for biology as a whole, and consequently underlies understanding of metazoan physiology and pathology, the consequences of this conceptual advance have been both profound and far-reaching. In the immune system, for example, assimilation of active cell death by apoptosis has had very important implications for the mechanisms underlying the development and selection of both T-lymphocytes and B-lymphocytes (reviewed by Rathmell and Thompson, 2002; Williams, 1994). Both positive and negative selection within the thymus, for example, rely on cellular self-destruction through apoptosis for the elimination of harmful (autoreactive) and useless (unable to recognize self-MHC) thymocytes during development. Similarly clonal deletion in the periphery relies on apoptosis for specific down-regulation of the immune response (Rathmell and Thompson, 2002). Clearly failure in the control or execution of apoptosis can underlie immune pathologies, which involve inappropriate persistence of cells with autoreactive antigen receptors.

The impact of the concept of apoptosis on our understanding of the mechanisms of cell-mediated cytotoxicity has been just as important. The observation that cytotoxic T-lymphocytes, as well as other cytotoxic cell types, induce target cells to self-destruct, sometimes through the target cell surface receptor Fas (reviewed by Krammer, 2000) rather than simply destroying them, has provided crucial avenues of investigation into potential pathological effects of failure of target cell apoptosis. These consequences may be particularly important when apoptosis suppression is induced by viral infection. This provides a continuing evolutionary battlefield between, on the one hand, the host and its efforts to eliminate virally infected cells by apoptosis, and, on the other hand, the virus and its interest in keeping the host cell alive long enough to produce more infectious virus (reviewed by Cuconati and White, 2002; Hasnain *et al.*, 2003). Failure of the immune system to block the spread of viral infection by induction of infected cell apoptosis can therefore result in a significant increase in the spread of the virus through the body and consequently increase the pathological consequences of infection.

Apoptosis therefore serves many different purposes in the immune system, which can generally be divided into those involved in developmental processes and those involved in protection against infection. It is not possible to state definitively the original purpose served by such active cell death when it first appeared during evolution, but it is clear that protection against infection by selective elimination of infected cells would even have benefits for unicellular organisms, in situations where they exist in populations of genetically highly related cells. The accumulating evidence supporting the existence of apoptosis, or some similar process, in protozoa also suggests that it may not be restricted to metazoans (reviewed by Welburn *et al.*, 1997).

In this chapter we evaluate the possible roles that apoptosis may have in the host–parasite interaction in non-mammalian animals (*Table 1*). Inevitably such an evaluation is restricted to areas where appropriate research has been undertaken and, as such, this review concentrates on those interactions that are of medical, agricultural or economic importance. In addition, the authors do not wish, at this stage, to contribute to the debate on the biological definition of a parasite. Such a definition has proven controversial although MacInnis in 1976 suggested that parasitism is an association in which 'one partner, the parasite, of a pair of interacting species, is dependent upon a minimum of one gene or its products from the other interacting species, defined as the host, for its survival'. For the purpose of this review therefore viruses, bacteria and all eucaryote infections are considered as parasites since many infections appear to adopt a similar relationship with the apoptotic process. Extensive studies have been undertaken on the relationship between infection and mammalian hosts, particularly humans. Such information has served as an invaluable starting point to assess the interaction between parasites and apoptosis in non-mammalian hosts.

Table 1 *Roles of apoptosis in infection*

Effect of apoptosis	Reference
1. Killing of the host: effects on population dynamics	Caputo *et al.* (2002)
2. Dissemination of infection	Hong *et al.* (1999a,b); Bjorklund *et al.* (1997)
3. Related to infectivity and virulence	Clem *et al.* (1994) Zhang *et al.* (1999); Sakai *et al.* (1999); Ishikawa *et al.* (2000)
4. Decrease killing mechanisms of host	Bayles *et al.* (1998); Wesson *et al.* (2000)
5. Liberation of pathogen from infected cells	Ojcius *et al.* (1999)
6. Decrease host fecundity	Hopwood *et al.* (2001)
7. Relationship to immunological privileged site	Griffith *et al.* (1996)
8. Limiting parasite migration	Jung *et al.* (2000); Murakawa *et al.* (2001)
9. Limiting pathogen population	Barcinski and DosReis (1999)
10. Killing of the pathogen	Bishop *et al.* (2002); Jos-Jaso-Friedman *et al.* (2000); Hagen *et al.* (1998)
11. Re-modelling of host for mutual benefit for both 'parasite' and host	Foster *et al.* (2000)

2 Basic concepts of apoptosis

Although analysis of apoptosis in *Drosophila* has illuminated elements of the control of cell death such as Reaper, Grim and Hid (reviewed by Abrams, 1999), not previously seen in other systems, it has been the genetic analysis of the lower metazoan, nematode *Caenorhabditis elegans*, which has played the most crucial role in analysis of programmed cell death. This was recognized recently by the award of the Nobel Prize to Bob Horvitz, together with Sydney Brenner and John Sulston, for their *C. elegans* studies including the first extensive genetic analysis of programmed cell death in any organism. *C. elegans* development proceeds, in genetically identical nematodes, by an entirely predictable process of cell division and development. This embryonic programme of development also includes the death of about 100 cells in the same reproducible manner. Horvitz and colleagues produced and identified *C. elegans* mutants in which this predictable developmental cell death failed to occur (reviewed by Ellis *et al.*, 1991). This was in itself a watershed in studies on cell death, since, prior to this work, cell death was not often viewed as a genetically programmed, active process, and the possibility of cells 'failing to die' would have been very rarely considered. The reprieved cells are perfectly viable and persist through the lifetime of the nematode. Such failure of cell death does not prevent a viable nematode being produced, in contrast to more complex organisms such as *Drosophila* and mice, where such failure of cell death is not compatible with development and long-term survival of the organism.

The genetic analysis of *C. elegans* mutants showing deficiencies in cell death allowed Horvitz and colleagues to identify several groups of genes which play important roles in cell death, and to group them into the different stages of the process. These stages are signalling, cellular self-destruction, removal of the dead cell by phagocytosis, and DNA degradation (reviewed by Ellis *et al.*, 1991). This was of great importance as the first genetic framework produced for a cell death programme in any organism, and provided a clear precedent for genetically programmed cell death by active cellular self-destruction at a time when the active gene-dependent nature of apoptosis in mammalian cells was not widely accepted (Williams *et al.*, 1992. Even more important, it gradually became clear that apoptosis in mammalian cells followed the same general pattern. This was presumably because of a common evolutionary origin for the processes. The framework established for *C. elegans* has thus proved invaluable in designing experiments exploring the molecular mechanisms of mammalian cell apoptosis.

Three of the *C. elegans* genes, in particular, were of great importance in the analysis of apoptosis in both nematodes and mammals. These are components of the central control mechanism of cell death, i.e. *ced*-3, *ced*-4 and *ced*-9 (Ellis *et al.*, 1991). *C. elegans* in which *ced*-3 or *ced*-4 have been inactivated by mutation illustrate the effects of failure of developmental cell death, since both these genes are required in the cell death process. In both cases, mammalian homologues have now been identified which also play important roles in apoptosis. In the case of *ced*-3, at least 12 mammalian homologues have been identified and the family encoded by these genes, the caspase family, makes up the biochemical core of the apoptotic process.

Caspases (cysteine aspartate-specific proteases; Alnemri *et al.*, 1996) are present in inactive pro-enzyme form in nucleated cells and are generally activated by cleavage. They are site-specific proteases, which recognize sequences of three or four amino acids, always ending in aspartate. In this way, rather than causing indiscriminate degradation of cellular proteins, caspases act on particular targets to cause the biochemical and morphological changes characteristic of apoptosis. The cleavage of targets is effectively irreversible, as is appropriate for the elimination of unwanted or potentially harmful cells. Some of the most important targets for caspases are other caspases, so that the response is amplified in a cascade and thus shows some similarities to complement activation or blood-clotting. Caspase 3, for example, is activated during apoptosis by upstream caspases in response to many different stimuli, whereas other caspases, acting upstream in particular signalling pathways, act to initiate the cascade.

Many of the targets for proteolytic cleavage have been identified in the past 10 years, and their cleavage is clearly related to the biochemistry and morphology of apoptosis. Caspase cleavage of ICAD (inhibitor of caspase-activated deoxyribonuclease) for example, releases CAD (caspase activated deoxyribonuclease), which in turn degrades genomic DNA. This activation of a DNAase results in the introduction of a very large number of breaks into the genomic DNA and is another important irreversible event, which is characteristic of many cells undergoing apoptosis (Enari *et al.*, 1998; Sakahira *et al.*, 1998).

APAF1 (apoptotic protease activating factor 1) has been identified as the mammalian homologue of Ced-4 (Li *et al.*, 1997). Functionally there are similarities between APAF1 and Ced-4, particularly that they both act to activate Ced-3/caspases. However, there are also important differences: Ced-4 appears to interact directly with the apoptosis inhibitor Ced-9, but this does not appear to be the case for APAF1 and the mammalian Ced-9 homologue Bcl-2 (Hausmann *et al.*, 2000). In general, the correspondence between the *C. elegans* pathway and the mammalian apoptosis pathways are not exact, with the mammalian apoptosis pathways displaying substantially more complexity: Ced-3 in *C. elegans* has a whole family of homologues, the caspases, in mammalian cells; similarly, Ced-9 in *C. elegans* has a whole family of homologues in mammalian cells, i.e. the Bcl-2 family.

Bcl-2 itself was isolated as the product of a candidate oncogene, before the significance of apoptosis control had been recognized, because of its presence at the t(14,18) chromosomal translocation characteristic of follicular lymphoma (e.g. Chenlevy *et al.*, 1989). The demonstration that, unlike other previously isolated oncogene products, Bcl-2 promoted cell survival rather than proliferation, was a critical development in the molecular analysis of apoptosis (Vaux *et al.*, 1988). Subsequent isolation of several other members of the family demonstrated that these mammalian Ced-9 homologues could act either as inhibitors (e.g. Bcl-2 and Bcl-x) or inducers of apoptosis, and were often localized to intracellular membranes, particularly on mitochondria (Zamzami *et al.*, 1998). Bcl-2 family members form heterodimers and some small but important proteins, containing only one of the 4 BH (Bcl-2 homology) domains, act to transduce pro-apoptotic signals to the larger Bcl-2 family proteins on the mitochondria (reviewed by Adams and Cory, 2001).

The localization of several Bcl-2 family members to mitochondrial membranes has helped to focus attention on the role of this organelle in apoptosis – and in many cases it is now clear that mitochondria do play a very important role. The observation that cytochrome c was released from mitochondria during apoptosis induced by many stimuli reinforced this idea, especially since it was clear that the cytochrome c itself was an active player in the formation, with APAF1 and caspase 9, of the apoptosome which initiates the caspase cascade (Liu *et al.*, 1996).

The exact role of Bcl-2 family members in controlling apoptosis has been investigated for many years, with most recent work focusing on mitochondria. Several mechanisms of action have been suggested (reviewed by Hengartner, 2000). Bcl-2 family members may directly participate in formation of pores in the outer mitochondrial membrane which might either allow cytochrome c to pass, on apoptosis, or might act as an ion channel to produce changes in membrane polarization which later lead to release of proteins. Such pores could be produced by oligomerization of Bcl-2 family proteins and/or by interaction of Bcl-2 family members with other proteins in the mitochondrial membranes, such as VDAC and the adenosine nucleotide transporter.

Several of the genes identified in *C. elegans* as being important in cell death encode proteins involved in the phagocytosis of apoptotic cells (Ellis *et al.*, 1991), reflecting the importance of cell disposal which is physiologically controlled and which results in minimal damage to the cells remaining. The biochemical changes in apoptosis include modifications of the cell membrane which mark the cell for disposal and which are recognized by both 'professional' phagocytes, such as macrophages, and also by surrounding cells which do not otherwise actively phagocytose other cells. Changes to the surface of apoptotic cells include the exposure of phosphatidylserine, which is found only on the inner leaflet of the plasma membrane of healthy cells as well as the appearance, or unmasking, of cell surface proteins (reviewed by Savill and Fadok, 2000).

3 Apoptosis in lower vertebrates

Although there have been extensive studies on the apoptotic process in mammals (see *Figure 2*) and invertebrates and its role in cellular pathology and development, very little is known in comparison about apoptosis in ectothermic vertebrates such as teleost fish and amphibians. In recent years there has been an increased interest in the form and function of apoptosis in these lower vertebrates. In fish, apoptosis has been detected in several tissues, for example, reproductive organs (Janz *et al.*, 2001), the retina (Kunz *et al.*, 1994) and the gills (Wendelaar Bonga and van de Mrij, 1989). In studies carried out by Caputo *et al.* (2002) in the paedomorphic goby, *Aphia minuta*, apoptosis was noted in the intestinal enterocytes and was implicated in the adult mortality that occurs immediately after the breeding season. The authors suggested that the significance of this observation lay in the fact that it was the first recorded case in which the apoptotic process was associated with natural host mortality and had implications when considering population dynamics. Information obtained from studies on fish has also revealed that controlled cell death is involved in normal cell differentiation (Rojo and Gonzalez, 1999; Soutschek and Zupapnec, 1995), and in response to stress and pollutants. For example, Julliard *et al.* (1993) suggested that exposure to copper induced apoptosis in the olfactory system of rainbow trout (*Oncorhynchus mykiss*), Toomey *et al.* (2001) noted that 2.3.7.8-tetrachlorodibenzo-p-dioxin increased apoptosis in several organs in the embryos of *Fundulus heteroclitus* and Fischer and Dietrich (2000) observed that toxins produced from cyanobacterium, which are associated with fish kills, was associated with apoptosis in hepatocytes and renal tissue of carp (*Cyprinus carpio*). Microcystin-LR, a commonly encountered

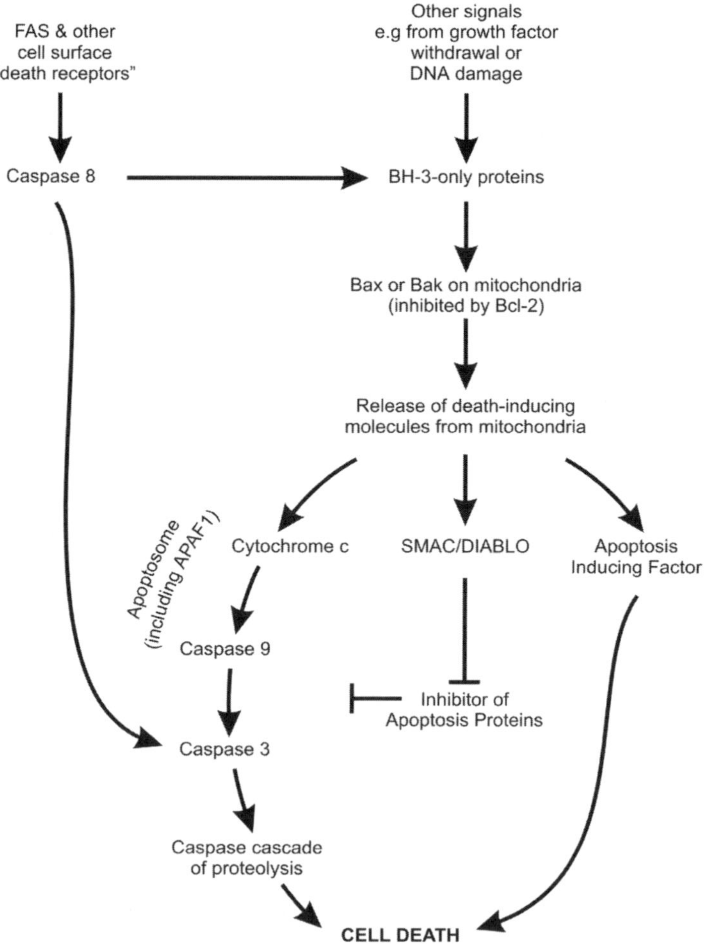

Figure 2 *Outline of apoptosis control in mammalian cells.*

hepatotoxin in China, and cadmium have also been noted to induce apoptosis in carp and rainbow trout hepatocytes (Li *et al.*, 2001; Risso-de Faverney *et al.*, 2001). Other environmental perturbations such as temperature have also been associated with apoptosis in fish tissues (Nolan *et al.*, 2000a) as has consumption of toxicants (Kamunde *et al.*, 2001; Lundebye *et al.*, 1999) and the administration of drugs (Gogal *et al.*, 1999). The association of apoptosis with environmental conditions is of course not unique to fish and has been described in many phyla. For example, Micic *et al.* (2001) noted tri-n-butyltin induced apoptosis in the gill tissue of the blue mussel *Mytilus galloprovincialis* and indeed several authors (Piechotta *et al.*, 1999; Sweet *et al.*, 1999) have suggested that apoptosis might even be used as a biomarker of environmental quality. This relationship between pollution and apoptosis has important implications when considering the effects of another stressor, infection, on this association (see Hoole *et al.*, 2003 for review).

With regards to the host–parasite interaction, recent studies have revealed that apoptosis is an important component in the immune system of lower vertebrates.

Several studies in amphibians have shown that apoptosis is a common occurrence in the lymphocyte component of the immune response. Grant *et al.* (1995) first reported the time course of mitogen-induced apoptosis in the thymic and splenic lymphocytes of adult *Xenopus laevis*. Later studies on the splenocytes of this amphibian by Mangurian *et al.* (1998) indicated that these cells express Fas, which possibly shared both structural and functional homologies with that occurring in humans. In addition, McMahan *et al.* (2000) revealed that PMA, which sometimes 'rescues' mammalian cells from apoptosis, stimulates controlled cell death in the splenocytes of *Xenopus*. The availability and extensive knowledge of amphibian metamorphosis has meant that this group of animals has served as a useful biological model to investigate the role of apoptosis during morphogenesis of the immune system. Ducoroy *et al.* (1999) noted that the enterotoxin from the bacterium *Staphylococcus aureus*, and glucocorticoid treatment induced apoptosis in the thymocytes of the larvae of the urodele amphibian, the axolotl. Studies by Grant *et al.* (1998) revealed that in *Xenopus*, apoptosis occurred in the lymphocytes of the thymus and spleen during metamorphosis although the levels differed between the organs and the metamorphic stage. They also noted that the apoptosis observed may not be correlated to plasma glucocorticoid titres. The association of apoptosis and these stress hormones has been extensively studied in fish (e.g. Espelid *et al.*, 1996; Verburg-van Kemenade *et al.*, 1999; Weyts *et al.*, 1997) and has implications when considering the association between parasites and apoptosis (see later). Studies on the sequencing of genes associated with the apoptotic process in lower vertebrates have been somewhat limited. For example, Yabu *et al.* (2001) cloned and characterized caspase 3 in zebrafish which showed 60% identity with that occurring in *Xenopus*, chicken and mammals. Recently caspase 6 from rainbow trout has been sequenced and has been shown to be expressed particularly in pronephric cells after exposure to cortisol and LPS and during confinement of the fish (Laing *et al.*, 2001). In addition, we have partially sequenced and cloned Bcl2 and APAF1 from the pronephric cells of carp, *Cyprinus carpio*.

In amphibians, as in mammals, apoptosis involves the translocation of phosphatidylserine from the inner leaflet of the cell membrane to the surface. This is utilized by macrophages as a recognition system in both amphibians and mammals and would thus appear to be conserved during vertebrate evolution (Nera *et al.*, 2000). Indeed, recent studies on carp, *Cyprinus carpio*, have also indicated that apoptosis in fish leucocytes (*Figure 3*) is associated with exposure of phosphatidylserine on their outer membrane (Saeij *et al.*, 2003; Saha *et al.*, 2003). It is of interest that studies on lower vertebrates such as fish have also indicated that targets of the immune response may be killed by apoptosis. Initial studies carried out by Meseguer *et al.* (1996) noted that leucocytes of sea bass (*Dicentrarchus labrax*) and gilthead seabream (*Sparus aurata*) were able to destroy tumour cells by apoptosis and necrosis. Greenlee *et al.* (1991) also noted that non-specific cytotoxic cells can induce apoptosis in rainbow trout (*Oncorhynchus mykiss*). Since these initial observations there have been extensive studies on the association between cytotoxic cells in fish and apoptosis (see Evans *et al.*, 2001; Shen *et al.*, 2000 for reviews). Many of these studies have concentrated on these cells in tilapia and catfish and recent investigations by Bishop *et al.* (2002) have suggested that one killing mechanism involved the release of Fas ligand from activated non-specific cytotoxic cells and the presence of Fas receptor on the target cells. This may have interesting implications when considering the killing of eucaryotic infections in animals.

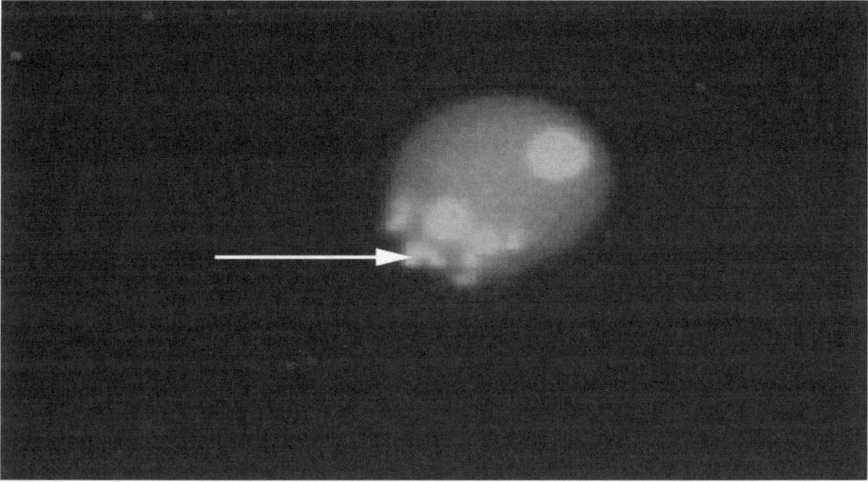

Figure 3 *Pronephric leucocyte of carp* Cyprinus carpio *undergoing cortisol induced apoptosis. Note apoptotic bodies (arrow). Stained with acridine orange.*

4 Association between viral infections and apoptosis

4.1 *Viruses and higher vertebrates*

The intricate association between the virus and the physiology and reproductive capacity of its host cell has resulted in the extensive use of this group of pathogens as a tool to study the detailed mechanisms of the apoptotic process. In addition, the medical and veterinary importance of several viruses have stimulated research into the interaction between the virion and apoptosis. Lilesi (1997) highlighted the fact that such investigations have resulted in the presence of apparently opposing hypotheses based on whether the host utilizes apoptosis as an anti-viral defence mechanism or the virus modulates the immune response to maintain an infection. However, it is clear that both hypotheses may be correct, depending on the specific viral infection. Infection of vertebrate hosts with a virus initiates a series of physiological events that either leads to the production of an effective immune response or the establishment of suitable conditions that allow continued and extensive viral replication. The immune response generated involves both innate and specific components, both of which incorporate a cellular aspect. In innate immunity this cellular component includes macrophages, natural killer cells (NK) and granulocytes with associated immunoreactive chemicals such as complement and cytokines. The latter are also involved in the co-ordination of an effective specific response which incorporates T- and B-lymphocytes and the production of specific antibodies. The interdependence and complexity of the physiological and immunological responses to viral infection means that the viron has a wide range of possible avenues open to it to manipulate the viral–host interaction in the virus's favour. One such avenue, the effect of the virus pathogen on the apoptotic process in host cells, has been extensively studied in mammals. The interaction between the virus and the apoptotic pathway can be manifested with two diametrically opposing outcomes which share one aim, that of providing a favourable environment for viral replication and dissemination. The viral proteins can either have an anti-apoptotic role or a stimulatory function that is dependent upon the target cell. For example, studies on the measles virus by Auwaerter *et al.* (1996) have revealed that a depressed immune response was not mediated by viral infection of thymocytes but originated from infected myelomonocytic cells and epithelial cells in the thymus. Indeed, this indirect 'bystander' action of virus action on apoptosis has also been described in HIV infections (Finkel *et al.*, 1995; Su *et al.*, 1995). Viral induction and inhibition of apoptosis can be targeted at a range of key points within the apoptotic process. Several viruses are thought to mediate an effect on apoptosis by directly interacting with cell surface receptors and their associated signalling pathways. For example, attachment proteins of the reovirus σ1, which are also used to determine the specificity of the viral strain, induce apoptosis when attached to the cell surface of mouse cells (Tyler *et al.*, 1995). This was further substantiated by Ramsey-Ewing and Moss (1998) who utilized mutant viral constructs that were unable to perform the early stages of transcription but were still able to induce apoptosis in mammalian cells, i.e. ovary cells of Chinese hamsters. In addition, extensive studies have been carried out on the interaction of HIV and CD4+ immunocompetent cells (Gougeon and Montagnier, 1993) in an attempt to explain the selectivity of apoptosis-mediated cell death in bystander cells whilst infected cells are apparently resistant to apoptosis. Everett and McFadden (1999) proposed a model in which the gp120 viral protein binds to cell surface receptors which trigger apoptosis. In infected cells it was postulated that expression of viral genes, e.g. *tat*, may induce resistance. It would appear that cell receptor/virus induction of apoptosis is not limited to mammalian cells. Brojatsch *et al.* (1996) and Adkins *et al.* (1997) noted that avian leucosis/sarcoma retrovirus uses a member of the TNF-receptor superfamily to induce apoptosis in avian cells. The interaction with the cell death receptor (Fas/TNF) and associated pathways is however a complex issue since several viruses, e.g. adenovirus, Epstein–Barr virus, Molluscum contagiosum virus, are thought to mediate their anti-apoptotic properties partly through effects on this route (see Everett and McFadden, 1999).

The importance of the *bcl2* family of genes as a control point within the apoptotic pathway has meant that it is a prime target for manipulation by viral infections, e.g. adenovirus, African swine fever virus and human herpesvirus 8 which produce anti-apoptotic proteins. For example, the latent membrane protein (LMP)-1 of Epstein–Barr virus-infected B-lymphocytes protects the cells from apoptosis by inducing the expression of *bcl2* (Henderson *et al.*, 1991). However, it should be noted that viral proteins can have a range of effects on gene expression. The oncoprotein of human T cell leukaemia virus 1 termed 'Tax' can induce up-regulation of the apoptosis-suppressing gene A20 (Laherty *et al.*, 1993) and yet in Jurkat cells the protein induces apoptosis (Chilichlia *et al.*, 1995). The variation in this response is thought to be dependent on the duration of the viral protein–cell interaction. One of the roles of Bcl2 may be the regulation of the calcium pump in the endoplasmic reticulum (ER) (Kuo *et al.*, 1998). This is utilized by the virus to up-regulate protein synthesis and this significant increase in ER activity may induce apoptosis.

One group of host proteins is a particular target for viruses. Caspases are activated by a variety of 'death signals', pro-caspases being cleaved at specific aspartic acid residues resulting in the production of two subunits. These subunits, acting as a tetramer, carry out the protein cleavage, which produces the typical apoptotic morphological changes in target cells. Several viruses, e.g. African swine fever virus and poxvirus produce gene products that inhibit the function of caspases. For example in cowpox the product termed cytokine response modifier A (CRMA), which is a serpin-like molecule, inhibits several caspases, e.g. 3, 6 and 8 in mammals but, interestingly, it also blocks apoptosis caused by synthesis of reaper in *Drosophilia* (see Villa *et al.*, 1997).

4.2 *Viruses and invertebrates*

The fact that viruses can also affect the apoptotic process in lower animals is perhaps most elegantly shown in studies involving baculoviruses. These large DNA viruses occur widely in the environment and infect a range of arthropods, particularly members of the Phyla Uniramia and Crustacea. Their association with insect pests has meant there has been considerable interest in their relationship with apoptosis as a mechanism of biological control (Narayanan, 1998). In addition, since baculoviruses can only replicate in certain insect cells, yet appear to be able to infect any kind of cells, there has been considerable interest in their use as vectors in human gene therapy and as tools to study the apoptotic process in general. As in the studies on mammals highlighted above, this group of viruses is associated with both anti- and pro-apoptotic activity. To date two anti-apoptotic gene families, *p35* and *iap* (inhibitor of apoptosis protein), have been identified. The *p35* gene was first identified in 1987 by Friesen and Miller from *Autographa californica* multicapsid nucleopolyhedrovirus (ACMNV) and subsequently found to act as a direct inhibitor of caspases (Bump *et al.*, 1995). Xue and Horvitz (1995) also noted that p35 inhibited the cell death protease, Ced3, in the nematode *Caenorhabditis elegans*. In ACMNV-infected cells the p35 appeared after 8–12 hours post-infection and accumulated for up to 36 hours post-infection. Such p35 was located in the cytosol and early expression of the gene was required to inhibit virus-induced apoptosis (Hershberger *et al.*, 1994). Homologous genes to p35 have been located in *Bombyx mori* NPV and *Spodoptora littoralis* NPV and also prevent viral-induced apoptosis (Kamita *et al.*, 1993).

IAPs (inhibitors of apoptosis proteins; reviewed by Salvesen and Duckett, 2002) were initially described in baculoviruses but IAP homologues are now known from insects, birds and mammals. IAPs are metalloproteins which contain one or more zinc-binding motifs called the BIR (baculovirus IAP repeat) and the RING (Really Interesting New Gene) domain. The BIR areas have been shown to bind to a wide range of apoptotic proteins, e.g. *Drosophila* Reaper, Grim, Doom and Hid (Miller, 1999). In addition, it has been suggested that IAPs may also act by inhibiting caspase activity (Deveraux and Reed, 1999; Gogal *et al.*, 2000; Wang *et al.*, 1999).

In comparison to the above, our knowledge of virus–host interactions in other animals is somewhat limited. Studies on a range of lower animals have indicated that one important role of apoptosis in viral infections is to aid the virus during the replication and dissemination process. For example, recent histological, cytochemical and ultrastructural studies by Khanobdee *et al.* (2002) have revealed that infection of the giant tiger shrimp, *Penaeus monodon*, with yellow head virus is associated with apoptosis in the lymphoid organs. The authors suggest that this may be associated with mortality. In previous studies on *P. vannamei* juveniles chronically infected with Taura syndrome virus (TSV), apoptosis was detected in lymphoid organ spheroid cells.

The authors, Hasson *et al.* (1999), suggested three possible outcomes to an TSV infection: (1) the virus may continue to replicate in the cells; (2) it may be eliminated by the cells; or (3) a persistent infection may occur with a balance between the elimination and replication of the virus.

4.3 *Viruses and lower vertebrates*

Studies on the relationship between viruses and apoptosis in ectothermic animals have mainly concentrated on fish, particularly in association with viruses of economical and/or ecological importance. Infectious pancreatic necrosis virus (IPNV) is a member of the Birnaviridae, a family of viruses that includes *Drosophila* X virus and infectious bursa disease virus (IBDV) of domestic fowl. IPNV is the causative agent of the disease of the same name which is endemic in most parts of Northern America, Europe and the Far East. This lethal disease is particularly important in hatchery situations in the culture of both salmonids and some species of non-salmonid fish. The virus induces a pathological response and damage in the intestine and pancreas of the affected fish, and carrier fish release the virus into the water body via various secretions. Recent studies have suggested that apoptosis may play an important role within the virus–host interaction. Hong *et al.* (1999a) utilized green fluorescent protein from the Cnidarian, *Aequorea victoria*, to visualize cellular changes in Chinook salmon embryo cells (CHSE-214) infected with IPNV. The initial response of these cells, up to 6 hours post-infection, was to produce cellular changes similar to those associated with apoptosis (Hong *et al.*, 1998). However, in later stages of the infection (greater than 6 hours), non-typical apoptotic changes were noted. For example, formation of membrane blebs which were released from the cell membrane and holes, 0.39–0.78 μm in diameter occurred in the cell membrane. The apoptosis process was confirmed using electrophoretic analysis of DNA fragmentation. The authors suggested that apoptosis was eventually replaced by necrosis. Such cellular disruption may have assisted in viral dissemination. Subsequent studies by Hong *et al.* (1999b) revealed that the virus may mediate its effect by altering apoptotic gene expression. *Mcl-1*, a member of the *Bcl2* family of genes, which was first identified in human myeloid leukaemia cells, is thought to inhibit apoptosis induced by the overexpression of *c-myc*. The authors noted that *mcl-1* expression was reduced in cells infected with IPN during the first 8 hours of infection. These results contrast to studies in mammals where Epstein–Barr virus-transforming protein LMP1 induces a rapid but transient stimulation of *mcl-1* in B cells (Wang *et al.*, 1996). Recently Hong and Wu (2002) have extended these studies to other apoptotic genes and have revealed that expression of *bad*, a *bcl2*-related family member, which promotes apoptosis, is up-regulated in IPNV-infected cells. The use of tyrosine kinase inhibitors, e.g. genistein suggested that the viral effects may be mediated via a viral receptor which triggered *bad* expression via a tyrosine kinase pathway.

It is of interest that Essabauer and Ahne (2002) also noted that in an epithelioma carp papulosum cell line (EPC) infected with epizootic haematopoietic necrosis virus (EHNV), apoptosis may involve activation of a protein kinase-dependent pathway. This iridovirus, which is pathogenic in several fish species, induces apoptosis, as detected using annexin-V labelling, in infected cell lines 12 hours post-infection. Pycnotic nuclei occurred after 18 hours of infection. Previous studies by Bjorklund *et al.* (1997) have also revealed that when EPC cell lines were infected with the rhabdovirus spring viraemia in carp virus (SVCV), which causes a lethal disease associated with peritonitis and inflammation in carp and other fish species, apoptosis occurred, as detected using electron and confocal microscopy and DNA fragmentation. This apoptosis was correlated with an increase is virus titre and was inhibited using acid cysteine proteinase inhibitors. Chiou *et al.* (2000) noted that the rhabdovirus matrix protein from infectious haematopoietic necrosis virus (IHNV), which induces a pathogenic disease primarily in salmonid fish, is associated with a reduction in host transcription and induction of apoptosis in CHSE-214 and EPC cell lines.

4.4 *Conclusion*

As indicated above, there appears to be compelling evidence that, as in mammals, in virus-induced apoptosis in lower vertebrates, the virus can be the beneficiary of the programmed cell death which assists in viral replication and dissemination. Although the possible anti-viral properties of apoptosis in lower animals have been suspected, extensive evidence to support this hypothesis is surprisingly absent. The best evidence comes from studies on the interaction between baculoviruses and insects and in particular the use of viral mutants that lack the anti-apoptotic gene, *p35*. For example, p35 mutant *Autographa californica* multicapsid nucleopolyhedrovirus (ACMNV) allows apoptosis in SF-21 cells derived from *Spodoptera frugiperda* which subsequently drastically reduces the yield and infectivity of the virus (Clem and Miller, 1993; Hershberger *et al.*, 1992). Further studies carried out by Clem *et al.* (1994) in which recombinant virus carrying IAP was used instead of p35 and subsequently restored the infectivity, confirm the hypothesis that infectivity is related to apoptosis.

5 Association between bacterial infections and apoptosis

As with viral infections the importance of bacterial pathogens, particularly with several medically problematic diseases, has led to a considerable amount of interest in the role of apoptosis within the bacteria–host interaction. Several extensive reviews have been produced in recent years which eloquently describe in detail how bacteria and host might compete to utilize the apoptotic process to their own benefits (see Gao and Abu Kwaik, 2000a,b; Monack and Falkow, 2000). However it is worth briefly considering the basic principles involved. Many of these medically important species of bacteria belong to genera that have important counterparts that infect lower animals. Indeed some, such as *Aeromonas hydrophila*, because of their ubiquitous occurrence, in this case in freshwater particularly with a high organic content, can be problematic in a range of animals. These studies have recently stimulated interest in the role of apoptosis in bacterial infections of lower vertebrates and invertebrates.

Several medically important bacterial parasites, e.g. *Salmonella*, *Yersinia*, *Mycobacterium*, *Shigella*, *Neisseria* and *Staphylococcus* have adopted a range of strategies by which they manipulate the apoptotic mechanisms in their hosts. Such effects can either be direct or indirect, the latter usually involving the production of a bacterial factor or toxin. In addition, whether the bacteria occur intracellularly or extracellularly may affect the strategy adopted. The Gram-positive coccus, *Staphylococcus aureus* causes a number of diseases in humans including endocarditis, brain abscess, arthritis, gastroenteritis as well as being associated with wound infections. The bacterium infects the host cells and escapes into the cytoplasm from the phagosome utilizing the pore-forming staphylococcal α-toxin. Bayles and co-workers (Bayles *et al.*, 1998; Wesson *et al.*, 2000) have shown that this toxin induces apoptosis in epithelial cells mediated through activation of caspases 3 and 8. This association of a bacterium with the phagocytotic process and the phagolysosomes gives the parasite an opportunity to affect the normal killing mechanisms associated with the system. During the maturation of the phagosomes there is a change in markers which is prevented in *Salmonella* and Mycobacteria infections and may be associated with the apoptotic process. *Salmonella* also induces apoptosis in macrophages and although it might be assumed that such an effect may immunocomprise the host, Gao and Abu Kwaik (2000a) proposed an alternative hypothesis in which they suggested that apoptosis induction in macrophages may contribute to the anti-bacterial immune response in bystander dendritic cells. Such cells, it was proposed, would migrate to the lymphoid organs and stimulate T cells. In another bacterium that infects macrophages, *Legionella pneumophila*, the induction of apoptosis is associated with activation of caspase-3 and not caspase-1 (Gao and Abu Kwaik, 1999, 2000c). This bacterium, which occurs widely in water, can also invade and kill free-living amoebae, however death does not apparently occur via an apoptotic mechanism. This implies that the parasite–host interaction in terms of apoptosis may be unique to that individual association or at least to the phyla it involves.

Recent studies have highlighted that the association between apoptosis and the bacterium–host interaction is not limited to higher animals. It is interesting to speculate that studies primarily on human infections but involving bacterial species that have a ubiquitous occurrence may be extrapolated to lower animals. For example *Aeromonas hydrophila* is associated with gastrointestinal infections in humans and haemorrhagic septicaemia in fish (Hoole *et al.*, 2001; Roberts and Shepherd, 1986). Studies by Falcon *et al.* (2001) indicate that a cytotoxic entertoxin produced by the bacterial species induces apoptosis in human intestinal cells in culture and Chopra *et al.* (2000) proposed that an aerolysin-related cytotoxic entertoxin (Act) up-regulated production of Bcl-2 in macrophages. Several studies have been carried out on other bacterial infections in fish. For example, Huang *et al.* (2000) carried out *in vitro* and *in vivo* experiments on the potential of *Staphylococcus epidermidis* and its secretions produced in culture, to induce apoptosis in tilapia, *Oreochromis aureus*. Apoptosis was detected in lymphocytes and macrophages of the spleen and kidney and apoptotic figures were occasionally detected in the brain, liver, gonad, mesentery, stomach, intestine and skeletal muscle of infected fish. Other studies on tilapia by Evans and co-workers have concentrated on apoptosis in relationship to non-specific cytotoxic cells and infection with *Streptococcus iniae*. Flow cytometric analysis by Evans *et al.* (2000) revealed that non-specific cytotoxic cells from the spleen, pronephros and peripheral blood expressed cytosolic but not membrane Fas ligand. The expression of this and associated proteins, i.e. CAS and FADD were increased in non-specific cytotoxic cells obtained from the peripheral blood and exposed to fish serum, suggesting that cytokine-like factors increased the cytotoxicity of non-specific cytotoxic cells by stimulating proteins associated with apoptosis. Later studies by these workers (Taylor *et al.*, 2001) showed that different isolates of *S. iniae* regulated the anti-bacterial activity of non-specific cytotoxic cells by affecting levels of apoptosis.

The concept that in the host–parasite interaction the apoptotic process either benefits the parasite or the host, but not both, has recently been questioned by studies carried out on sepiolid squids. Studies by Foster and co-workers on the co-operative association between the luminous bacterium *Vibrio fischeri* and the epithelial cells of the light organs of Hawaiian squid *Euprymna scolopes* have revealed that the bacterium induces apoptosis in these cells (Foster and McFall-Ngai, 1998), possibly involving lipopolysaccharide (LPS) (Foster *et al.*, 2000), which is associated with essential remodelling of the host tissue. Apoptosis may thus be considered as an essential component in this relationship between the bacterium and the invertebrate.

6 Association between apoptosis and protozoan and metazoan infections

6.1 *Introduction*

The mechanisms by which viral and bacterial pathogens regulate the apoptotic pathway in both vertebrates and invertebrates have been well characterized and the possible biological benefit(s) either to the host or the pathogen extensively discussed. In contrast, the interaction between the apoptotic process and eucaryotic parasites has only recently received attention. Indeed, it is only in the last few years that the molecular components in this interaction have been studied. Such studies have mainly been directed towards an understanding of the association of apoptosis with parasites that are of medical importance and thus, by inference, infect mammals. However, the involvement of lower animals, e.g. insect vectors within the life cycle of several such parasites, e.g. *Plasmodium* sp., *Trypanosoma* sp., has recently focused the attention of biologists on the role and involvement of apoptosis in the parasite interaction in lower animals. In addition, the increasing importance of unicellular and multicellular parasites in the culture of lower vertebrates, e.g. fish, has added to the urgency to understand the complexity of the interaction between parasitization by eucaryotes and apoptosis. The introduction of a eucaryotic parasite, as opposed to the viruses and procaryotic bacteria, also introduces another dimension into the role of apoptosis in infections, that of apoptosis within the pathogen. The understanding of apoptosis in this parasite–host interaction not only has implications for our understanding of the pathology involved, which may lead to the development of new approaches to disease control, but also will contribute to the debate on the evolutionary origin and role of apoptosis in animals. This is perhaps exemplified by a consideration of cytochrome c which appears to serve two contrasting roles within a cell, i.e. that of being a crucial component of the respiratory chain and at the same time playing an important part in the apoptotic process. The answer to this apparently diverse role of cytochrome c probably lies in the proposed symbiotic origin of mitochondria (Blackstone and Green, 1999).

6.2 *Apoptosis and parasitic protozoa*

In 2001 several extensive reviews were published that considered the involvement of apoptosis in the protozoan–host interaction. All of these reviews (e.g. DosReis and Barcinski, 2001; Heussler *et al.*, 2001; Luder *et al.*, 2001), however, consider the extensive research carried out on the role of apoptosis in protozoan infections of mammals, particularly humans. In contrast, very little is known about how the apoptotic process relates to protozoan infections in lower animals. However, a consideration of the protozoan–mammalian interaction may be useful in determining the mechanisms by which the parasite manipulates the apoptotic process. Studies on apoptosis in relation to the interactions between mammals and their protozoan parasites have mainly been associated with intracellular pathogens of medical and veterinary importance. For example, *Plasmodium* sp., the causative organism of malaria, has a complex life cycle within its vertebrate host which involves intracellular parasitization of both liver parenchyma cells and erythrocytes; *Trypanosoma* sp. such as *T. cruzi*, which induces the disease termed 'Chagas' disease' in Latin America has an infective stage in the vertebrate blood which invades various host cells including immunocompetent cells, the macrophages. Indeed, *Theileria parva*, which is responsible for East Coast fever in cattle, also replicates within lymphocytes. *Toxoplasma gondii*, an intracellular parasite which infects a range of vertebrate definitive hosts including many mammalian and bird species is a ubiquitous opportunistic pathogen of particular importance in fetuses and immunosuppressed patients, as is *Cryptosporidium parvum* which causes diarrhoea.

Many of these parasites adopt several mechanisms to manipulate the apoptotic process as a means of reducing the immune response or creating an internal environment that increases their chances of survival. For example, Toure-Baldo *et al.* (1995, 1996) noted that *Plasmodium falciparum* induces apoptosis in various immune cells of its host and in chronic infections may alter lymphocyte physiology. One particular aspect of the apoptotic mechanism which appears to be a common target in all the above protozoan parasites is Fas and its associated ligand (FasL). For example, Matsumoto *et al.* (2000) observed that in Macaca monkeys infected with *P. coatneyi* that apoptosis occurred in immunocompetent cells and was associated with an increase in soluble FasL in serum. Liesenfeld *et al.* (1997) also noted that a similar relationship between expression of Fas/FasL and apoptosis in Peyer's patches occurred in *Toxoplasma* infections, an observation that was supported by the studies of Hu *et al.* (1999) who correlated pathology and apoptosis in *Toxoplasma*-infected eyes with Fas/FasL expression in the retina.

Protozoan parasites of mammals are also known to inhibit apoptosis and utilise a number of target sites within the apoptotic pathway. Nuclear factor kappa B (NFκB), a transcription factor that regulates genes encoding apoptosis inhibitor proteins, is activated by *Theileria parva* and thus plays an important role in protecting infected T cells against apoptosis (Heussler *et al.*, 1999). Recent studies by Lau *et al.* (2001) noted that the parasite *Plasmodium yoelii* induced an increase in expression of glutamylcysteine synthase which is known to regulate activation of NFκB (Manna *et al.*, 1999). The crucial role that caspases play in the apoptotic process also means that this group of enzymes is a prime target for manipulation by protozoan parasites. Studies by Ojcius *et al.* (1999) revealed that *Cryptosporidium parvum* induced apoptosis in intestinal cell lines and this could be inhibited using caspase-inhibitors. The authors suggested that the parasite may induce apoptosis which aids its exit from the infected cell or the apoptotic process may lead to parasite liberation. In contrast, Nakajima-Shimada *et al.* (2000) noted that in HeLa cells infected with *Trypanosoma cruzi* there was a reduction in caspase 8 which they suggested was associated with inhibition of the earliest steps of death receptor-mediated apoptosis. In addition, Heussler *et al.* (2001)

reported work carried out by Zeynalzedegan that indicated that *Toxoplasma gondii* also inhibited the activation of caspase 3. Several workers have noted that the evolutionary conserved family of heat shock proteins (HSP) can inhibit the apoptotic pathway either by affecting caspase activity and/or the release of cytochrome c (Mosser *et al.*, 2000). Indeed, Heussler *et al.* (2001) suggested that HSPs may have co-evolved with apoptosis in metazoans. Recent studies (Ishikawa *et al.*, 2000; Sakai *et al.*, 1999; Zhang *et al.*, 1999) have also suggested that parasitic protozoa may induce HSP expression which leads to a reduction in the apoptotic process and that this may be correlated with parasitic virulence of *Plasmodium yoelii*, *Leishmania major* and *Trypanosoma cruzi* respectively.

The involvement of a vector in the life cycle of these medically and veterinary important parasitic protozoa has recently led to considerable interest in the host–parasite interactions in their invertebrate host. Much of this study has concentrated on the interaction between *Plasmodium* and its mosquito vector. The anopheline mosquito, when feeding on a malaria-infected host, acquires gametocytes of the *Plasmodium* parasite with the blood meal. Gametes produced are fertilized in the blood bolus within the insect midgut and the ookinetes, which develop from the zygotes, penetrate the midgut wall. The ookinete transforms into an oocyst below the basal lamina of the gut and forms sporozoites, which invade the salivary glands and pass into the mammalian host when the mosquito next feeds. The invasion of the insect midgut by the ookinete has been the focus of recent research. Previous studies by Shahabuddin and Pimenta (1998) revealed that when chicken malaria *P. gallinaceum* infects *Aedes aegypti* the parasite invades special epithelial cells termed 'Ross cells' which have high levels of vesicular ATPase. This specificity would imply the existence of some parasite recognition mechanism. Recently however, Han *et al.* (2000) have proposed an additional strategy for ookinete invasion of the midgut cells which incorporates apoptosis. The theory called 'the time bomb theory' has been based on studies utilizing *Plasmodium berghei* and *Anopheles stephensi*. In this host–parasite interaction the ookinete does not appear to exhibit the specificity seen in *P. gallinaceum* and invades normal midgut epithelial cells which do not have high levels of vesicular ATPase. However, the invasion process is associated with high levels of nitric oxide synthase, which result in genomic DNA fragmentation and cell death and thus trigger the time bomb. The parasite may have, however, adopted a survival strategy since the ookinete is capable of extensive lateral movement which may remove it from the source of the toxic chemicals. The authors propose that this survival technique is very efficient, as 95% of the parasitic invasion survive, associated with areas of damage cells. Clearly the role of apoptosis in this theory requires further elucidation. Recently however, there have been investigations which link *Plasmodium* infections in anopheles mosquitos with apoptosis in the reproductive tract of the vector. Studies on oogenesis in *Drosophila melanogaster* have revealed that apoptosis occurs in nurse cells in the reproductive follicles (Foley and Cooley, 1998) and that expression of *reaper* and *hid* genes occurs in follicle cells (Cho and Nagoshi, 1999). Investigations carried out by Hopwood *et al.* (2001) have revealed a significant correlation between follicle resorption and apoptosis in the follicle cells of *Anopheles stephensi* infected with *Plasmodium yoelli nigeriensis*. This follicle resorption occurs when the ookinetes are invading the midgut and involves caspases as inhibition of these enzymes with z-VAD.fmk (benzyloxycarbonyl-Val-Ala-Asp (Ome) fluoromethylketone) stops resorption. These effects, which appear to be mediated through apoptosis, have interesting implications in the host–parasite interaction as reduction in fecundity appears to be a common phenomenon in parasitized molluscs and arthropods (see Hurd, 1990 for review). This may be beneficial to the parasite as resources that would have been utilized in egg production in the host are diverted away from this function and may be utilized to assist in parasite or host survival. Koella (1999) highlighted that this may be a common strategy adopted by parasites which exploit the trade-off between longevity of the host and its reproduction.

6.3 *Apoptosis and parasitic metazoa*

The role of apoptosis in the parasite-induced alteration of the host internal environment which may result in increased survival of the infection has also been studied in metazoan parasites that infect mammals. Several studies on the blood fluke *Schistosoma mansoni* have revealed an intricate relationship between the pathology induced, particularly by the parasite eggs, and apoptosis in immunocompetent cells. Rumbley *et al.* (1998) noted an increase in apoptosis in granuloma T cells but not splenic T cells suggesting that there were regional/organ differences in modulation of the T cell repertoire by the parasite. Fallon *et al.* (1998) noted that an increase in apoptosis in CD4+ and CD8+ splenic T cells may be correlated to the onset of egg laying by the parasite and a Th-2 type response. Such studies indicate that apoptosis may play an important role in the chronic response to schistosomiasis. This differential effect on a leucocyte type also occurs in other parasitic infections, for example, rats infected with the nematode *Trichinella spiralis*. Gon *et al.* (1997) noted that eosinophils from infected animals were more resistant to apoptosis than those from non-infected controls. In addition, an increase in eosinophils was noted in the intestine at the time the worms had begun to migrate. The association between apoptosis in intestinal cells and parasitic gut infections has also been observed in other host–parasite interactions. Hyoh *et al.* (1999) observed that acute infection of rats with the nematode *Nippostrongylus brasiliensis* induced apoptosis in the epithelial cells of the villi of the small intestine. Starke and Oaks (1999) also observed that when the tapeworm, *Hymenolepis diminuta*, was removed from rats using Praziquantel, the mucosal mast cell numbers approached those levels found in non-infected control animals after 4 weeks post-drug treatment. This coincided with a decline in the apoptosis in this cell type. The authors suggested that apoptosis may play an important role in the involution of parasite-induced mastocytosis.

In contrast to the above, there have been limited studies that have investigated the relationship between metazoan parasites and apoptosis in lower vertebrates. Morley and Hoole (1997) observed that homogenates and excretory products of the tapeworm *Khawia sinensis* induced suppression of blastogenesis in the pronephric, splenic and thymic lymphocytes of carp (*Cyprinus carpio*). Attempts to correlate this reduction in leucocyte activity with parasite-induced *in vitro* apoptosis produced inconclusive results. This supported unpublished observations by these workers which indicated that there was no significant difference in apoptosis in the pronephric and splenic leucocytes of carp infected with *K. sinensis* for 6 weeks. In contrast, unpublished studies by the authors in association with ML Roberts (Roberts, 1997) have revealed that in carp infected with 500 cercariae of the blood fluke, *Sanguinicola inermis*, a significant increase in apoptosis ($P < 0.05$) occurred in pronephric cells 32 days after infection. It may be, therefore, that the apoptotic response in the host not only depends on the parasite species but also its location and that of the organ affected. Indeed, Griffith *et al.* (1996) noted that in immune privileged sites such as the eye, Fas-mediated apoptosis in lymphocytes prevented a local inflammatory response and also induced a systemic immuno-tolerance to the antigen which had been applied to the eye. This has interesting implications when considering those parasitic infections of lower vertebrates, for example *Diplostomum* spp. and *Tyrodelphys* spp., that infect the eyes of a wide range of fish species. Recent studies by Jung and co-workers (Jung *et al.*, 2000; Murakawa *et al.*, 2001) have revealed that infections of the marine fish, Chub mackerel, with the larval stage of *Anisakis simplex* induces the production of a substance that can induce apoptosis in a dose-dependent manner in mammalian tumour cells. This substance was found to be a flavoprotein and was considered to be a novel reticulo-plasmin. It was suggested that apoptosis could be induced by two mechanisms, both of which utilized caspase 9 cleavage and cytochrome c viz. a rapid system that was mediated through hydrogen peroxide and a relatively delayed mechanism mediated via L-lysine deprivation. The authors also detected the apoptosis-inducing protein in the capsules that formed around the larval nematode and suggested that this protein may function to prevent parasite migration to the host tissue and thus impede infection.

6.4 *Apoptosis within the parasite*

One interesting aspect when considering the relationship between apoptosis and the host–eucaryotic parasite interaction is apoptosis occurring within the parasite. It is feasible that death of parasites through an apoptotic mechanism can be mediated either through adverse effects that occur in the host and/or may be self-inflicted by the parasite as part of its normal development. The latter has important implications when considering the evolutionary origins of apoptosis. Apoptosis has a clear developmental and physiological function in multicellular animals where it serves to remove unwanted cells. Intuitively it might be considered that controlled cell death would be of little functional value to a unicellular animal such as a parasitic protozoan. However, this assumes the individuality of the single protozoan is sacrosanct and this is not the case since many parasitic protozoa include a clonal expansion stage, e.g. the oocyst in malaria (Tibayrenc *et al.*, 1990). Over recent years there have been several studies that have indicated that protozoan parasites in particular, e.g. *Trypanosoma cruzi, Trypanosoma brucei, Plasmodium falciparum* and *Leishmania* undergo apoptosis (see Barcinski and DosReis, 1999; DosReis and Barcinski, 2001; Luder *et al.* 2001 for a review). In some of these parasites, e.g. leishmania, apoptotic death is thought to be Ca^{2+}-dependent and involves the mitochondrial fraction (see DosReis and Barcinski, 2001) and recent studies by Arnoult *et al.* (2002) have revealed that in *L. major* apoptosis involves cysteine proteinase activation and the release of cytochrome c from the mitochondria. Several recent studies have considered parasite apoptosis in an insect vector. For example Al-Olayan *et al.* (2002) noted that apoptosis occurred in the ookinetes of *Plasmodium berghei* and that more than 50% of this mosquito midgut stage die by controlled cell death prior to the parasite invasion of the gut wall. When parasite apoptosis was inhibited using the caspase-inhibitor Z-VAD.fmk, the infection intensity was doubled.

Barciniski and DosReis (1999) speculated on the functional significance of apoptosis within parasitic protozoans. Briefly the following was proposed:

1. Apoptosis may regulate parasite numbers to the mutual benefit of the parasite population and their host. For example, in *Trypanosoma brucei* in tsetse fly, parasite numbers remain constant, which may allow the more efficient use of proline used by parasite and host as an energy source.
2. Apoptotic parasites may selectively interfere with the host immune response. In mammalian hosts the antigens produced by apoptotic protozoan parasites may induce a CD4+ Th-2 cell response, a parasite-protective response in the host.

Such proposals may also be applicable when considering the role of apoptosis in metazoan parasites although studies on the functional role of apoptosis in the embryogenesis and development in more complex parasites is lacking. It is proposed however that apoptosis is likely to play a role in such parasites since, on an evolutionary scale, apoptosis has been implicated in the control of cell numbers in diploblastic animals. For example, Cikala *et al.* (1999) described apoptosis in the Cnidarian, *Hydra vulgaris*. When the polyps are starved, subjected to temperature shock or treated with colchicine, cell death morphologically indistinguishable from apoptosis occurred. In addition, cell death was associated with two genes showing a strong homology to the caspase 3 family.

As indicated above, apoptosis in parasites may be induced as part of the natural anti-parasitic killing mechanism of the host. Jaso-Friedmann *et al.* (2000) investigated apoptosis in *Tetrahymena* spp. This genus of protozoa has certain representatives that are opportunistic pathogens in teleosts (e.g. Hoffman *et al.*, 1975). In addition, *Tetrahymena thermophilia* and *T. pyriformis* have been used in vaccine development against the ciliated protozoan pathogen *Ichthyophthirius multifiliis* the causative organism of the disease 'Ich' or white spot in many fish species (Clark *et al.*, 1995; Gaertig *et al.*, 1999). The *Tetrahymena* genus is also a model system in the study of the phylogenetic development of signalling proteins. Studies by Graves *et al.* (1985) have previously shown that non-specific cytotoxic cells in fish were involved in the lysis of parasitic protozoans. In the studies of Jaso-Friedmann *et al.* (2000) it was revealed that *Tetrahymena* spp. express the Fas receptor on their membrane and that the Fas ligand–receptor interaction in teleosts may be important in the anti-parasitic innate immunity mediated by apoptosis induced by non-specific cytotoxic cells. This host-induced apoptosis in parasites may not be restricted to vertebrate hosts. Hagen *et al.* (1998) noted that death of the microfilarial stage of the nematode *Onchocerca ochengi* in its blackfly host *Simulium damnosum* was mediated through apoptosis. Survival of the parasite was significantly increased by the injection of the caspase inhibitor boc-D.fmk, and serine protease inhibitors.

7 Apoptosis and stress

The details given above indicate the possible direct interaction between infection with viral, bacteria, protozoan and metazoan parasites and apoptosis either within the invading organism or the infected host. It should not be forgotten however that infection can also be considered as a stressor and thus can affect apoptosis via an indirect route mediated through the stress response. As in mammals this response in lower vertebrates involves hormonal, physiological and behavioural changes. The hormonal changes, which are usually indicative of stress, are associated with increases in catecholamines and corticosteroids (Wendelaar Bonga, 1997). Several studies in a range of vertebrates have indicated that parasites are associated with the induction of stress in the host as manifested in an association with corticosteroids. For example, Fleming (1997, 1998) noted that in lambs and dairy cows, *Bos taurus*, inoculation with the nematodes *Haemonchus contortus* and *Ostertagia ostertagi* respectively resulted in a significant increase in cortisol levels compared to uninfected controls. The author suggested that in the *Ostertagia ostertagi–Bos taurus* interaction the increase in cortisol levels noted were correlated with the emergence of the young adult parasites from the gastric glands. Although Grutter and Pankhurst (2000) noted that infection of the coral reef fish, *Hemigymnus melapterus* with gnathiid isopods did not result in an increase in host plasma cortisol levels, we still propose that the parasite might induce apoptosis in lower vertebrates via an indirect route mediated through corticosteroids, as several workers have noted an association between parasitism and cortisol levels in lower vertebrates. For example in fish, Sures *at al.* (2001) noted that infection of the European eel, *Anguilla anguilla*, with the nematode *Anguillicola crassus* induced an increase in plasma cortisol levels which was associated with the time of larval development and the appearance of the adult parasites. Further studies carried out on ectoparasites of fish have also revealed that that parasitization was associated with an increase in cortisol levels, e.g. *Gyrodactylus derjavini* infections in rainbow trout (*Oncorhynchus mykiss*) (Stoltze and Buchmann, 2001) and *Lepeophtheirus salmonis* infections of wild sea trout (*Salmo trutta*) (Poole *et al.*, 2000). Administration of corticosteroids may also affect parasite burden. Harris *et al.* (2000) noted that implantation of hydrocortisone acetate into Arctic charr (*Salvelinus alpinus*) and brook trout (*S. frontalis*) significantly increased the population of the monogenean, *Gyrodactylus salaris* although the response of Arctic charr was affected by the differences in resistance between individual fish. Wendelaar Bonga and co-workers (Nolan *et al.*, 2000b; Ruane *et al.*, 1999) have also noted a relationship between the administration of cortisol in food fed to rainbow trout and the reduced establishment of the branchurian ectoparasitic crustacean *Argulus foliaceus*.

Additional support for our proposal that parasites might indirectly affect apoptosis in the host via a stress response can be obtained from the numerous studies that have related apoptosis in immune cells of higher vertebrates to levels of corticosteroids (e.g. Ayala *et al.*, 1995). Indeed, over recent years there have been several investigations in fish that have shown that corticosteroids may induce apoptosis in leucocytes. Initial studies carried out by Alford *et al.* (1994) revealed that confinement-induced stress reduced the level of apoptosis in the peripheral leucocytes of channel catfish, *Ictalurus punctatus*, although the role played by cortisol was unclear. However, Weyts *et al.* (1997) demonstrated for the first time that apoptosis was an immune regulatory mechanism in fish and that the mechanism was subject to control via glucocorticosteroids. Using *Cyprinus carpio* subjected to handling stress it was ascertained that *in vitro* cortisol inhibits proliferation of peripheral blood lymphocytes. In unstimulated cells apoptosis was induced due to the absence of stimulating signals or 'neglect' as the authors referred to this occurrence whilst mitogen (LPS or PHA)-stimulated lymphocytes were sensitive to cortisol-induced apoptosis. Verburg-van Kemenade *et al.* (1999)

observed that the B-lymphocytes of *C. carpio* also differed in their sensitivity to cortisol-induced apoptosis depending upon their organ of origin. Several studies have also indicted that corticosteroid analogues such as dexamethasone can induce apoptosis in the immune cells of fish (Evans *et al.*, 2001; Hoole *et al.*, 2003; Walsh *et al.*, 2002). An essential component in understanding the relationship between cortisol and apoptosis is the interaction and distribution of glucocorticoid receptors. Studies in fish have revealed that cortisol or its analogues down-regulate the glucocorticoid receptors in the liver (Lee *et al.*, 1992 Pottinger, 1990), brain (Lee *et al.*, 1992), gill (Shrimpton and Randall, 1994) and erythrocytes (Pottinger and Brierley, 1997), whilst *in vivo* treatment of coho salmon with cortisol increased the numbers of glucocorticoid receptors in splenic and pronephric leucocytes. Weyts *et al.* (1998) suggested that the binding of cortisol to the peripheral blood leucocytes of *Cyprinus carpio* exhibited saturation kinetics.

Unfortunately, although a clear route by which parasitic infection could induce apoptosis via a stress response is obvious, studies that directly correlated infection with cortisol biochemistry and the induction of apoptosis in immune cells are very limited. It is only recently that studies carried out by Saeij *et al.* (2003) have attempted to ascertain the link between stress induced by handling of carp and their increased susceptibility to the protozoan parasite, *Trypanoplasma borreli*. Cortisol suppressed *T. borreli*-induced expression of a range of immune factors such as interleukin-1β, serum amyloid A, tumour necrosis factor-α and inducible nitric oxide synthase. In addition, cortisol induced apoptosis in activated peripheral blood leucocytes which was manifested in a disruption in mitochondrial transmembrane potential and a reduction in the levels of glutathione.

8 Summary

It is clear that the roles of apoptosis in the interactions between the parasite and their non-mammalian hosts are multifaceted and highly dependent on individual associations between the two organisms involved. Whilst there are instances where both organisms appear to gain from the apoptotic mechanism induced, in the majority of cases apoptosis appears to favour only one of the parties. In the instances when the parasite benefits, the apoptosis has been related to infectivity and virulence, an interruption of the killing mechanism of the host, and liberation of the pathogen. However, there are occasions where the apoptotic process benefits the host, as controlled cell death has been associated with limiting the pathogen population, parasite migration within the host and, in some instances, actually killing the invading organism. Apoptosis thus appears to play several fundamental roles within the host–parasite relationship which is ultimately reflected in an effect on the host population either mediated through an alteration in host fecundity or reduction in host numbers. The next decade promises to be both exciting and productive with respect to our knowledge of the relationship between apoptosis in non-mammalian animals and infection. Over the last few years the information obtained from studies on the apoptotic process in mammals and invertebrates (i.e. *C. elegans* and *Drosophila*) have been effectively used to increase our understanding of the apoptotic process in other animals such as insects, fish and amphibians. Such knowledge has paved the way for extensive studies on the effect of infections to be carried out.

Acknowledgements

This work was supported in part by the European Community's Improving Human Potential Programme under contract (HPRN-CT-2001-00214), (PARITY).

References

Abrams, J.M. (1999) An emerging blueprint for apoptosis in Drosophila. *Trends in Cell Biol.* **9:** 435-440.

Adams, J.M. and Cory, S. (2001) Life-or-death decisions by the Bcl-2 protein family. *Trends in Biochem. Sci.* **26:** 61-66.

Adkins, H.B., Brojatsch, J., Naughton, J., Rolls, M.M., Pesola, J.M. and Yong, J.A.T. (1997) Identification of a cellular receptor for subgroup E avian leukosis virus. *Proc. Natl Acad. Sci. USA* **94:** 11617–11622.

Alford, P.B., Tomasso, J.R., Bodine, A.B. and Kendall, C. (1994) Apoptotic death of peripheral leukocytes in channel catfish: effects of confinement-induced stress. *J. Aqua. Animal Health* **6:** 64–69.

Alnemri, E.S., Livingston, D.J., Nicholson, D.W., Salvesen, G., Thornberry, N.A., Wong, W.W. and Yuan, J.Y. (1996) Human ICE/CED-3 protease nomenclature. *Cell* **87:** 171–173

Al-Olayan, E.B., Williams, G.T. and Hurd, H. (2002) Apoptosis in the malaria protozoan, *Plasmodium berghei*: a possible mechanism for limiting intensity of infection in mosquito. *Inter. J. Parasitol.* **32:** 1133–1143.

Arnoult, D., Akarid, K., Grodet, A., Petit, P.X., Estaquier, J. and Ameisen, J.C. (2002) On the evolution of programmed cell death: apoptosis of the unicellular eukaryote *Leishmania major* involves cysteine proteinase activation and mitochondrion permeabilization. *Cell Death Different.* **9:** 65–81.

Auwaerter, P.G., Kaneshima, H., McCune, J.M., Wiegand, D. and Griffin, D. E. (1996) Measles virus infection of thymic epithelium in the SCID-hu mouse leads to thymocyte apoptosis. *J. Virol.* **70:** 3734–3740.

Ayala, A., Herdon, C.D., Lehman, D.L., Demaso, C.M., Ayala, C.A. and Chaudry, I.H. (1995) The induction of accelerated thymic programmed cell-death during polymicrobial sepsis-control by corticosteroids but not tumor-necrosis-factor. *Shock* **3:** 259–267.

Barcinski, M.A. and DosReis, G.A. (1999) Apoptosis in parasites and parasite-induced apoptosis in the host immune system: a new approach to parasitic diseases. *Brazil. J. Med. Biol. Res.* **32:** 395–401.

Barde, Y.A. (1989) Trophic factors and neuronal survival. *Neuron* **2:** 1525-1534.

Bayles, K.W., Wesson, C.A., Liou, L.E., Fox, L.K., Bohach, G.A. and Trumble W.R. (1998) Intracellular *Staphylococcus aureus* escapes the endosome and induces apoptosis in epithelial cells. *Infect. Immun.* **66:** 336–342.

Bishop, G.R., Taylor, S., Jaso-Friedmann, L. and Evans, D.L. (2002) Mechanisms of nonspecific cytotoxic cell regulation of apoptosis: cytokine-like activity of Fas ligand. *Fish Shellfish Immunol.* **13:** 47–67.

Bjorklund, H.V., Johansson, T.R. and Rinne, A. (1997) Rhabdovirus-induced apoptosis in a fish cell line is inhibited by a human endogenous acid cysteine proteinase inhibitor. *J. Virology* **71:** 5658–5662.

Blackstone, N.W. and Green, D.G. (1999) The evolution of a mechanism of cell suicide. *BioEssays* **21** : 84-88.

Brojatsch, J., Naughton, J., Rolls, M.M., Zingler, K. and Young, A.T. (1996) CAR1, a TNFR-related protein, is a cellular receptor for cytopathic avian leukosis-sarcoma viruses and mediates apoptosis. *Cell* **87:** 845–855.

Bump, N.J., Hackett, M., Hugunin, M., *et al.* (1995) Inhibition of ICE family proteases by baculovirus antiapoptotic protein p35. *Science* **269:** 1885–1888.

Caputo, V., Candi, G., Arneri, E., La Mesa, M., Cinti, C., Provinciali, M., Nisi Cerioni, P. and Gregorini, A. (2002) Short lifespan and apoptosis in *Aphia minuta*. *J. Fish Biol.* **60:** 775–779.

Chenlevy, Z., Nourse, J. and Cleary, M.L. (1989) The *bcl-2* candidate proto-oncogene product is a 24-kilodalton integral-membrane protein highly expressed in lymphoid-cell lines and lymphomas carrying the t(14,18) translocation. *Mol. Cell. Biol.* **9:** 701-710.

Chilichilia, K., Moldenhauer, G., Daniel, P.T., Busslinger, M., Gazzolo, L., Schirrmacher, V. and Khazaie, K. (1995) Immediate effects of reversible HTLV-1 tax function: T cell activation and apoptosis. *Oncogene* **10:** 269–277.

Chiou, P.P., Kim, C.H., Ormonde, P. and Leong J. C. (2000) Infectious hematopoietic necrosis virus matrix protein inhibits host-directed gene expression and induces morphological changes of apoptosis in cell cultures. *J. Virol.* **74:** 7619–7627.

Cho, S-H. and Nagoshi, R.N. (1999) Induction of apoptosis in the germline and follicle layer of *Drosophila* egg chambers. *Mech. Develop.* **88:** 159–172.

Chopra, A.K., Xu, X., Ribardo, D., Gonzalez, M., Kuhl, K., Peterson, J.W. and Houston, C.W. (2000) The cytotoxic entertoxin of *Aeromonas hydrophila* induces proinflammatory cytokine production and activates arachidonic acid metabolism in macrophages. *Infect. Immun.* **68:** 2808–2818.

Cikula, M., Wilm, B., Hobmayer, E., Bottger, A. and David, C.N. (1999) Identification of caspases and apoptosis in the simple metazoan *Hydra*. *Current Biol.* **9:** 959–962.

Clark, T.G., Lin, T.L. and Dickerson, H.W. (1995) Surface immobilization antigens of *Ichthyophthirius multifiliis*: their role in protective immunity. *Ann. Rev. Fish Dis.* **5:** 113–131.

Clem, R.J. and Miller, L.K. (1993) Apoptosis reduces both the *in vitro* replication and the *in vivo* infectivity of a baculovirus. *J. Virol.* **67:** 3730–3738.

Clem, R.J., Robson, M. and Miller, L.K. (1994) Influence of infection route on the infectivity of baculovirus mutants lacking the apoptosis-inhibiting gene p35 and the adjacent gene p94. *J. Virol.* **68:** 6759–6762.

Cuconati, A. and White, E. (2002) Viral homologs of BCL-2: role of apoptosis in the regulation of virus infection. *Genes Develop.* **16:** 2465-2478.

Deveraux, Q.L. and Reed, J.C. (1999) IAP family proteins – suppressors of apoptosis. *Genes Develop.* **13:** 589–597.

DosReis, G.A. and Barcinski, M.A. (2001) Apoptosis and parasitism: from the parasite to the host immune response. *Advances Parasitol.* **49:** 133–161.

Ducoroy, P., Lesourd, M., Padros, M.R. and Tournefier, A. (1999) Natural and induced apoptosis during lymphocyte development in the axolotl. *Develop. Comp. Immunol.* **23:** 241–252.

Ellis, R.E., Yuan, J.Y. and Horvitz, H.R. (1991) Mechanisms and functions of cell-death. *Ann. Rev. Cell Biol.* **7:** 663-698.

Enari, M., Sakahira, H., Yokoyama, H., Okawa, K., Iwamatsu, A. and Nagata, S. (1998) A caspase-activated DNase that degrades DNA during apoptosis, and its inhibitor ICAD. *Nature* **391:** 43-50.

Espelid, S., Lokken, G.B., Sterio, K. and Bogwald, J. (1996) Effects of cortisol and stress on the immune system in Atlantic salmon (*Salmo salar* L.). *Fish Shellfish Immunol.* **6:** 95–110.

Essabauer, S. and Ahne, W. (2002) The epizootic haematopoietic necrosis virus (Iridoviridae) induces apoptosis *in vitro*. *J. Vet. Med. B Infect. Dis. Vet. Public Health* **49:** 25–30.

Evans, D.L., Taylor, S.L., Leary, J.H., Bishop, G.R., Eldar, A. and Jaso-Friedmann, L. (2000) *In vivo* activation of *Tilapia* nonspecific cytotoxic cells by *Streptococcus iniae* and amplification with apoptosis regulatory factor(s). *Fish Shellfish Immunol.* **10:** 419–434.

Evans, D.L., Leary, J.H. and Jaso-Friedmann, L. (2001) Nonspecific cytotoxic cells and innate immunity: regulation by programmed cell death. *Develop. Comp. Immunol.* **25:** 791–805.

Everett, H. and McFadden, G. (1999) Apoptosis: an innate immune response to virus infection. *Trends Microbiol.* **7:** 160–165.

Falcon, R.M., Carvalho, H.F., Joazeiro, P.P., Gatti, M.S. and Yano, Y. (2001) Induction of apoptosis in HT29 human intestinal epithelial cells by cytotoxic enterotoxin of *Aeromonas hydrophilia*. *Biochem. Cell Biol.* **79:** 525–531.

Fallon, P.G., Smith, P. and Dunne, D.W. (1998) Type 1 and type 2 cytokine-producing mouse CD^{4+} and CD^{8+} T cells in acute *Schistosoma mansoni* infection. *Euro. J. Immunol.* **28:** 1408–1416.

Fischer, W.J. and Dietrich, D.R. (2000) Pathological and biochemical characterization of microcystin-induced hepatopancreas and kidney damage in carp (*Cyprinus carpio*). *Toxicol. Appl. Pharmacol.* **164:** 73–81.

Fleming, M.W. (1997) Cortisol as an indicator of severity of parasitic infections of *Haemonchus contortus* in lambs (*Ovis aries*). *Comp. Biochem. Physiol. B Biochem. Mol. Biol.* **116:** 41–44.

Fleming, M.W. (1998) Experimental inoculations with *Ostertagia ostertagi* or exposure to artificial illumination alter peripheral cortisol in dairy calves (*Bos taurus*). *Comp. Biochem. Physiol. A Mol. Integ. Physiol.* **119:** 315–319.

Foley, K. and Cooley, L. (1998) Apoptosis in late stage *Drosophila* nurse cells does not require genes within the H99 deficiency. *Development* **125:** 1025–1082.

Foster, J.S., Apicella, M.A. and McFall-Ngai, M.J. (2000) *Vibrio fischeri* lipopolysaccharide induces developmental apoptosis, but not complete morphogenesis, of *Euprymna scolopes* symbiotic light organ. *Develop. Biol.* **226:** 242–254.

Foster, J.S. and McFull-Ngai, M.J. (1998) Induction of apoptosis by cooperative bacteria in the morphogenesis of host epithelial tissues. *Develop. Genes Evol.* **208:** 295–303.

Friesen, P.D. and Miller, L.K. (1987) Divergent transcription of early 35 and 94-kilodalton protein genes encoded by the Hind-III K genome fragment of the baculovirus *Autographa californica* nuclear polyhedrosis virus. *J. Virol.* **61:** 2264–2272.

Funkel, T.H., Tudor-Williams, G., Banda, N.K., Cotton, M.F., Curiel, T., Monks, C., Baba, T.W., Ruprecht, R.M. and Kupter, A. (1995) Apoptosis occurs predominately in bystander cells and not in productively infected cells of HIV- and SIV-infected lymph nodes. *Nat. Med.* **1:** 129–134.

Gaertig, J., Gao, Y., Tishgarten, T., Clark, T.G. and Dickerson, H.W. (1999) Surface display of a parasite antigen in the ciliate *Tetrahymena thermophila*. *Nature Biotechnol.* **17:** 462–465.

Gao, L-Y. and Abu Kwaik, Y. (1999) Activation of caspase-3 during *Legionella pneumophilia*-induced apoptosis. *Infection Immunity* **67:** 4738–4746.

Gao, L-Y. and Abu Kwaik, Y. (2000a) Hijacking of apoptotic pathways by bacterial pathogens. *Microbes Infect.* **2:** 1705–1719.

Gao, L-Y. and Abu Kwaik, Y. (2000b) The modulation of host cell apoptosis by intracellular bacterial pathogens. *Trends Microbiol.* **8:** 306–313.

Gao, L-Y. and Abu Kwaik, Y. (2000c) The mechanism of killing and existing the protozoan host *Acanthamoeba polyphaga* by *Legionella pneumophila*. *Environ. Microbiol.* 2: 79–90.

Gogal, L., McCall, K., Agapite, J., Hartwieg, E. and Steller, H. (2000) Induction of apoptosis by *Drosophila* reaper, hid, and grim through inhibition of IAP function. *EMBO J.* **19:** 589–597.

Gogal, R.M., Smith, B.J., Robertson, J.L., Smith, S.A. and Holladay, S.D. (1999) Tilapia (*Orecochromis niloticus*) dosed with azathioprine display immune effects similar to those seen in mammals, including apoptosis. *Vet. Immunol. Immunopathol.* **68:** 211–229.

Gon, S, Saito, S., Takeda, Y., Miyata, H., Takatsu, K. and Sendo, F. (1997) Apoptosis and *in vivo* distribution and clearance of eosinophils in normal and *Trichinella*-infected rats. *J. Leukocyte Biol.* **62:** 309–317.

Gougeon, M-L. and Montagnier, L. (1993) Apoptosis in AIDS. *Science* **260:** 1269–1270.

Grant, P., Clothier, R.H., Johnson, R.O. and Rubens, L.N. (1998) *In situ* lymphocyte apoptosis in larval *Xenopus laevis*, the South African clawed toad. *Develop. Comp. Immunol.* **22:** 449–455.

Grant, P., Clothier, R.H., Johnson, R.O., Schott and Ruben, L.N. (1995) The time course, localization and quantitation of T- and B-cell mitogen-driven apoptosis *in vivo. Immunol. Lett.* **47:** 227–231.

Graves, S.S., Evans, D.L. and Dawe, D.L. (1985) Antiprotozoan activity of non-specific cytotoxic cells (NCC) from channel catfish (*Ictalurus punctatus*). *J. Immunol.* **134:** 78–85.

Greenlee, A.R., Brown, R.A. and Ristow, S.S. (1991) Nonspecific cytotoxic-cells of rainbow trout (*Oncorhynchus mykiss*) kill YCA-1 Target by both necrotic and apoptotic mechanisms. *Develop. Comp. Immunol.* **3:** 153–164.

Griffith, T.S., Yu, X., Herndon, J.M., Green, D.R. and Ferguson, T.A. (1996) CD95-induced apoptosis of lymphocytes in an immune privileged site induces immunological tolerance. *Immunity* **5:** 7–16.

Grutter, A.S. and Pankhurst, N.W. (2000) The effects of capture, handling, confinement and ectoparasite load on plasma levels of cortisol, glucose and lactate in the coral reef fish *Hemigymnus melapterus*. *J. Fish Biol.* **57:** 391–401.

Hagen, H.E., Klager, S.L. and Williams, G.T. (1998) Is apoptosis involved in mechanisms to eliminate *Onchocera ochengi* during *Simulium damnosum* immune response? *Trop. Med. Intern. Health* **3:** 945–950.

Han, Y.S., Thompson, J., Kafatos, F.C. and Barillas-Murray, C. (2000) Molecular interactions between *Anopheles stephensi* midgut cells and *Plasmodium berghei*: the time bomb theory of ookinete invasion of mosquitos. *EMBO J.* **19:** 6030–6040.

Harris, P.D., Soleng, A. and Bakke, T.A. (2000) Increased susceptibility of salmonids to the monogenean *Gyrodactylus salaris* following administration of hydrocortisone acetate. *Parasitology* **120:** 57–64.

Hasnain, S.E., Begum, R., Ramaiah, K.V.A., Sahdev, S., Shajil, E.M., Taneja, T.K., Mohan, M., Athar, M., Sah, N.K. and Krishnaveni, M. (2003) Host–pathogen interactions during apoptosis. *J. Biosci.* **28:** 349–358.

Hasson, K.W., Lightner, D.V., Mohney, L.L., Redman, R.M. and White, B.M. (1999) Role of lymphoid organ spheroids in chronic Taura syndrome virus (TSV) infections in *Penaeus vannamei. Dis. Aqua. Organ.* **38:** 93–105.

Hausmann, G., O'Reilly, L.A., van Driel, R., Beaumont, J.C., Strasser, A., Adams, J.M. and Huang, D.C.S. (2000) Pro-apoptotic apoptosis protease-activating factor 1 (Apaf-1) has a cytoplasmic localization distinct from Bcl-2 or Bcl-x(L). *J. Cell Biol.* **149:** 623-633

Henderson, S., Rowe, M., Gregory, C., Croom-Carter, D., Wang, F., Longnecker, R., Keiff, E. and Rickinson, A. (1991) Induction of bcl-2 expression by Epstein-Barr virus latent membrane protein 1 protects infected b cells from programmed cell death. *Cell* **65:** 1107–1115.

Hengartner, M.O. (2000) The biochemistry of apoptosis. *Nature* **407:** 770-776.

Hershberger, P.A., Dickson, J.A. and Friesen, P.D. (1992) Site-specific mutagenesis of the 35-kilodalton protein gene encoded by *Autographa californica* nuclear polyhedrosis virus: cell line-specific effects on virus replication. *J. Virol.* **67:** 3730–3738.

Hershberger P.A., LaCount, D.J. and Friesen, P.D. (1994) The apoptotic suppressor P35 is required early during baculovirus replication and is targeted to the cytosol of infected cells. *J. Virol.* **68:** 3467–3477.

Heussler, V.T., Kuenzi, P. and Rottenberg, S. (2001) Inhibition of apoptosis by intracellular protozoan parasites. *Intern. J. Parasitol.* **31:** 1166–1176.

Heussler, V.T., Machado, J., Fernandez, P.C., Botteron, C., Chen, C-G., Pearse, M.J. and Dobbelaere, D.A.E. (1999) The intracellular parasite *Theileria parva* protects infected T cells from apoptosis. *Proc. Natl Acad. Sci. USA* **96:** 7312–7317.

Hoffman, G.L., Landolt, M., Camper, J.E., Coats, D.W., Stookey, J.L. and Burek, J.D. (1975) A disease of freshwater fishes caused by *Tetrahymena corlissi* Thompson 1955 and a key for identification of holotrich ciliates of freshwater fishes. *J. Parasitol.* **61:** 217–223.

Hong, J.R. and Wu, J.L. (2002) Induction of apoptotic death in cells via bad gene expression by infectious pancreatic necrosis virus infection. *Cell Death Different.* **9:** 113–124.

Hong, J.R., Tai, L.L., Hsu, Y.L. and Wu, J.L. (1998) Apoptosis precedes necrosis of fish cell line by infectious pancreatic necrosis. *Virology* **250:** 76–84.

Hong, J.R., Lin T.L., Yang, J.Y., Hsu, Y.L. and Wu, J.L. (1999a) Dynamics of nontypical apopototic morphological changes visualized by green fluorescent protein in living cells with Infectious Pancreatic Necrosis virus infection. *J. Virol.* **73:** 5056–5063.

Hong, J.R., Hsu, Y.L. and Wu, J.L. (1999b) Infectious pancreatic necrosis virus induces apoptosis due to down-regulation of survival factor Mcl-1 protein expression in a fish cell line. *Virus Res.* **63:** 75–83.

Hoole, D., Bucke, D., Burgess, P. and Wellby, I. (2001) *Diseases of Carp and other Cyprinid Fishes.* Blackwell Science, Oxford, UK.

Hoole, D. Lewis, J.W., Schuwerack, P.M.M., Chakravarthy, C. and Cartwright, J.R. (2003) Inflammatory interactions in fish exposed to pollutants and parasites: a role for apoptosis and C reactive protein. *Parasitology* **126:** S71-S85.

Hopwood, J.A., Ahmed, A.M., Polwart, A., Williams, G.T. and Hurd, H. (2001) Malaria-induced apoptosis in mosquito ovaries: a mechanism to control vector egg production. *J. Exp. Biol.* **204:** 2773–2780.

Hu, M.S., Schwartzman, J.D., Yeaman. G.R., Collins, J., Seguin, R. Khan, I.A. and Kasper, L.H. (1999) Fas–FasL interaction involved in pathogenesis of ocular toxoplamosis in mice. *Infection Immunity* **67:** 928–935.

Huang, S-L., Liao, I-C. and Chen, S-N. (2000) Induction of apoptosis in tilapia, *Oreochromis aureus* Steindachner, and in T0–2 cells by *Staphylococcus epidermidis*. *J. Fish Dis.* **23:** 363–368.

Hurd, H. (1990) Physiological and behavioural interactions between parasites and invertebrate hosts. In: Baker, J.R. and Muller, R. (eds) *Advances in Parasitology*, **29**, pp. 271–317. Academic Press, London.

Hyoh, Y., Nishida, M., Tegoshi, T., Yamada, M., Uchikawa, R., Matsuda, S. and Arizono, N. (1999) Enhancement of apoptosis with loss of cellular adherence in the villus epithelium of the small intestine after infection with the nematode *Nippostrongylus brasiliensis* in rats. *Parasitology* **119:** 199–207.

Ishikawa, H., Hisaeda, H., Taniguchi, M., *et al.* (2000) CD4(+)v (alpha)14 NKT cells play a crucial role in an early stage of protective immunity against infection with *Leishmania major*. *Intern. Immunol.* **12:** 1267–1274.

Janz, D.M., McMaster, M.E., Weber, L.P., Munkittrick, K.R. and van der Kraak, G. (2001) Recovery of ovary size, follicle cell apoptosis, and HSP70 expression in fish exposed to bleached pulp mill effluent. *Can. J. Fish. Aqua. Sci.* **58:** 620–625.

Jaso-Friedmann, L., Leary, J.H. and Evans, D.L. (2000) Role of nonspecific cytotoxic cells in the induction of programmed cell death of pathogenic protozoans: participation of the Fas ligand–Fas receptor system. *Exp. Parasitol.* **96:** 75–88.

Julliard, A.K., Saucier, D. and Astic, L. (1993) Effects of chronic low level copper exposure on ultrastructure of the olfactory system in rainbow trout. *Histol. Histopathol.* **8:** 665—672.

Jung, S.-K., Mai, A., Iwamaoto, M., Arizono, N., Fujimoto, D., Sakamaki, K. and Yonehara, S. (2000) Purification and cloning of an apoptosis-inducing protein derived from fish infected with *Anisakis simplex*, a causative nematode of human anisakiasis. *J. Immunol.* **165:** 1491–1497.

Kamita, S.G., Majima, K. and Maeda, S. (1993) Identification and characterisation of the p35 gene of *Bombyx mori* nuclear polyhedrosis virus that prevents virus-induced apoptosis. *J. Virol.* **67:** 455–463.

Kamunde, C.N., Grosell, M., Lott, J.N.A. and Wood, C.M. (2001) Copper metabolism and gut morphology in rainbow trout (*Oncorhynchus mykiss*) during chronic sublethal dietary copper exposure. *Can. J. Fish. Aqua. Sci.* **58:** 293–305.

Kerr, J.F.R., Wyllie, A.H. and Currie, A.R. (1972) Apoptosis: a basic biological phenomenon with wide-ranging implications in tissue kinetics. *Brit. J. Cancer* **26** : 239-257.

Khanobdee, K., Soowannayan, C., Flegel, T.W. Ubol, S. and Withyachumnarnkul, B. (2002) Evidence for apoptosis correlated with mortality in the giant black tiger shrimp, *Penaeus monodon* infected with yellow head virus. *Dis. Aqua. Organ.* **48:** 79–90.

Koella, J.C. (1999) An evolutionary view of the interactions between anopheline mosquitos and malaria parasite life cycles. *Microbes Infect.* **1:** 303–308.

Krammer, P.H. (2000) CD95's deadly mission in the immune system. *Nature* **407:** 789-795.

Kunz, Y.W., Wildenburge, G., Goodrich, L. and Callaghan, E. (1994) The fate of ultraviolet receptors in the retina of the Atlantic Salmon (*Salmo salar*). *Vision Res.* **34:** 1375–1383.

Kuo T.H., Kim, H.R.C., Zhu, L.P., Yu Y.J., Lin, H.M. and Tsang, M (1998) Modulation of endoplasmic reticulum calcium pump by Bcl-2. *Oncogene* **17:** 1903–1910.

Laherty, C. D., Perkins, N.D. and Dixit, V.M. (1993) Human T cell leukemia virus 1 Tax and phorbol 12-myristate 13-acetate induce expression of the A20 zinc finger protein by distinct mechanisms involving nuclear factor-κB. *J. Biol. Chem.* **268:** 5032–5039.

Laing, K.J., Holland, J., Bonilla, S., Cunningham, C. and Secombes, C.J. (2001) Cloning and sequencing of caspase 6 in rainbow trout, *Oncorhynchus mykiss*, and analysis of its expression under conditions known to induce apoptosis. *Develop. Comp. Immunol.* **25:** 303–312.

Lau, A.O., Sacci, J.B. and Azad, A.F. (2001) Host responses to *Plasmodium yoeli* hepatic stages: a paradigm in host–parasite interaction. *J. Immunol.* **166:** 1945–1950.

Lee, P.C., Goodrich, M., Struve, M., Yoon, H.L. and Weber, D. (1992) Liver and brain glucocorticoid receptor in rainbow trout, *Oncorhynchus mykiss*: down-regulation by dexamethasone. *Gen. Comp. Endocrinol.* **87:** 222–231.

Li, X., Liu, Y. and Song, L. (2001) Cytological alterations in isolated hepatocytes from common carp (*Cyprinus carpio* L.) exposed to microcystin-LR. *Environ. Toxicol.* **16:** 517–522.

Li, P., Nijhawan, D., Budihardjo, I., Srinivasula, S.M., Ahmad, M., Alnemri, E.S. and Wang, X.D. (1997) Cytochrome c and dATP-dependent formation of Apaf-1/caspase-9 complex initiates an apoptotic protease cascade. *Cell* **91:** 479-489.

Liles, W.C. (1997) Apoptosis – role in infection and inflammation. *Curr. Opin. Infect. Dis.* **10:** 165–170.

Liesenfeld, O., Kosek, J.C. and Suzuki, Y. (1997) Gamma interferon induces Fas-dependent apoptosis of Peyer's patch T cells in mice following peroral infection with *Toxoplasma gondii*. *Infection Immunity* **65:** 4682–4689.

Liu, X.S., Kim, C.N., Yang, J., Jemmerson, R. and Wang, X.D. (1996) Induction of apoptotic programme in cell-free extracts: Requirement for dATP and cytochrome c. *Cell* **86:** 147–157.

Luder, C.G.K., Gross, U. and Lopes, M.F. (2001) Intracellular protozoan parasites and apoptosis: diverse strategies to modulate parasite–host interactions. *Trends Parasitol.* **17:** 480–486.

Lundebye, A-K., Berntssen, M.H.G., Wendelaar Bonga, S.E. and Maage, A. (1999) Biochemical and physiological responses in Atlantic salmon (*Salmo salar*) following dietary exposure to copper and cadmium. *Marine Pollution Bulletin* **39:** 137–144.

Luder, C.G.K., Gross, U. and Lopes, M.F. (2001) Intracellular protozoan parasites and apoptosis: diverse strategies to modulate parasite–host interactions. *Trends Parasitol.* **17:** 480–486.

MacInnis, A.J. (1976) How parasites find their host: some thoughts on the inception of host–parasite integration. In: Kennedy, C.R. (ed.) *Ecological Aspects of Parasitology*, pp. 3–20. North-Holland, Amsterdam.

Mangurian, C., Johnson, R.O., McMahan, R., Clothier, R.H. and Ruben, L.N. (1998) Expression of a Fas-like proapoptotic molecule on the lymphocytes of *Xenopus laevis*. *Immunol. Lett.* **64:** 31–38.

Manna, S.K., Kuo, M.T. and Aggarwal, B.B. (1999) Overexpression of gamma-glutamylcysteine synthetase suppresses tumor necrosis factor-kappa B and activator protein-1. *Oncogene* **18:** 4371–4382.

Matsumoto, J., Kawai, S., Terao., K., Kirinoki, M., Yasutomi, Y., Aikawa, M. and Matsuda, H. (2000) Malaria infection induces rapid elevation of the soluble Fas ligand level in serum and subsequent T lymphocytopenia possible factors responsible for the differences in susceptibility of two species of *Macaca* monkeys to *Plasmodium coatneyi* infection. *Infection Immunity* **68:** 1183–1188.

McMahan, R., Johnson, R.O., Rubens, L.N. and Clothier, R.H. (2000) Apoptosis and the cell cycle in *Xenopus laevis*: PHA and PMA exposure of splenocytes. *Immunol. Lett.* **70:** 179–183.

Meseguer, J., Esteban, M.A. and Malero, V. (1996) Nonspecific cell-mediated cytotoxicity in the sea water teleosts (*Sparus aurata* and *Dicentrarchus labrax*). Ultrastructural study of target cell death mechanism. *Anatomy Anatom. Record* **4:** 499–505.

Micic, M., Bihari, N., Labura, Z., Muller, W.E.G. and Batel, R. (2001) Induction of apoptosis in the blue mussel *Mytilus galloprovincialis* by tri-n-butyltin chloride. *Aqua. Toxicol.* **55:** 61–73.

Miller, L.K. (1999) An exegesis of IAPs: salvation and surprises from BIR motifs. *Trends Cell Biol.* **9:** 323–328.

Monack, D. and Falkow, S. (2000) Apoptosis as a common bacterial virulence strategy. *Intern. J. Med. Microbiol.* **290:** 7–13.

Morley, N.J. and Hoole, D. (1997) The *in vitro* effect of *Khawia sinensis* on leucocyte activity in carp (*Cyprinus carpio*). *J. Helminth.* **71:** 47–52.

Mosser, D.D., Caron, A.W., Bourget, L., Meriin, A.B., Sherman, M.Y., Morimoto, R.I. and Massie, B. (2000) The chaperone function of hsp70 is required for protection against stress-induced apoptosis. *Mol. Cell. Biol.* **20:** 7146–7159.

Murakawa, M., Jung, S-K., Lijima, K. and Yonehara, S. (2001) Apoptosis-inducing protein, AIP, from parasite-infected fish induces apoptosis in mammalian cells by two different molecular mechanisms. *Cell Death Different.* **8:** 298–307.

Nakajima-Shimada, J., Zou, C., Takagi, M., Umeda, M., Nara, T. and Aoki, T. (2000) Inhibition of Fas-mediated apoptosis by *Trypanosoma cruzi* infection. *Biochim. Biophys. Acta* **1475:** 175–183.

Narayanan, K. (1998) Apoptosis: Its role in microbial control in insect pests. *Curr. Sci.* **75:** 114–122.

Nera, M.S., Vanderbeek, G., Johnson, R.O., Ruben, L.N. and Clothier, R.H. (2000). Phosphatidylserine expression on apoptotic lymphocytes of *Xenopus laevis*, the South African clawed toad, as a signal for macrophage recognition. *Develop. Comp. Immunol.* **24:** 641–652.

Nolan, D.T., Hadderingh, R.H., Spanings, F.A.T., Jenner, H.A. and Wendelaar Bonga, S.E. (2000) Acute temperature elevation in tap and Rhine water affects skin and gill epithelia, hydromineral balance, and gill Na^+/K^+-ATPase activity of brown trout (*Salmo trutta*) smolts. *Can. J. Fish. Aqua. Sci.* **57:** 708–718.

Nolan, D.T., van der Salm, A.L. and Wendelaar Bonga, S.E. (2000) The host–parasite relationship between the rainbow trout (*Oncorhynchus mykiss*) and the ectoparasite *Argulus foliaceus* (Crustacea: Branchiura): epithelial mucous cell response, cortisol and factors which may influence parasite establishment. *Contrib. Zool.* **69:** 57–63.

Ojcius, D.M., Perfettini, J-L., Bonnin, A. and Laurent, F. (1999) Caspase-dependent apoptosis during infection with *Cryptsporidium parvum*. *Microbes Infection* **1:** 1163–1168.

Piechotta, G., Lacorn, M., Lang, T., Kammann, U., Simat, T., Jenke, H-S. and Steinhart, H. (1999) Apoptosis in dab (*Limanda limanda*) as possible new biomarker for anthropogenic stress. *Ecotoxicol. Environment. Safety* **42:** 50–56.

Poole, W.R., Nolan, D. and Tully, O. (2000) Modelling the effects of capture and sea lice (*Lepeophtheirus salmonis* [Kroyer]) infestation on the cortisol stress response in trout. *Aqua. Res.* **31:** 835–841.

Pottinger, T.G. (1990) The effect of stress and exogenous cortisol on receptor-like binding of cortisol in the liver of rainbow trout, *Oncorhynchus mykiss*. *Gen. Comp. Endocrinol.* **78:** 194–203.

Pottinger T.G. and Brierley, I. (1997) A putative cortisol receptor in the rainbow trout erythrocyte: stress prevents starvation-induced increases in specific binding of cortisol. *J. Experiment. Biol.* **200:** 2035–2043.

Ramsey-Ewing, A. and Moss, B. (1998) Apoptosis induced by a postbinding step of vaccinia virus entry into Chinese hamster ovary cells. *Virology* **242:** 138–149.

Rathmell, J.C. and Thompson, C.B. (2002) Pathways of apoptosis in lymphocyte development, homeostasis, and disease. *Cell* **109:** S97-S107

Risso-de Faverny, C., Devaux, A., Lafaurie, M., Girard, J.P., Bailly, B. and Rahmani, R. (2001). Cadmium induces apoptosis and genotoxicity in rainbow trout hepatocytes through generation of reactive oxygene species. *Aqua. Toxicol.* **53:** 65–76.

Roberts, M.L. (1997) The immune response of carp (*Cyprinus carpio* L.) to the blood fluke *Sanguinicola inermis* Plehn, 1905 (Trematoda: Sanguinicolidae). Ph.D. Thesis, Keele University, UK.

Roberts, R.J. and Shepherd, C.J. (1986) *Handbook of Trout and Salmon Diseases,* 2nd Edn. Blackwell Science, Oxford, UK.

Rojo, C. and Gonzalez, E. (1999) Ontogeny and apoptosis of chloride cells in the gill epithelium of newly hatched rainbow trout. *Acta Zool.* **80:** 11-13.

Ruane, N.M., Nolan, D.T., Rotllant, J., Tort, L., Balm, P.H.M. and Wendelaar Bonga, S.E. (1999) Modulation of the response of rainbow trout (*Oncorhynchus mykiss* Walbaum) to confinement, by an ectoparasitic (*Argulus foliaceus* L.) infestation and cortisol feeding. *Fish Physiol. Biochem.* **20:** 43–51.

Rumbley, C.A., Zekavat, S.A., Sugaya, H., Perrin, P.J., Ramadan, M.A. and Phillips, S.M. (1998). The schistosome granuloma: characterization of lymphocyte migration, activation, and cytokine production. *J. Immunol.* **161:** 4129–4137.

Saeij, J.P.J., Verburg-van Kemenade, L.B.M. van Muiswinkel, W. B. and Wiegertjes, G.F. (2003) Daily handling stress reduces resistance of carp to *Trypanoplasma borreli*: *in vitro* modulatory effects of cortisol on leukocyte function and apoptosis. *Develop. Comp. Immunol.* **27:** 233–245.

Saha, N.R., Usami, T. and Suzuki, Y. (2003) A double staining flow cytometric assay for the detection of steroid induced apoptotic leucocytes in common carp (*Cyprinus carpio*). *Develop. Comp. Immunol.* **27:** 351–363.

Sakahira, H., Enari, M., and Nagata, S. (1998) Cleavage of CAD inhibitor in CAD activation and DNA degradation during apoptosis. *Nature* **391:** 96-99.

Sakai, T., Hisaeda, H., Ishikawa, H., Maekawa, Y., Zhang, M., Nakao, Y., Takeuchi, T., Matsumoto, K., Good, R.A. and Himeno, K. (1999) Expression and role of heat-shock protein 65 (HSP65) in macrophages during *Trypansoma cruzi* infection: involvement of HSP65 in prevention of apoptosis of macrophages. *Microbes Infection* **1:** 419–427.

Salvesen, G.S. and Duckett, C.J. (2002) IAP proteins blocking the road to death's door. *Nature Rev. Mol. Cell Biol.* **3:** 401–410.

Saunders, T.W. (1966) Death in embryonic systems. *Science* **154:** 604–612.

Savill, J. and Fadok,V. (2000) Corpse clearance defines the meaning of cell death. *Nature* **407:** 784-788.

Shahabuddin, M. and Pimenta, P.F. (1998) *Plasmodium gallinaceum* preferentially invades vesicular ATPase-expressing cells in *Aedes aegypti* midgut. *Proc. Natl Acad. Sci. USA* **95:** 3385–3389.

Shen, L., Stuge, T.B., Zhou H., *et al.* (2000) Channel catfish cytotoxic cells: a mini-review. *Develop. Comp. Immunol.* **26:** 141–149.

Shrimpton, J.M. and Randall, D.J. (1994) Downregulation of corticosteriod receptors in gills of coho salmon due to stress and cortisol treatment. *Am. J. Physiol.* **267:** R432–R438.

Soutschek, J. and Zupapnc, G.D.H. (1995) Apoptosis as a regulator of cell proliferation in the posterion/prepacemaker nucleus of adult gymnotiform fish, *Apteronotus leptorhynchus*. *Neurosci. Lett.* **202:** 133–136.

Starke, W.A. and Oaks, J.A. (1999) *Hymenolepis diminuta*: Praziquantel removal of adult tapeworms is followed by apoptotic down-regulation of mucosal mastocytosis. *Exp. Parasitol.* **92:** 171–181.

Stoltze, K. and Buchmann, K. (2001) Effect of *Gyrodactylus derjavini* infections on cortisol production in rainbow trout fry. *J. Helminth.* **75:** 291–294.

Su, L., Kaneshima, H., Bonyhadi, M., Salimi, S., Kraft, D. Rabin, L and McCune, J.M. (1995) HIV-1-induced thymocyte depletion if associated with indirect cytopathogenicity and infection of progenitor cells *in vitro*. *Immunity* **2:** 25–36.

Sures, B., Knopf, K. and Kloas, W. (2001) Induction of stress by the swimbladder nematode *Anguillicola crassus* in European eels, *Anguilla anguilla*, after repeat experimental infection. *Parasitology* **123:** 179–184.

Sweet, L.I., Passino-Reader, D.R., Meier, P.G. and Omann, G.M. (1999) Xenobiotic-induced apoptosis: significance and potential application as a general biomarker of response. *Biomarkers* **4:** 237–253.

Taylor, S.L., Jaso-Friedmann, L., Allison, A.B., Eldar, A. and Evans, D.L. (2001) *Streptococcus iniae* inhibition of apoptosis of nonspecific cytotoxic cells: a mechanism of activation of innate immunity in teleosts. *Dis. Aqua. Organ.* **46:** 15–21.

Tibayrenc, M., Kjellberg, F. and Ayala, F. (1990) A clonal theory of parasitic protozoa: the population structures of *Entamoeba, Giarda, Leishmania, Naegleria, Plasmodium, Trichomonas* and *Trypanosoma* and their medical taxonomical consequences. *Proc. Natl Acad. Sci. USA* **87:** 2414–2418.

Toure-Baldo, A., Sarthou, J.L. and Roussilhon, C. (1995) Acute *Plasmodium falciparum* infection is associated with increased percentage of apoptosis of apoptotic cells. *Immunol. Lett.* **46:** 59–62.

Toure-Baldo, A., Sarthou, J.L., Aribot, G., Michel, P., Trape, J.F., Rogier, C. and Roussilhon, C. (1996) *Plasmodium falciparum* induces apoptosis in human mononuclear cells. *Infection Immunity* **64:** 744–750.

Toomey, B.H., Bello, S., Hahn, M.E., Cantrell, S., Wright, P., Tillitt, D.E. and Di Giulio, R.T. (2001) 2,3,7,8-Tetrachlorodibenzo-p-dioxin induces apoptotic cell death and cytochrome P4501A expression in developing *Fundulus heteroclitus* embryos. *Aqua. Toxicol.* **53:** 127–138.

Tyler, K.L., Squier, M.K., Rodgers, S.E., Schneider, B.E., Oberhaus, S.M., Gridina, T.A., Cohen, J.J. and Dermody, T.S. (1995) Differences in the capacity of reovirus strains to induce apoptosis are determined by the viral attachment protein sigma 1. *J. Virol.* **69:** 6972–6979.

Vaux, D.L., Cory, S. and Adams, J.M (1988) Bcl-2 gene promotes hematopoietic-cell survival and cooperates with c-myc to immortalize pre-B-cells. *Nature* **335 :** 440-442.

Verburg-van Kemenade, B.M.L., Nowak, B., Engelsma, M.Y. and Weyts, F.A.A. (1999) Differential effects of cortisol on apoptosis and proliferation of carp B-lymphocytes from head kidney, spleen and blood. *Fish Shellfish Immunol.* **9:** 405–415.

Villa, P., Kaufmann, S.H. and Earnshaw, W.C. (1997) Caspases and caspase inhibitors. *Trends Biochem. Sci.* **22:** 388–393.

Walsh, C.J., Wyffels, J.T., Bodine, A.B. and Luer, C.A. (2002) Dexamethasone-induced apoptosis in immune cells from peripheral circulation and lymphomyeloid tissues of juvenile clearnose skates, *Raja eglanteria. Develop. Comp. Immunol.* **26:** 623–633.

Wang, S.L., Hawkins, C.J., Yoo, S.J., Muller, H.-A.J. and Hay, B.A. (1999) The *Drosophila* caspase inhibitor DIAP1 is essiential for cell survival and is negatively regulated by HID. *Cell* **98:** 453–463.

Wang, S., Rowe, M. and Lundgren, E. (1996) Expression of the Epstein Barr virus transforming protein LMP1 causes a rapid and transient stimulation of the Bcl-2 homologue Mcl-1 levels in B-cell lines. *Cancer Res.* **56:** 4610–4613.

Welburn, S.C., Barcinski, M.A. and Williams, G.T. (1997) Programmed cell death in Trypanosomatids. *Parasitology Today* **13:** 22-26.

Wendelaar Bonga, S.E. (1997) The stress response in fish. *Physiol. Rev.* **77:** 591–625.

Wendelaar Bonga, S.E. and van der Meij, C.M.J. (1989) Degeneration and death, by apoptosis and necrosis, of the pavement and chloride cells in the gills of the teleost *Oreochromis mossambicus. Cell Tissue Res.* **255:** 235–243.

Wesson, C.A., Deringer, J., Liou, L.E., Bayles, K.W., Bohach, G.A. and Trumble W.R. (2000) Apoptosis induced by *Staphylococcus aureus* in epithelial cells utilizes a mechanism involving caspase 8 and 3. *Infection Immunity* **68:** 2998–3001.

Weyts, F.A.A., Verburg-van Kemenade, B.M.L. and Flik, G. (1998) Characterisation of glucocorticoid receptors in peripheral blood leukocytes of carp, *Cyprinus carpio* L. *Gen. Comp. Endocrinol.* **111:** 1–8.

Weyts, F.A.A., Verburg-van Kemenade, B.M.L., Flik, G., Lambert, J.G.D. and Wendelaar Bonga, S.E. (1997) Conservation of apoptosis as an immune regulatory mechanism: Effects of cortisol and cortisone on carp lymphocytes. *Brain Behav. Immun.* **11:** 95–105.

Williams, G.T. (1994) Apoptosis in the immune-system. *J. Path.* **173**: 1-4.

Williams, G.T., Smith, C.A., McCarthy, N.J. and Grimes, E.A. (1992) Apoptosis: Final control point in cell biology. *Trends Cell Biol.* **2:** 263-267.

Wyllie, A.H., Kerr, J.F.R. and Currie, A.R. (1980) Cell death: The significance of apoptosis. *Int. Rev Cytol.* **68:** 251-306.

Xue D. and Horvits, H.R. (1995) Inhibition of the *Caenorhabditis elegans* cell-death CED-3 by a CED-3 cleavage site in baculovirus p35 protein. *Nature* **377:** 248–251.

Yabu, T., Kishi, S., Okazaki, T. and Yamashita, M. (2001) Characterization of zebrafish caspase-3 and induction of apoptosis through ceramide generation in fish fathead minnow tailbud cells and zebrafish embryo. *Biochem. J.* **360:** 39–47.

Zamzami, N., Brenner, C., Marzo, I., Susin, S.A. and Kroemer (1998) Subcellular and submitochrondrial localization of BCL-2-like oncoproteins. *Oncogene* **16:** 2265–2282.

Zhang, M., Hisaeda, H., Sakai, T., Ishikawa, H., Hao, Y.P., Nakano, Y., Ito, Y. and Himeno, K. (1999) Macrophages expressing heat-shock protein 65 play an essiential role in protection of mice infected with *Plasmodium yoelli*. *Immunology* **97:** 611–665.

3

Thionine-positive cells in relation to parasites

Michael E. Nielsen, Thomas Lindenstrøm, Jens Sigh and Kurt Buchmann

1 Definition of thionine-positive cells

Thionine-positive cells can be defined as cells with the ability to be metachromatically stained with the basic aniline dye thionine. The cells within this definition comprise a broad array of cells and include cell types such as mast cells, goblet cells, basophils, eosinophils and cells from amyloid, cartilage and mucus tissues. In the following, emphasis will only be put on two types of thionine-positive cells that probably have evolved from the same ancient type of cells. The cell types that will be addressed here are mast cells and basophils, which both are involved in host defences against parasites (Galli *et al.*, 1984). The fundamental knowledge of the structure and function of both mast cells and basophils is based on work in mammals, especially humans. Thus, the majority of the information in this chapter originates from work with mammalian species, however references to other vertebrates and invertebrates will be included.

The mast cells and basophils were first described in the late 19th century by the German physician and biochemist Paul Ehrlich. He described the mast cells and the basophils as large, distinctively stained cells containing basophilic granules. The mast cells and basophils can be histologically recognized by their ability to have their granules metachromatically stained with thionine. Only a few tissue elements display metachromasia, changing in the tissues from blue (the usual orthochromatic form) to purplish red or reddish purple. Thus, thionine and other cationic dyes, called metachromatics, are of considerable value in the study of specific elements of connective tissue. Among other metachromatic dyes the most commonly used are toluidine blue O, methylene blue, gallocyanin, and pinacyanole (Schubert and Hamerman, 1956). The cytoplasmic granules of these cells contain many biologically active substances, which contribute to their roles in host defence against parasites (Galli *et al.*, 1984). The roles of mast cells in the expression of host resistance to certain parasites are dependent on factors such as parasite species and the infection site (Costa *et al.*, 1997; Galli and Dvorak, 1995; Gill, 1986; Neitz *et al.*, 1993).

Host–Parasite Interactions, edited by G. F. Wiegertjes & G. Flik.

1.1 *Origin and structure of thionine-positive cells*

The thionine-positive mast cells and basophils both derive from CD34$^+$ haematopoietic progenitor cells. Basophils, which have a quite short life span, complete their differentiation in the bone marrow, circulate in the bloodstream and are not generally found in connective tissues (Wardlaw *et al.*, 1994). However, they can transmigrate through endothelial layers and subsequently invade inflamed tissue, a process that is controlled by the particular expression of cell surface recognition molecules. Current evidence indicates that mature basophils are terminally differentiated (Cheng *et al.*, 2001). In relation to mast cells, basophils are generally characterized by their blood circulation, their distinctly segmented nuclei and the expression of a specific basophilic antigen (Bsp-1) (Cheng *et al.*, 2001). In addition, they express a wide range of integrins, cell adhesion molecules, chemokine receptors and other surface molecules (Bochner and Schleimer, 2001). With respect to basophils, there is as yet no unique adhesion molecule identified on the surface of these cells, though relative levels of many of these surface structures differ from mast cells (Bochner and Schleimer, 2001).

Mast cells are longer-lived and arise from mast cell-committed precursors, which as agranular cells circulate in the bloodstream (Valent *et al.*, 1992), transverse the vascular space and upon entry into tissues differentiate and mature (Cheng *et al.*, 2001). Although mast cells are recognized in peripheral tissues, little is known about the phenotype of mast cell precursors, their fate from the bone marrow to the tissues, the migration and homing processes and the factors and adhesion molecules that affect those processes. Mast cell progenitors have in addition to the bone marrow been found in the umbilical cord blood as well as in fetal liver (Irani *et al.*, 1992; Ishikaza *et al.*, 1992). Phenotypic identification of immature mast cells has been hampered by the low frequency of these cells in circulation and the lack of specific surface markers in the early stages of development.

A number of studies indicate that interactions with collagen, laminin, fibronectin and vitronectin may be important for tissue localization of mast cells (Bochner and Schleimer, 2001). These interactions are mediated through various surface recognition molecules of which many are shared with basophils. However, in addition mast cells express c-kit (CD117 – ligand for SCF (stem cell factor) as well as the low-affinity IgG receptor FcγRII (CD32). Furthermore, mast cells are chymase- and tryptase-positive as well as positive for carboxypeptidase A (CPA) (Cheng *et al.*, 2001). As basophils and mast cells are different morphologically, functionally and biochemically (*Table 1*) they have been considered to be developmentally unrelated (Cheng *et al.*, 2001). However, this viewpoint has been challenged various times and the issue of uniformity and relatedness of these two cell types is still a matter of dispute (Cheng *et al.*, 2001). Recently it has been suggested that differentiated mast cells and basophils may not be fixed in a certain state. It is rather the microenvironment under which they are placed that is of importance; both to the appearance (morphologically and biochemically) and to the possibility of these cells to reversibly alter their expression of serine proteases (Cheng *et al.*, 2001). This is based on studies where metachromatic cells with features of both mast cells and basophils could be developed *in vitro* upon cultivation with conditioned media. Metachromatic FcϵRI$^+$/Try$^+$/Chy$^+$ cell populations could be developed *in vitro* from human bone marrow cells from normal donors in the presence of SCF and conditioned media (Kirschenbaum *et al.*, 1992; Li *et al.*, 1996, 1998).

However, these cells had features of basophils in terms of morphology. The different expression of integrins and cell adhesion molecules on these two cells (e.g. the presence of β2 integrins and VCAM-1 in basophils) has been interpreted as determinants of cell lineages, but according to Cheng *et al.* (2001), this is probably more a requirement for diapedesis of basophils into inflamed tissue.

Table 1 *Basic characteristics of mast cells and basophils*

Cell features	Mast cells	Basophils
Haematopoietic progenitor cell	$CD34^+$	$CD34^+$ (differentiation to CFU-Ba)
Site of differentiation	Extramedullary tissue (a few in marrow)	Bone marrow
Nucleus	Round to oval	Long, variably segmented
Circulation in blood	Rarely recognized	Present at low concentrations
Life span in tissue	Weeks to months	Days, possibly weeks
Mitotic potential	Present	Not reported
Major mediator of differentiation and proliferation	SCF	IL-3
Metachromatic staining with basic dyes	At acid and neutral pH	At acid but not neutral pH
Chloracetate activity	Positive	Usually negative (except cats)
Tryptase activity	Positive	Negative

CFU-Ba (colony-forming unit, basophil), CD34+ (stem cell with CD34 antigen), IL (interleukin), SCF (stem cell factor).
(Modified after Scott and Stockham, 2000.)

Cytologically, mast cells are classified as large round cells. Individual mast cells typically resemble mononuclear cells and contain a variable amount of cytoplasm filled with metachromatic granules. This definition should, however, be broadened to include cells with multilobulated nuclei (Gurish *et al.*, 1997). Basophils have a characteristic bilobed nucleus, which is easily recognizable with electron microscopy. Granules of both mast cells and basophils are membrane-bound and filled with a closely packed, electron-dense material (Wheater *et al.*, 1979). The composition of the mast cell granules is similar to basophil granules, but there is no evidence that mast cells are merely basophils resident in connective tissues, although both cell types appear to have similar actions and may degranulate in response to similar stimuli, e.g. parasites (Wheater *et al.*, 1979).

Within tissues, non-granulated mast cells continue to differentiate, develop granules, and may subsequently have a lifespan ranging from weeks to months. The basophils are subsequently distributed into the bloodstream where they develop granules; in contrast to the mast cells they reside in the bloodstream except during tissue inflammation. Mast cells are normally present in small numbers in the connective tissue of all organs, but particularly in the dermal layer of skin around blood vessels and nerves (*Figure 1*), and are identified by their cytoplasmic granules (*Figure 2*). Mast cells have been considered the tissue equivalent of the circulating basophil, but although there is evidence that they arise from common precursors in the bone marrow, there is no clear evidence that mature basophils are able to differentiate into mast cells. The two cell types are readily distinguished by their morphology on light microscopy and the presence of chloroacetate esterase activity in mast cells (Scott and Stockham, 2000).

1.2 *Chemical constituents of mast cells and basophils*

Both mast cells and basophils have cytoplasmatic granules which contain a wide range of molecules (*Table 2*). Histamine is a highly characteristic substance of mast cells and basophils, as it is the only mammalian cell type in which it is produced. Additionally, a

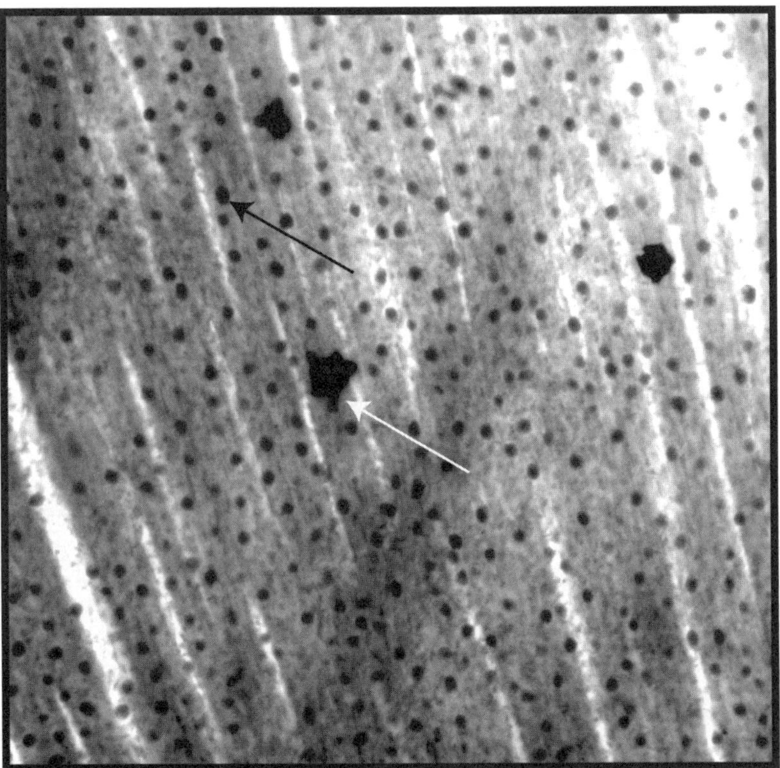

Figure 1 *The skin from ventral body surface of a brown trout stained with 0.1% thionine in absolute ethanol (pH 4). The sampled fish were fixed in absolute ethanol for 2 weeks. Fins and skin were dissected and stained in 0.1% thionine in absolute ethanol (pH 4) for 15 minutes, destained in absolute ethanol for 2 minutes and mounted in Depex. It is essential to use absolute ethanol for fixing and staining in order to avoid loss of granular staining (Reite and Evensen, 1994). Thionine-positive cells appear as metachromatically stained purple granulated cells (black arrow) on the surrounding pale blue tissue. The thionine-positive cells are numerously scattered throughout the body surface and fins and are located superficially. A melanophore is indicated by a white arrow. (200 x magnification).*

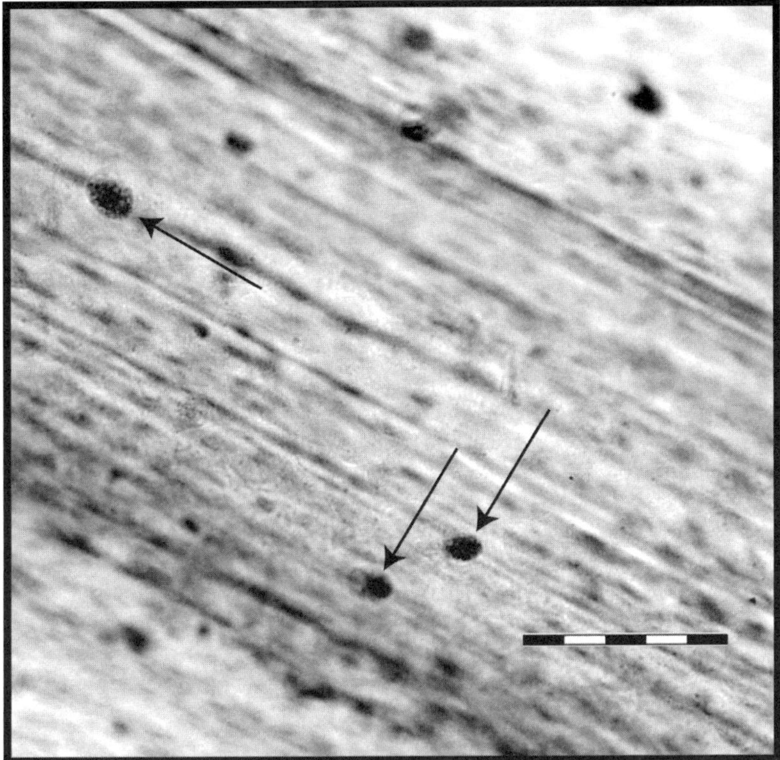

Figure 2 *The pectoral fin of a brown trout stained with 0.1% thionine in absolute ethanol (pH 4). The thionine-positive cells indicated by the black arrows are clearly granulated. The densities of thionine-positive cells in the fins are lower than on the body surface, and decline towards the outer edges of the fins. Bar = 50 μm.*

wide range of other molecules such as proteases, cytokines, chemotactic factors, serotonin and proteoglycans are abundant in mast cells and basophils. In particular, the glycosaminoglycans heparin and heparan sulphate are considered to be important markers of mast cell function.

The membrane-bound secretory granules present in mast cells and basophils can be released into the extracellular space and thereby mediate important physiological and pharmacological processes (Anwar *et al.*, 1980; Austen and Orange, 1975; Lagunoff, 1972; Lagunoff and Pritzl, 1976). Mast cell secretory products include biogenic amines like histamine and serotonin, proteoglycans such as heparin, numerous enzymes, arachidonic acid products like prostaglandins and leukotrienes, and many cytokines including chemokines that attract eosinophils (Scott and Stockham, 2000) (see *Table* 2).

2 Thionine-positive cells and anti-parasitic immunity

Both mast cells and basophils contain preformed mediators, which are granule associated, and these are accountable for the metachromasia upon staining with the cationic dyes previously described. Most probably, this particular staining property is primarily due to the presence of proteoglycans (e.g. chondroitin sulphate and heparin)

Table 2 *Constituents of mast cells and basophils*

Constituents of cytoplasmatic granules	Mast cells	Basophils
Major cell products		
Biogenic amines	Histamine, adenosine, serotonin	Histamine, serotonin
Enzymes	α-tryptase, β-tryptase, chymase, carboxy peptidase, cathepsin G, acid hydrolases A_2, amionpeptidase and hexosaminodase	Neutral proteases with bradykinin-generating activity, elastase, β-glucoronidase and cathepsin G-like enzyme
Proteoglycans	Heparin and chrondroitin sulphate	Heparin, chrondroitin sulphate and dermatan sulphate
Other	—	Major basic protein
Mediators synthesized ***de novo*** (* also stored as preformed)		
Arachidonic acid products	PGD_2, LTC_4, LTB_4, LTD_4, LTE_4, TxB_2 and PGE_2	LTC_4
Cytokines and chemokines	SCF*, TNF-α*, IL-1α, IL-1β, IL-3, IL-4, IL-5, IL-6, IL-8*, IL-10, IL-13, IL-16, GM-CSF, TGF-β, MCP-1, MIP-1α, MIP-1β, bFGF, RANTES, endothelin and lymphotactin	IL-4, IL-8*, IL-13, TNF-α* and MIP-1α
Other	PAF	PAF

bFGF (basic fibroblast growth factor), CFU-Ba (colony-forming unit, basophil), GM-CSF (granulocyte-macrophage colony-stimulating factor), IL (interleukin), LTB_4 (leukotriene B_4), LTC_4 (leukotriene C_4), LTD_4 (leukotriene D_4), LTE_4 (leukotriene E_4), MCP (monocyte chemotactic protein), MIP (macrophage inflammatory protein), PAF (platelet-activating factor), PGD_2 (prostaglandin D_2), PGE_2 (prostaglandin E_2), RANTES (regulated on activation, normal T-cell expressed and secreted), SCF (stem cell factor), TNF (tumour necrosis factor) and TGF-β (transforming growth factor-β), TxB_2 (thromboxane B_2). (Modified after Scott and Stockham, 2000.)

in these granules (Galli *et al.*, 2002). The biological function of these proteoglycans is not fully understood – although it has recently been described that heparin in mice is required for the normal packaging of certain neutral proteases in mast cell cytoplasmatic granules (Humphries *et al.*, 1999). Mast cells display a large degree of heterogeneity in regard to the content of such proteoglycans and/or proteases depending on source of species or tissues or developmental stage (Galli *et al.*, 2002). Mast cell heterogeneity in mammals has also been described in relation to their profile of cytokines and suggests specific biologic functions for the different mast cell subsets (Bradding *et al.*, 1995). Although homogeneity of cell populations has been considered to be a feature of basophils (Abraham and Arock, 1998), these cells have also recently been demonstrated to exhibit some phenotypic heterogeneity (Galli *et al.*, 2002).

2.1 *The involvement of mast cells and basophils in innate immunity against parasites*

The involvement of both mast cells and basophils in anti-parasitic immunity has been acknowledged for decades, with parasite infections quite often leading to increased levels of circulating basophils, a marked increase in IgE levels in serum and increased numbers of mast cells and/or basophils in infected tissues (Galli *et al.*, 2002). The contribution of basophils and mast cells to adaptive immune responses is already comprehensively recognized based on their IgE-mediated interactions with parasites. However, the importance of basophils and not least mast cells in innate immunity towards parasitic infections is increasingly being recognized and supported by various studies (Coelho-Castro *et al.*, 2002; Issekutz *et al.*, 2001; Knight *et al.*, 2000).

The distribution of mast cells in skin mucosal surfaces as well as around blood vessels makes them ideally placed for immune surveillance and for the early interaction with pathogens at the portals of entry (Abraham and Arock, 1998; Galli *et al.*, 1999). From the intensive work on the involvement of these cells in allergic diseases, it is clear that they are endowed with the intrinsic capacity of mobilizing a rapid and vigorous inflammatory response in affected hosts (Galli *et al.*, 1999). Besides being important effector cells in innate immunity, it is probably this role of initialization and orchestration, which has attained greatest attention in unravelling the role of mast cells in innate responses against parasites.

2.2 *Recognition of pathogens and receptors involved*

Innate recognition by non-opsonin-dependent mechanisms

Basophils have in addition to mast cells been shown to directly recognize various pathogens in a serum-independent way. Contrasting the situation for mast cells, it is currently not known whether basophils upon such opsonin-independent binding have the ability to phagocytose and kill microbes (Abraham and Arock, 1998). The best-documented example of a direct activation is probably by enterobacteria such as *E. coli*, *Klebsiella pneumonia* and *Enterobacter cloacae* (Malaviya *et al.*, 1994). This innate and serum-independent recognition is accomplished by a GPI-anchored receptor molecule (CD48) on mast cells, which recognizes a mannose-binding lectin on a type 1 fimbriae (FimH) present on such enterobacteria (Galli *et al.*, 2002). Upon recognition, the mast cells were activated and thus induced to degranulate and exhibit bactericidal activity (Malaviya *et al.*, 1994). Amongst the Gram-positive bacteria, mast cells have likewise been documented to bind strongly to *Staphylococcus aureus* and *Streptococcus faecalis* (Abraham and Arock, 1998). Regarding parasitic organisms, mast cells have been shown to associate to live promastigote forms of both *Leishmania major* and *L. infantum* without the presence of serum components. This association subsequently induces release of preformed mediators (Bidri *et al.*, 1997). The precise mechanism behind this interaction is however not fully elucidated (Abraham and Arock, 1998; Bidri *et al.*, 1997). A recognition mechanism analogous to the

Gram-negative enterobacteria has recently been described in the parasite *Schistosoma mansoni* (Coelho-Castelo *et al.*, 2002). Here, like the situation for FimH, a mannose-binding lectin most likely involving only one carbohydrate-binding site was observed to be responsible for serum-independent mast cell activation. This process could be abolished by addition of D-mannose in a dose-dependent way. An approximately 60 kDa protein was identified in both the cercariae and tegumental surface of adult worms. The mast cell ligand to this Sm60 is, however, still unknown (Coelho-Castelo *et al.*, 2002). *Trichinella spiralis* have also recently been shown to contain antigens (TSL-1), which in a serum-independent way can alter the cytokine profile of cultured mast cells (Arizmendi *et al.*, 2001). Thus, evidence for the presence of innate, non-opsonin-dependent recognition mechanisms in mast cells is emerging and it is therefore conceivable that pattern recognition receptors are present and involved. Interestingly, a number of Toll-like receptors (TLR2, TLR4 and TLR6) have recently been identified on bone marrow-derived mast cells (McCurdy *et al.*, 2001). Cytokine production upon stimulation with LPS was shown to be completely dependent on the presence of a functional TLR4 (McCurdy *et al.*, 2001). Consequently, mice containing a point mutation in TLR4 were completely refractory to LPS stimulation of mast cells in terms of cytokine responses (McCurdy *et al.*, 2001).

Some parasites have been described to produce superantigens, which can crosslink the high-affinity IgE receptors (FcϵRI) (either directly or indirectly by receptor bound IgE) in a non-specific manner. Thus, basophils and mast cells can be activated in non-immune individuals by a *Necator americanus* lectin, which can crosslink receptor-bound IgE through carbohydrate moieties on the Fc-portion (Falcone *et al.*, 2001).

Innate recognition by opsonin dependent mechanisms

Recognition of pathogens by mast cells can also be achieved after opsonization by soluble host components such as complement. Pathogens can be coated with the C3b fragment of complement and can subsequently be recognized by complement receptors present on the mast cell membrane (Abraham and Arock, 1998). The deposition of C3b can be achieved in an Ig-independent way either through the alternative, the lectin or by the classical pathway involving CRP (C-reactive protein) in replacement of specific antibodies. Mast cell adherences to schistosomula of *Schistosoma mansoni* have actually been reported to involve complement-mediated opsonisation (Sher, 1976). The importance of complement in innate recognition of pathogens by mast cells has been stressed in studies involving CLP (caecal ligation and puncture; a model of acute septic peritonitis) (Henz *et al.*, 2001). Thus, by exploiting complement C3-deficient mice, it has been shown that these animals exhibit markedly reduced mast cell activation during CLP leading to diminished levels of TNF-α in the peritoneum (Henz *et al.*, 2001). Furthermore, mice deficient in complement receptor 1 (CD35) or complement receptor 3 (CD11b/CD18) have both demonstrated impaired activation of peritoneal mast cells after CLP reflected by lowered histamine release, recruitment of neutrophils, bacterial clearance and survival (Prodeus *et al.*, 1997; Rosenkranz *et al.*, 1998). Complement components have also been shown to be involved in the direct activation of basophils and mast cells. Hence, the anaphylactic fragments of complement C3a and C5a serve as secretagogues on these cells through interaction with the C3aR and the C5aR (CD88), respectively (Bochner and Schleimer, 2001; Zwirner *et al.*, 1998). Likewise, other host-derived factors such as sub-fragments of fibrinogen and fibronectin have shown to exhibit similar activating capabilities (Abraham and Arock, 1998).

2.3 *Phagocytosis and microbicidal activity*

Phagocytosis and subsequent killing of microbes have not been described in basophils (Abraham and Arock, 1998). Mast cells, however, are known to be phagocytic to both micromolecular as well as particulate materials (Henz *et al.*, 2001). However, this ability decreases with the increased maturity of the cells and its significance in relation to other phagocytic cells is not well elucidated (Henz *et al.*, 2001). *In vitro* experiments have clearly shown that mast cells are capable of phagocytosis and killing of bacteria, although the phagocytic capacity/efficiency was significantly lower than 'traditional phagocytes' (Galli *et al.*, 1999). Mediator release (histamine, β-glucuronidase, LTB_4) as a result of phagocytosis occurs in a dose-dependent fashion. Opsonization with complement seems to increase the release of these mediators (Henz *et al.*, 2001). Still, the triggering of mediator release based on e.g. parasite stimuli, appears to induce either a slower release or in a smaller quantity than seen in 'classical' FcεRI-dependent activation by IgE cross-linking (Galli *et al.*, 1999).

Microbicidal activities of mast cell are dependent upon both oxidizing and non-oxidizing mechanisms. Thus, mast cell phagocytosis of $FimH^+$ *E. coli* through interaction with surface-bound CD48 has shown to induce oxidative burst as measured by chemoluminiscence assay (CL) (Abraham and Arock, 1998; Malaviya *et al.*, 1994). Viability of engulfed bacteria was highly affected by mast cell activation. In contrast, a $FimH^-$ *E. coli* mutant did not provoke the production of reactive oxygen metabolites. Superoxide anions rather than hydrogen peroxide appeared to be the oxygen species involved, as the oxidative burst reaction could be completely inhibited by superoxide dismutase but not by catalase, a known quencher of hydrogen peroxide (Malaviya *et al.*, 1994). Reactive oxygen intermediates have the ability to directly affect parasites adversely, although it must be recalled that such factors could also have a detrimental impact on host tissues. Non-oxidizing microbicidal activity includes acidification of phagocytic vacuoles and fusion of lysosomal granules to the phagosome (Abraham and Arock, 1998). For parasites refractory to phagocytosis, highly reactive mast cell amines and serine proteases with defined substrate specificities (Solivan *et al.*, 2002; Wong *et al.*, 2002) could likewise have harmful effects on parasite structures. McKean and Pritchard (1989) have described digestive action of a mast cell protease on cuticular collagens of *Necator americanus*, which could give an indication of such a function. Still, a variety of microbicidal activities displayed by mast cells remain to be characterized (Abraham and Arock, 1998).

2.4 *Surface molecules of importance to chemotaxis*

Both basophils and mast cells are equipped with a multitude of adhesion molecules and receptors for chemotactic factors, which are important for their homing to various tissues as well as for migration and recruitment during infections with e.g. parasites (Bochner and Schleimer, 2001). Complement fragments C3a and C5a produced during complement activation can serve as potent chemoattractants through their respective complement receptors on basophils and mast cells and thereby attract these to the site of infection (Bochner and Schleimer, 2001). These thionine-positive cells are also endowed with certain seven transmembrane span chemokine receptors. Mature mast cells only express CCR3 but during differentiation they additionally express CXCR2 (IL-8 ligand), CXCR4 and CCR5 (Bochner and Schleimer, 2001). Mast cells express c-kit (CD117), which binds the kit ligand (KL); probably better known as SCF (stem cell factor). Besides being an essential cytokine for mast cell differentiation, SCF is known to be a selective chemoattractant for mast cells and can directly modulate mast cell adhesion (Bochner and Schleimer, 2001). Contrasting this, basophils express CCR1, CCR2, CCR3, CCR6, CXCR2 and CXCR4, probably reflecting the prerequisite of such surface structures for the recruitment and accumulation of basophilic cells at infected sites (Bochner and Schleimer, 2001). Likewise, basophils display surface expression of the IL-3 receptor (CD123, γc). IL-3 exposure of basophils directly augments their adhesiveness to endothelium and up-regulates certain integrins (Bochner and Schleimer, 2001). Integrins and cellular adhesion molecules are crucial for the tissue distribution, homing, migration and recruitment of both cell types. The surface expression of such migration-related molecules could be induced or altered by activation and exposure to chemokines and cytokines. Basophils and mast cells share many of these molecules such as the β1 (VLA) family of integrins and the intercellular adhesion molecules (ICAM 1–3; belonging to the immunoglobulin superfamily). Besides CD11c/CD18, which seems to be weakly expressed on mast cells from some tissues, the β2 integrins are not found on mast cells as opposed to basophils (Bochner and Schleimer, 2001). The importance of these surface molecules in anti-parasite responses has been stressed in studies, where monoclonal antibodies were used to block certain integrins during infection (Issekutz *et al.*, 2001). Antibody-mediated blockage in infected rats of α4β7 integrins, which specifically mediates adhesion to mucosal addressin cell adhesion molecules (MadCAM)-1, significantly inhibited the mast cell hyperplasia normally seen during *Nippostrongylus brasiliensis* infections (Issekutz *et al.*, 2001). The response of intestinal eosinophils was, however, not affected indicating the specific inhibition of mast cells. Neither was the initial degranulation of mast cells affected by the antibody treatment and the effect observed is therefore unlikely to be due to inhibition of the initial activation of resident mast cells. Although final clearance of worms was not counteracted, the expulsion process was markedly delayed (Issekutz *et al.*, 2001). Thus, despite of the presence of a pro-inflammatory reaction, the expression of integrins is necessary and important for the recruitment and migration of these thionine-positive cells.

2.5 *Parasite-induced mediator release and its significance*

The activation of mast cells and basophils, either directly by parasite-derived ligands or secretory/excretory products, by involvement of antigen-specific IgE crosslinking or by host-derived factors led to the release of mediators. These can either be preformed granule-associated substances or mediators synthesized *de novo*. The preformed mediators include histamine, heparin and serine proteases such as tryptase and chymase, the cyto- and chemokines SCF, IL-8 and TNF-α as well as VEGF (vascular endothelial cell growth factor) (Henz *et al.*, 2001). The preformed pool of granule-associated TNF-α and IL-8 seems to provide a crucial perquisite for one of the main functions of mast cells in innate immune responses – namely the initiation of inflammation and recruitment of neutrophils to the site of infection (Abraham and Arock, 1998; Henz *et al.*, 2001). Thus, by using mast-cell-deficient $Kit^{W}/Kit^{W V}$ mice it has been shown that impaired killing of bacteria was directly correlated with lowered recruitment of neutrophils most likely as a result of a diminished TNF-α driven attraction of these cells (Malaviya *et al.*, 1996). Furthermore, mast-cell-deficient mice exhibit reduced release of TNF-α following bacterial challenge (Henz *et al.*, 2001). TNF-α is a potent pro-inflammatory cytokine, which can activate endothelial cells and induce recruitment of leucocytes by inducing the expression of important chemokines such as IL-8 in an autocrine fashion. The contribution of preformed mast cell IL-8 to this recruitment has not been fully clarified, but due to its CXC signature, its chemotactic ability is directed accurately towards neutrophils (Henz *et al.*, 2001). Increased vascular permeability, in addition to endothelial activation, is a necessity for the access of plasma proteins to epithelial surfaces. Mast-cell-derived serine proteases seem to promote such leakage of plasma proteins probably by a proteolytic activity on tight junction proteins (Knight *et al.*, 2000), whereby leakage of plasma proteins can be allowed (Miller, 1996). Infection with many gastrointestinal nematodes leads to pronounced mast cell hyperplasia as well as differentiation and activation of mast cells (Nawa *et al.*, 1994). This mastocytosis is correlated with a systemic release of mast cell serine proteases, which, besides the probability of directly interfering with parasites as mentioned earlier, are essential for the induction and maintenance of increased vascular permeability (Knight *et al.*, 2000). The importance of such proteases for anti-parasitic responses has been investigated in transgenic mice lacking the mMCP-1 (murine mast cell protease) gene. The mMCP-1 is a β-chymase, which in contrast to mast cell tryptase (displaying trypsin specificity), is a chymotrypsin-like serine protease. Mice lacking the mMCP-1 gene expelled *Trichinella spiralis* at a reduced level in both primary and secondary infections. However, expulsion of *Nippostrongylus brasiliensis*, a non-invasive gastrointestinal nematode in contrast to *Trichinella spiralis*, did not differ between $mMCP\text{-}1^{-/-}$ mice and their wild-type littermates (Knight *et al.*, 2000). Although infection with *N. brasiliensis* induces mastocytosis, its clearance in the mouse seems to be independent of mast cells (Nawa *et al.*, 1994), which could explain the lack of influence of mMCP-1. The initial establishment and survival of *T. spiralis* larvae was not affected by the presence of mMCP-1, and it was only during the phase of worm loss that significant differences in worm burdens could be detected (Knight *et al.*, 2000). While mast cell chymases are clearly involved in expulsion of some nematode infections, the mechanisms by which it operates still remain obscure (Knight *et al.*, 2000). Interestingly, a potent chymotrypsin/elastase inhibitor has recently been isolated from *Trichuris suis*, a burrowing gastrointestinal parasite of pigs (Rhoads *et al.*, 2000). This inhibitor, designated TsCEI, highly inhibits the hydrolysis of peptide substrates by chymase (Rhoads *et al.*, 2000). As TsCEI is actively secreted, it indicates that the target for this inhibitor might be host proteases rather than endogenous parasite proteases. Thereby, it could provide a mechanism through which the parasites may interfere with mast cell mediators and by this means significantly contribute to their survival in the host (Rhoads *et al.*, 2000). TsCEI has been assigned to a distinct family of small serine protease inhibitors designated 'smapins', of which 14 members have hitherto been described from parasites (Zang and Maizels, 2001). Out of these, two smapins from *Ascaris suum* also have inhibitory activity on chymotrypsin and could therefore potentially interfere with mast cell chymases in a way like TsCEI (Zang and Maizels, 2001). Finally, mast cell chymase has been further implicated as a central modulator of inflammation by its ability to directly activate interleukin-1β (Mitzutani *et al.*, 1991). IL-1β is produced as a precursor molecule that is cleaved to give a mature peptide. In contrast to IL-1α, only the mature peptide of IL-1β has biological activity and the cleavage of its precursor is normally achieved by a very specific protease called interleukin-converting enzyme (ICE) or caspase-1 (Dinarello, 1997). By cleavage of the IL-1β precursor, mast cell chymase secreted upon parasite stimulation can initiate inflammation by activating this pro-inflammatory cytokine.

Upon activation, mast cells and basophils are stimulated to *de novo* synthesis of a range of pharmacologically active substances. These include lipid metabolites, which are derived from arachidonic acid by either the cyclo-oxygenase pathway leading to the production of prostaglandins and thromboxanes or the lipoxygenase pathway leading to the production of leukotrienes (*Table 2*) (Ikai, 1999). Their effect ranges from immune modulation (e.g. PGE_2) to chemotaxis (e.g. LTB_4) and muscle contraction (e.g. LTC_4); all of which will facilitate expulsion of, for example, gastrointestinal parasites (Henz *et al.*, 2001). Furthermore, activation of mast cells and basophils induces *de novo* synthesis of a number of chemokines of both CXC (IL-8), CC (MIP-1α, MIP-1β, MCP-1, RANTES) and C (lymphotactin) designation (*Table 2*), which will attract various leucocytes to the site of infection. Finally, the expression of a vast array of cytokines is induced as a result of activation of mast cells and basophils (*Table 2*). Contrasting the situation from mast cells, no significant production of pro-inflammatory factors such as TNF-α, IL-1, IL-5, IL-6, or GM-CSF has been reported from basophils (Falcone *et al.*, 2001). Thus, basophils seem more restricted, but are known to release IL-4 and IL-13 upon activation (Galli *et al.*, 2002). The disease status and micro-environmental context (e.g. influence of other cytokines and growth factors) under which mast cells are influenced seem to be of high importance as to which cytokines these cells emit (Henz *et al.*, 2001). Furthermore, the various tissue types of mast cells differ considerably in their profile of secreted cytokines (Henz *et al.*, 2001).

2.6 *Mast cells as antigen-presenting cells*

The knowledge that mast cells display phagocytic activity has prompted the question, whether they also have the ability to serve as antigen-presenting cells (APC) and thereby initiate a specific response. Mast cells can process antigens and subsequently display these on both MHC I and MHC II. This is not the case for basophils (Abraham and Arock, 1998). The significance of this mast cell feature *in vivo* and in relation to professional APC has, however, not yet been established (Galli et el., 1999). MHC class I has for a long time been known to be present on mast cells (Henz *et al.*, 2001). The situation for MHC II is more complex – generally, no or low levels of baseline expression of MHC II can be observed in resting mast cells. However, MHC II can be up-regulated as seen on mast cells isolated from infected tissue or during exposure to specific stimuli (Henz *et al.*, 2001). MHC can also be up-regulated upon stimulation with LPS as well as INF-γ, TNF-α and IL-4. Generally, the expression is dependent on the maturation state as well as activation state of the mast cells. Mast cells can migrate towards lymphocytes, as they are equipped with integrins and chemokine receptors. Migration of mast cells towards lymph nodes has actually been observed, although the chemotactic factor(s) responsible for this migration is not known (Henz *et al.*, 2001). However, IL-8 is speculated to be involved due to the presence of CXCRI and CXCRII receptors on mast cells (Henz *et al.*, 2001). Furthermore, mast cells seem to be equipped with most, if not all, basic prerequisites for interacting with lymphocytes such as displaying various important co-stimulating surface receptors. These include CD43 (leukosialin), CD48, CD154 (also known as CD40L), CD40, CD80 and CD86 (members of the B7 family) (Galli *et al.*, 2002; Ghannadan *et al.*, 1998; Henz *et al.*, 2001). However, the B7 receptors (CD80 and CD86) have only been studied in the context of antigen presentation (Henz *et al.*, 2001) and could not be detected on mast cells isolated from juvenile foreskin or adult mammary skin exploiting a panel of mAb against defined CD antigens (Ghannadan *et al.*, 1998). Although evidence is accumulating showing that mast cells can actually act as antigen-presenting cells, their capacity to do so seems restricted in relation to specialized APCs and remains to be clarified *in vivo* (Henz *et al.*, 2001).

2.7 *Mast cells and basophils in adaptive immunity against parasites*

Parasite infections are quite often associated with increased levels of circulating basophils and eosinophils, followed by highly augmented levels of IgE (Galli *et al.*, 2002). Parasite clearance will often be associated with increased numbers of thionine-positive cells in affected tissues. This is revealed by the infiltration of basophils in inflamed skin during infection with many ectoparasitic arthropods (Falcone *et al.*, 2001; Galli *et al.*, 2002) or the marked mastocytosis during infection with gastrointestinal nematodes (Galli *et al.*, 2002). In relation to the latter, mast-cell-deficient mice exhibit highly delayed expulsion of *Strongyloides ratti* infections (Nawa *et al.*, 1994). The initial activation of mast cells leads to a degranulation process, which manifests itself as an initial drop in the density of mucosal mast cells during the earlier phases of infection (Issekutz *et al.*, 2001). By which mechanism(s) this initial activation and degranulation is based has not been fully elucidated (Issekutz *et al.*, 2001), but it is plausible that pattern recognition receptors, as discussed earlier, might be involved. This activation initiates the release of inflammatory mediators (including cytokines – *Table 2*) and increased mucosal permeability, which facilitates egress of plasma proteins and diapedesis of myeloid and lymphoid cells (Miller, 1996). Although the cytokines released by mast cells are highly dependent on the type of mast cells studied, under most circumstances, cytokine patterns elicited by these cells are of a Th_2 profile (*Table 2*) (Henz *et al.*, 2001). Thus, under the influence of such Th_2 cytokines in the milieu of the APC and the naive, undifferentiated T cell in peripheral lymphoid tissue, the specific response will be biased towards a Th_2 response (Miller, 1996). However, mast cells stimulated with PMA have interestingly been shown to secrete INF-γ, which suggests that they are also able to induce Th_1 responses depending on the type of stimulus they are exposed to and therefore could be involved in the Th_1/Th_2 dichotomy (Henz *et al.*, 2001). Th_2 cells will release IL-4 and -5, which are responsible for the IgE isotype switching and IL-3, -4, -9, -10 and -13 accounting for the mast cell hyperplasia seen in helminth infections (Miller, 1996). It is generally accepted that the local mast cell hyperplasia observed during the expulsion phase of helminth infections is dependent on T cells and IL-3 (Issekutz *et al.*, 2001). This is based upon studies using nude, hypothymic mice, which have been shown to be unable to expel infections with *Nippostrongylus brasiliensis*, *Strongyloides ratti* and *S. venezuelensis* (Nawa *et al.*, 1994). Furthermore, it has been shown that *N. brasiliensis*-associated intestinal mastocytosis can be adoptively transferred by sIg^- immune thoracic duct lymphocytes (Nawa *et al.*, 1994). The significance of IL-3 for intestinal mastocytosis has been deciphered by administration of anti-IL-3 antibodies, which suppress helminth-induced mast cell hyperplasia (Issekutz *et al.*, 2001). In addition, repeated administration of IL-3 to hypothymic mice has shown to enable T-cell-deficient hosts to expel *S. ratti* from the intestine with an associated mastocytosis (Nawa *et al.*, 1994). Likewise, mice deficient in IL-3 by gene targeting in embryonic stem cells have provided further evidence that IL-3 contributes to increased numbers of tissue mast cells as well as enhanced basophil production and immunity toward infection with *S. venezuelensis* (Lantz *et al.*, 1998). Thus, there is convincing evidence to support the notion that mastocytosis and IgE up-regulation following infection with intestinal helminths is driven by cytokines of a Th_2 profile. The mast cell hyperplasia is, however, not the mucosal defence mechanism *per se*. Evidence for immune mechanisms involving mucus-producing goblet cells has been gathered in recent years. Thus, although *N. brasiliensis* truly induces mastocytosis, this event does not seem to be a prerequisite for clearance as it is for burrowing parasite species (Nawa *et al.*, 1994). Actually, mast-cell-deficient mice seem to expel these parasites more or less comparable to wild-type littermates (Nawa *et al.*, 1994). Therefore, it has been suggested that the expulsion of non-dwelling nematodes, such as *N. brasiliensis*, relies on a T-cell-dependent 'damage' of worm tegument and likewise a T-cell-dependent proliferation of goblet cells (Nawa *et al.*, 1994). Normal worms seem to release excretory/secretory substances, which at least to some extent can counterbalance the detrimental effects on the tegument exerted by the host (Nawa *et al.*, 1994). Both clearance of *N. brasiliensis* and goblet cell hyperplasia can be adoptively transferred by immune T cells (Nawa *et al.*, 1994). The actual expulsion seems to be achieved through T-cell-independent qualitative alterations of terminal sugars of mucins (Nawa *et al.*, 1994). Interestingly, IL-3 deficiency does not impair expulsion of *N. brasiliensis* despite significant reductions in mastocytosis and basophilia (Lantz *et al.*, 1998).

Secondary infections with intestinal helminths are generally cleared quicker in more exaggerated reactions than primary infections (Knight *et al.*, 2000; Miller, 1996). This is due to a type-I hypersensitivity reaction based on the presence of parasite-specific IgE generated during the priming infection (Miller, 1996). Cross-linking of specific IgE avidly bound to mast cell and basophil FcεRI receptors induces a degranulation and mediator release, which is exaggerated and occurs faster than observed by innate recognition mechanisms (Galli *et al.*, 1999).

Basophils have been shown to be important players in immunity against various ectoparasitic infections – especially against ticks (Falcone *et al.*, 2001). During primary infestation, the feeding activity of arthropod ectoparasites may cause degranulation and mediator release from skin thionine-positive cells (Galli *et al.*, 1999). A multitude of cytokines and growth factors seem to have a marked effect on basophils, whereas their effect on mast cells is limited. These include IL-1, IL-3, IL-5, NGF-β and GM-CSF. Furthermore, basophils only contain small quantities of preformed proteases or toxic proteins (with the exception of MBP and lysophospholipase) (Falcone *et al.*, 2001). Activated basophils rapidly produce and release large quantities of IL-4 and IL-13, which is highly exaggerated in sensitized animals upon crosslinking of receptor-bound IgE (Falcone *et al.*, 2001). These are major Th_2-type cytokines, which might be crucial for the support of Th_0 cell subset differentiation to the Th_2 phenotype (Falcone *et al.*, 2001). Specific immunity can be initiated in draining lymph nodes by involvement of Langerhans cells as APCs and leads to production of parasite-specific Ig. Subsequent exposure will cause rapid rejection by a specialized form of delayed-type hypersensitivity (DTH) reaction called cutaneous basophil hypersensitivity (CBH) (Falcone *et al.*, 2001). Upon crosslinking of receptor-bound IgE, basophils rapidly produce and release large quantities of IL-4 and IL-13 – major Th_2-type cytokines (Falcone *et al.*, 2001). In the CBH type of acquired immunity, basophils are a major component and seem to be required for full expression of immune resistance to infestation of the skin in guinea pigs by larval ixodid ticks of the species *Amblyomma americanum* (Falcone *et al.*, 2001; Galli *et al.*, 1999). Likewise, sensitized guinea pigs infested by *Rhiphicephalus appendiculatus* rejected the ectoparasites vigorously by involvement of a marked accumulation of basophils in skin as well as a marked systemic basophilia (Falcone *et al.*, 2001). By similar mechanisms, challenge infestations with the tsetse fly *Glossina morsitans* were rejected in the same experimental animal system (Falcone *et al.*, 2001). In mice, where mast cells in skin predominate rather than basophils, expression of IgE-dependent immune resistance to cutaneous feeding by larval *Haemaphysalis longicornis* ticks seems to be dependent on mast cells (Galli *et al.*, 1999).

Conclusively, both types of thionine-positive cells can be regarded not only as important effector cells in innate immunity and for the initiation and orchestration of adaptive responses. They clearly also play a crucial role in acquired immunity towards various parasitic infections. However, their significance in host defence and not least the mechanisms by which they are enrolled have just begun to be firmly enlightened.

3 Thionine-positive cells in lower vertebrates

The main research efforts on putative mast cells have been based on their function in a range of mammalian species. However, the relatively few studies on various cell types, which were conducted with lower vertebrates and invertebrates have indicated the existence of corresponding cells in these distantly related animal groups. The distribution of these studies among systematic groups is, however, not balanced. Investigations on fish cells dominate whereas data for amphibian, reptilian and avian mast cell equivalents are poorly represented. No studies have yet characterized differences between circulating basophils and sedentary tissue mast cells in fish and these types are generally supposed to be interrelated also in fish.

3.1 *Evolution of thionine-positive cells*

It is believed that an ancestral cell type of the lympho-haematopoietic system existed among primitive animals such as invertebrates. The function of these cells was probably phagocytic and was able to recognize foreign antigens, and thereby increase engulfment. From this primitive ancestor several cell lineages were derived, which resulted in various categories such as macrophages, granulocytes, eosinophils, mast cells and lymphocytes.

From mammalian studies, histamine has been indicated as an important constituent of mast cells. However, additional molecules such as proteases, cytokines, chemotactic factors, serotonin and proteoglycans are abundant in mast cells. Glycosaminoglycans heparin and heparan sulphate, have been found in molluscs and arthropods and are generally believed to bind serotonin and histamine. Recently, the clam *Mercenaria mercenaria* was found to harbour abundant mast-like cells containing numerous granules with heparin (Ulrich and Boon, 2001) thus suggesting the presence of mast cells at this early phylogenetic stage. Likewise, the tunicate *Styela plicata* contains heparin-filled cells in various tissues, including epithelia (Cavalcante *et al.*, 2000). During evolution, the above-mentioned diversification of the cell lineage must have occurred several times. It is therefore suggested that strict equivalence between these different cell types in various taxonomic groups do not occur. Thus, in mammals eosinophils and mast cells are considered different entities, whereas in teleosts eosinophilic granular cells (EGC) are also thionine positive and thus considered to be mast cell equivalents (Reite, 1998).

Mast cells are also known from amphibia, such as the toad *Bufo viridis*, in which their development is highly dependent on sex hormone exposure (Minucci *et al.*, 1994). This was also demonstrated in mast cells from reptiles. Thus, oestrogen treatment of the lizard *Podarcis sicula* resulted in a strong increase in mast cell density (Minucci *et al.*, 1995). Birds are also known to possess mast cell equivalents. Basophils are a common part of the avian blood picture (Hauptmanova *et al.*, 2002). A comparative study performed by Takaya and Yoshida (1997) on mast cells from fish, amphibians, reptiles and mammals showed a clear variation of elemental constituents among these groups; sulphur contents varied especially. Mast cells from goldfish intestine had a low content in comparison with other groups. Although not fully elucidated, these differences may point to functional dissimilarities among mast cells from different systematic entities.

Due to the fact that fish mast cells seem to correspond to EGCs, it is appropriate to emphasize that anti-parasitic mechanisms are also present in mammalian eosinophils (Meeusen and Balic, 2000). Apart from releasing a range of mediators eliciting inflammatory reactions, direct effector mechanisms aimed at killing invading parasites are known. It cannot be excluded that similar mechanisms exist in fish EGCs.

3.2 *Anti-parasitic mechanisms of thionine-positive cells in fish*

Early work on putative mast cells in fish showed that histamine, the above-mentioned important constituent of mammalian mast cells, is very rare in this teleost cell type (Ellis, 1982). This will indicate that their function in fish differs at least partly from their mammalian counterparts. In addition it was demonstrated that EGCs also have mast cell properties. Thus, if the EGCs are fixed in absolute alcohol the water-soluble basophilic granules are preserved and display metachromasia, whereas fixation with formalin destroys this stainability (Reite, 1998). This also suggests a special function and evolution of fish mast cells.

Inflammation

Increase in circulating basophils during intra-orbital inflammation of carp has been described by Suzuki and Hibiya (1986). In addition, during wound healing migration of basophils to the site of inflammation is seen (Iger and Abraham, 1990). It has been suggested that basophils are stored in the haematopoietic tissue and enter circulation due to inflammatory stimuli (Suzuki and Iida, 1992).

The EGCs have been studied in some detail by Matsuyama and Iida (1999). They demonstrated that these cells degranulated following treatment with foreign antigen and indicated the presence of neutrophil-attracting substances in the granules of EGCs. Further work on the substances contained in these cells showed a clear role in attraction of neutrophil cells and augmenting their diapedesis into inflammatory sites (Matsuyama and Iida, 2002). The permeability of the blood vessels during this process was described (Matsuyama and Iida, 2001). EGCs were obtained from tilapia (*Oreochromis niloticus*) by washing the body cavity and the swim bladder membrane and subsequent centrifuged on a Percoll gradient. Lysates from the isolated fish cells were then injected into the swim bladder of the fish. It was seen that the vascular permeability was enhanced markedly.

Further studies by Matsuyama and Iida (2002) elucidated the effects of tilapia mast cell lysates on neutrophil adherence to endothelial cells. Following treatment with the mast cell constituent, the neutrophils changed morphology and increased their binding to cultured vascular endothelial cells. This suggested that adhesion molecules became expressed following lysate treatment. These could be speculated to explain the molecular basis of endothelial–neutrophil binding. The adhesion molecules were not identified, but future studies should investigate if equivalents to intercellular adhesion molecule ICAM-1 and vascular adhesion molecule VCAM-1 are produced due to exposure to mast cell contents.

EGCs are prevalent in rainbow trout tissue, where the cells have been found in gills, skin and gut (Holland and Rowley, 1998). Inflamed gill tissue in association with degranulation of EGCs following infection of rainbow trout with *Renibacterium salmoninarum* was shown by Flano *et al.* (1996). The authors also demonstrated differentiation and proliferation of EGCs when exposed to *R. salmoninarum* and suggested that this process was cytokine mediated and not merely run directly by bacterial products.

Direct anti-parasitic effects

Studies on brown trout skin heavily infected with the skin parasitic flagellate *Ichthyobodo necator* showed an increased density of putative mast cells (Blackstock and Pickering, 1980). Salmonids infected with intestinal parasites such as the acanthocephalan *Acanthocephalus dirus* displayed strong accumulation of basophils at the infection site (Bullock, 1963). The above-mentioned studies did not demonstrate a direct anti-parasitic effect of EGCs but teleost mast cells will at least in some cases contain molecules with direct effect on invading pathogens.

Silphaduang and Noga (2001) isolated a number of peptides designated piscidins from various tissues in hybrid striped bass and demonstrated a marked anti-bacterial effect of the substances. Subsequently they demonstrated, by the use of immunohistochemistry, that the peptide antibiotics were localized in mast cells of the host. Gills, skin, gut and vessels in the viscera were found to be positive to the peptide antibiotics. This finding indicates that teleost mast cells do not only exert their action through induction of inflammatory reactions, but also through direct killing of pathogens.

Some indications of cytotoxic activity of EGCs from sea bass were found by Cammarata *et al.* (2000) and Vazzana *et al.* (2002). Tumour cells were killed *in vitro* by leucocytes and further work showed that the killing effect was associated with isolated EGCs.

Direct killing activity of EGCs on protozoan and metazoan parasites has not been demonstrated but these cells are clearly activated upon infections with endo-parasitic helminths. Thus, Dezfuli *et al.* (2000) described the presence of EGCs in reactive foci around *Raphidascaris acus* infecting pancreas tissue of minnows. Whether mast cells produce substances acting on ecto-parasites, such as monogeneans and ciliates, is unknown. However, it was shown by Sigh and Buchmann (2000) that thionine-positive cells in trout skin degranulated following infection with *Gyrodactylus derjavini* and *Ichthyophthirius multifiliis*. Further work should elucidate if this process is associated with release of anti-parasitic compounds and if the degranulation leads to increased vascular permeability and release of plasma proteins.

4 Concluding remarks

Mast cells and basophils were originally described as granulated cells displaying metachromasia, when stained by thionine. Subsequently, a number of characters besides the histology have been recognized, which has eased the identification of these cells. Still, prevailing knowledge on the origin and uniformity of mast cells and basophils is insufficient. Search for homologues of mast cells and basophils in other systematic entities than mammals has been limited. This is primarily due to the fact that metachromasia is not a unique character for these cells. Hence, the fact that EGCs in teleosts display metachromasia has lead to the suggestion that they are putative mast cells. Whether they are functionally identical needs to be clarified. Generally, the nomenclature of leucocytes from non-mammalian species is based on resemblance to mammalian counterparts rather than direct descriptions of developmental lineages of haematopoietic precursors. This could lead to misassumptions, as function is often deduced only from morphological similarities. Due to the lack of additional characters than metachromasia, functional studies of mast cells and basophils in invertebrates and lower vertebrates are few. The involvement of mast cells and basophils in allergic diseases and disorders has been thoroughly investigated and led to a huge knowledge about their specific involvement in inflammatory processes. In addition, a number of reports on the contribution of these cells in anti-parasitic responses have been presented, although the mechanisms are insufficiently described. However, for a long time the significance and involvement of mast cells and basophils in natural immunity to infectious diseases in general has been enigmatic. The elucidation of this aspect has essentially just been initiated by a few reports within recent years.

Acknowledgements

This work was supported in part by the European Community's Improving Human Potential Programme under contract (HPRN-CT-2001-00214), (PARITY).

References

Abraham, S.N. and Arock, M. (1998) Mast cells and basophils in innate immunity. *Sem. Immunol.* **10:** 373–381.

Anwar, A.R., McKean, J.R., Smithers, S.R. and Kay, A.B. (1980) Human eosinophil- and neutrophil-mediated killing of schistosomula of *Schistosoma mansoni in vitro.* I. Enhancement of complement-dependent damage by mast cell-derived mediators and formyl methionyl peptides. *J. Immunol.* **124:** 1122–1129.

Arizmendi, N., Yépez-Mulia, L., Cedillo-Rivera, R., Ortega-Pierres, M.G., Muñoz, O., Befus, D. and Enciso-Moreno, J.A. (2001) Interleukin mRNA changes in mast cells stimulated by TSL-1 antigens. *Parasite* **8:** 114–116.

Austen, K.F. and Orange, R.P. (1975) Bronchial asthma: The possible role of the chemical mediators of immediate hypersensitivity in the pathogenesis of subacute chronic disease. *Am. Rev. Respir. Dis.* **112:** 423–436.

Bidri, M., Vouldoukis, I., Mossalayi, M.D., Debre, P., Guillosson, J.J., Mazier, D. and Arock, M. (1997) Evidence for direct interaction between mast cells and *Leishmania* parasites. *Parasite Immunol.* **19:** 475–483.

Blackstock, N. and Pickering, A.D. (1980) Acidophilic granular cells in the epidermis of the brown trout, *Salmo trutta* L. *Cell Tissue Res.* **210:** 359–369.

Bochner, B.S. and Schleimer, R.P. (2001) Mast cells, basophils, and eosinophils: distinct but overlapping pathways for recruitment. *Immunological Rev.* **179:** 5–15.

Bradding, P., Okayama, Y., Church, M.K. and Holgate, S.T. (1995) Heterogeneity of human mast cells based on their cytokine content. *J. Immunol.* **155:** 297–307.

Bullock, W.L. (1963) Intestinal histology of some salmonid fishes with particular reference to the histopathology of acanthocephalan infections. *J. Morphol.* **112:** 23–43.

Cammarata, M., Vazzana, M., Cervello, M., Vincenzo, A. and Parrinello, N. (2000) Spontaneous cytotoxic activity of eosinophilic granule cells separated from the normal peritoneal cavity of *Dicentrarchus labrax. Fish Shellfish Immunol.* **10:** 143–154.

Cavalcante, M.C.M., Allodi, S., Valente, A.P., Straus, A.H., Takahashi, H.K., Mourao, P.A.S. and Pavao, M.S.G. (2000) Occurrence of heparin in the invertebrate *Styela plicata* (Tunicata) is restricted to cell layers facing the outside environment – an ancient role in defense. *J. Biol. Chem.* **275:** 36189–36196.

Cheng, Q.I., Lixin, L. and Krilis, S.A. (2001) Cells with features of mast cells and basophils. *ACI Int.* **13/5:** 197–203.

Coelho-Castelo, A.A.M., Panunto-Castelo, A., Moreno, A.N., Dias-Baruffi, M., Namur, M.C., Oliver, Roque-Barreira, M.-C. and Rodrigues, V. (2002) Sm60, a mannose-binding protein from *Schistosoma mansoni* with inflammatory property. *Int. J. Parasitol.* **32:** 1747–1754.

Costa J.J., Weller P.F. and Galli, S.J. (1997) The cells of the allergic response – Mast cells, basophils, and eosinophils. *J. Am. Med. Ass.* **278:** 1815–1822.

Dezfuli, B.S., Simoni, E., Rossi, R. and Manera, M. (2000) Rodlet cells and other inflammatory cells of *Phoxinus phoxinus* infected with *Raphidascaris acus* (Nematoda). *Dis. Aquat. Org.* **43:** 61–69.

Dinarello, C.A. (1997) Interleukin-1. *Cytokine Growth Factor Rev.* **4:** 253–265.

Ellis, A.E. (1982) Histamine, mast cells and hypersensitivity responses in fish. *Dev. Comp. Immunol.* **2:** 147–155.

Falcone, F.F., Prichard, D.I. and Gibbs, B.F. (2001) Do basophils play a role in immunity against parasites? *Trends Parasitol.* **17:** 126–129.

Flano, E., Lopez-Fierro, P., Razquin, B.E. and Villena, A. (1996) In vitro differentiation of eosinophilic granular cells in *Renibacterium salmoninarum*-infected gill cultures from rainbow trout. *Fish Shellfish Immunol.* **6:** 173–184.

Galli, S.J. and Dvorak, A.M. (1995) Production, biochemistry and function of basophils and mast cells. In: Beutler, E., Lichtman, M.A., Coller, B.S. and Kipps, T.J. (eds) *Williams Hematology*, 5th edn., pp. 805–810, McGraw-Hill, New York.

Galli, S.J., Dvorak, A.M. and Dvorak, H.F. (1984) Basophils and mast cells: Morphologic insights into their biology, secretory patterns and function. *Prog. Allergy* **34:** 1–141.

Galli, S.J., Maurer, M. and Lantz, C.S. (1999) Mast cells as sentinels of innate immunity. *Cur. Opin. Immunol.* **11:** 53–59.

Galli, S.J., Wedemeyer, J. and Tsai, M. (2002) Analyzing the roles of mast cells and basophils in host defence and other biological responses. *Int. J. Haematol.* **75:** 363–369.

Ghannadan, M., Baghestanian, M., Wimazal, F., Eisenmenger, M., Latal, D., Kargul, G., Walchshofer, S., Sillaber, C., Lechner, K. and Valent, P. (1998) Phenotypic characterization of human skin mast cells by combined staining with toluidine blue and CD antibodies. *J. Invest. Dermatol.* **111:** 689–695.

Gill, H.S. (1986) Kinetics of mast cell, basophil and eosinophils populations at ***hyalomma anatolicum anatolicum*** feeding sites on cattle and the acquisition of resistance. *Parasitology* **93:** 305–315.

Gurish, M.F., Friend, D.S., Webster, M., Ghildyal, N., Nicodemus, C.F. and Stevens, R. (1997) Mouse mast cells that possess segmented multi-lobular nuclei. *Blood* **90:** 382–390.

Hauptmanova, K., Literak, I. and Bartova, E. (2002) Haematology and leucocytozoonosis of great tits (*Parus major* L.) during winter. *Acta Vet. Brno.* **71:** 199–204.

Henz, B.M., Maurer, M., Lippert, U., Worm, M. and Babina, M. (2001) Mast cells as initiators of immunity and host defence. *Exp. Dermatol.* **10:** 1–10.

Holland, J.W. and Rowley, A.F. (1998) Studies on the eosinophilic granule cells in the gills of the rainbow trout, *Oncorhynchus mykiss. Comp. Biochem. Physiol. C* **120:** 321–328.

Humphries, D.E., Wong, G.W., Friend, D.S., Gurish, M.F., Qiu, W.T., Huang, C.F., Sharpe, A.H. and Stevens, R.L. (1999) Heparin is essential for the storage of specific granule proteases in mast cells. *Nature* **400:** 769–772.

Iger, Y. and Abraham, M. (1990) The process of skin healing in experimentally wounded carp. *J. Fish Biol.* **36:** 421–437.

Ikai, K. (1999) Psoriasis and the arachidonic acid cascade. *J. Dermatol. Sci.* **21:** 135–146.

Irani, A.A., Craig, S.S., Nilsson, G., Ishizaka, T. and Schwartz, L.B. (1992) Characterization of human mast cells developed in vitro from fetal liver cells co-cultured with murine 3T3 fibroblasts. *Immunology* **77:** 136–143.

Ishikaza, T., Furitsu, T., Mitsui, H., Takei, M., Koszloski, M.A., Irani, A-M.A., Scbwartz, L.B., Dvorak, A.M., Zsebo, K.M. and Gillis, S. (1992) Development of human mast cells from umbilical cord blood cells by human and murine c-kit ligand. *FASEB J.* **6:** A1403.

Issekutz, T.B., Palecanda, A., Kadela-Stolarz, U. and Marshall, J.S. (2001) Blockade of esther alpha-4 or beta-7 integrins selectively inhibits intestinal mast cell hyperplasia and worm expulsion in response to *Nippostrongylus brasiliensis* infection. *Eur. J. Immunol.* **31:** 860–868.

Kirschenbaum A.S., Goff J.P., Kessler S.W., Mican J.M., Zsebo, K.M. and Metcalfe, D.D. (1992) Effect of IL-3 and stem cell factor on the appearance of human basophils and mast cells from CD34+ pluripotent progenitor cells. *J. Immunol.* **148:** 772–777.

Knight, P.A., Wright, S.H., Lawrence, C.E., Paterson, Y.Y.W. and Miller, H.R.P. (2000) Delayed expulsion of the nematode *Trichinella spiralis* in mice lacking the mucosal mast cell-specific granule chymase, mouse mast cell protease-1. *J. Exp. Med.* **192:** 1849–1856.

Lagunoff, D. (1972) Contributions of electron microscopy to the study of mast cells. *J. Invest. Dermatol.* **58:** 296–311.

Lagunoff, D. and Pritzl, P. (1976) Characterization of rat mast cell granule proteins. *Arch. Biochem. Biophys.* **173:** 554–563.

Lantz, C.S., Boesiger, J., Song, C.H., Mach, N., Kobayashi, T., Mulligan, R.C., Nawa, Y., Dranoff, G. and Galli, S.J. (1998) Role for interleukin-3 in mast cell and basophil development and in immunity to parasites. *Nature* **392:** 90–93.

Li, L., Meng, W. and Krilis, S.A. (1996) Mast cell expressing chymase but not tryptase can be derived by culturing human progenitors in conditioned media obtained from a human mastocytosis cell strain with c-kit ligand. *J. Immunol.* **156:** 4839–4844.

Li, L., Li, Y., Reddel, S.W., Cherrian, M., Friend, D.S., Stevens, R.L., and Krilis, S.A. (1998) Identification of basophilic cells in the peripheral blood of asthma, allergy, and drug-reactive patients that express mast cell granule proteases. *J. Immunol.* **161:** 5079–5086.

Malaviya, R., Ikeda, T., Ross, E. and Abraham, N. (1996) Mast cell modulation of neutrophil influx and bacterial clearance at sites of infection through TNF-α. *Nature* **381:** 77–80.

Malaviya, R., Ross, E.A., MacGregor, J.I., Ikeda, T., Little, J.R., Jakschik, B.A. and Abraham, S.N. (1994) Mast cell phagocytosis of FimH-expressing enterobacteria. *J. Immunol.* **152:** 1907–1914.

Matsuyama, T. and Iida, T. (1999) Degranulation of eosinophilic granular cells with possible involvement in neutrophil migration to site of inflammation in tilapia. *Dev. Comp. Immunol.* **23:** 451–457.

Matsuyama, T. and Iida, T. (2001) Influence of tilapia mast cell lysate on vascular permeability. *Fish Shellfish Immunol.* **11:** 549–556.

Matsuyama, T. and Iida, T. (2002) Tilapia mast cell lysates enhance neutrophil adhesion to cultured vascular endothelial cells. *Fish Shellfish Immunol.* **13:** 243–250.

McCurdy, J.D., Lin, T.-J. and Marshall, J.S. (2001) Toll-like receptor 4-mediated activation of murine mast cells. *J. Leukocyte Biol.* **70:** 977–984.

McKean, P.G. and Pritchard, D.I. (1989) The action of a mast cell protease on the cuticular collagens of *Necator americanus*. *Parasite Immunol.* **11:** 293–298.

Meeusen, E.N.T. and Balic, A. (2000) Do eosinophils have a role in the killing of helminth parasites. *Parasitol. Today* **16:** 95–101.

Miller, H.R.P. (1996) Mucosal mast cells and the allergic response against nematode parasites. *Vet. Immunol. Immunopathol.* **54:** 331–336.

Minucci, S., Baccari, G.C. and Matteo, L.D. (1994) The effect of sex hormones on lipid content and mast cell number in the harderian gland of the female toad, *Bufo viridis*. *Tissue Cell* **26:** 797–805.

Minucci, S., Vitiello, I.I., Marmorino, C., Di Matteo, L. and Baccari, G.C. (1995) Mast cell–Leydig cell relationships in the testis of the lizard *Podarcis s. sicula* Raf: thermal manipulation, ethane 1, 2-dimethane sulphonate (EDS) and sex hormone treatment. *Zygote* **3:** 259–264.

Mitzutani, H., Schechter, N.M., Lazarus, G., Black, R.A. and Kupper, T.S. (1991) Rapid and specific conversion of precursor interleukin 1β (IL-1β) to an active IL-1 species by human mast cell chymase. *J. Exp. Med.* **174:** 821–825.

Nawa, Y., Ishikawa, N., Tsuchiya, K., Horii, Y., Abe, T., Khan, A.I., Bing-Shi, B., Itoh, H., Ide, H. and Uchiyama, F. (1994) Selective effector mechanisms for the expulsion of intestinal helminths. *Parasite Immunol.* **16:** 333–338.

Neitz, A.W.H., Grothe, R., Pawlas, S. and Groeneveld H.T. (1993) Investigations into lymphocyte transformation and histamine release by basophils in sheep repeatly infested with *Rhipicphalus evertsi evertsi* ticks. *Exp. Appl. Acrology* **17:** 551–559.

Prodeus, A.P., Zhou, X., Maurer, M., Galli, S.J. and Carrol, M.C. (1997) Impaired mast cell dependent natural immunity in complement C3-deficient mice. *Nature* **390B:** 172–175.

Reite, O.B. (1998) Mast cells/eosinophilic granule cells of teleostean fish: a review focusing on staining properties and functional responses. *Fish Shellfish Immunol.* **8:** 489–513.

Reite, O.B. and Evensen, Ø. (1994) Mast cells in the swimbladder of Atlantic salmon *Salmo salar*: histochemistry and responses to compound 48/40 and formalin-inactivated *Aeromonas salmonicida*. *Dis. Aquat. Org.* **20:** 95–100.

Rhoads, M.L., Fetterer, R.H., Hill, D.E. and Urban, F.J. (2000) *Trichuris suis*: A secretory chymotrypsin/elastase inhibitor with potential as an immunostimulator. *Exp. Parasitol.* **95:** 36–44.

Rosenkranz, A.R., Coxon, A., Maurer, M., Gurish, M.F., Austen, K.F., Friend, D.S., Galli, S.J. and Mayadas, T.N. (1998) Cutting edge: Impaired mast cell development and innate immunity in Mac-1 (CD11b/CD18, CR3)-deficient mice. *J. Immunol.* **161:** 6463–6467.

Schubert, M. and Hamerman, D. (1956) Metachromasia: chemical theory and histochemical use. *J. Histochem. Cytochem.* **4:** 159–189.

Scott, M.A. and Stockham, S.L. (2000) Basophils and mast cells. In: Feldman B.F., Zinkl, J.G. and Jain, N.C. (eds) *Schalm's Veterinary Hematology*, 5th edn., pp. 308–317. Lippincott Williams & Wilkins, New York.

Sher, A. (1976) Complement-dependent adherence of mast cells to schistosomula. *Nature* **263:** 334–336.

Sigh, J. and Buchmann, K. (2000) Associations between epidermal thionin-positive cells and skin parasitic infections in brown trout *Salmo trutta*. *Dis. Aquat. Org.* **41:** 135–139.

Silphaduang, U. and Noga, E.J. (2001) Peptide antibiotics in mast cells of fish. *Nature* **414:** 268–269.

Solivan, S., Selwood, T., Wang, Z.-M. and Schechter, N.M. (2002) Evidence for diversity of substrate specificity among members of the chymase family of serine proteases. *FEBS Letters* **512:** 133–138.

Suzuki, Y. and Hibiya, T. (1986) Dynamics of leukocytic inflammatory responses in carp. *Fish Pathol.* **23:** 179–184.

Suzuki, Y. and Iida, T. (1992) Fish granulocytes in the process of inflammation. *Annu. Rev. Fish Dis.* **2:** 149–160.

Takaya, K. and Yoshida, T. (1997) Comparative study of elemental constituents of vertebrate mast cell and basophil granules by quantitative X-ray microanalysis on fresh preparations. *J. Trace Microprobe Techn.* **15:** 669–672.

Ulrich, P.N. and Boon, J.K. (2001) The histological localization of heparin in the northern Quahog clam, *Mercenaria mercenaria*. *J. Inv. Pathol.* **78:** 155–159.

Valent, P., Spanblochl, E., Sperr, W.R., Sillaber, C., Zsebo, K.M., Agis, H., Strobl, H., Geissler, K., Bettelheim, P. and Lechner, K. (1992) Induction of differentiation of human mast cells from bone marrow and peripheral blood mononuclear cells by recombinant human stem cell factor/kit-ligand in long-term culture. *Blood* **80:** 2237–2245.

Vazzana, M., Cammarata, M., Cooper, E.L. and Parrinello, N. (2002) Confinement stress in sea bass (*Dicentrarchus labrax*) depresses peritoneal leukocyte cytotoxicity. *Aquaculture* **210:** 231–243.

Wardlaw, A.J., Walsh, G.M. and Symon, F.A. (1994) Mechanisms of eosinophils and basophil migration. *Allergy* **49:** 797–807.

Wheater, P.R., Burkitt, H.G. and Daniels, V.G. (1979) *Functional Histology*, 1st. edn. Churchill Livingstone, New York.

Wong, G.W., Foster, P.S., Yasuda, S., *et al.* (2002) Biochemical and functional characterization of human transmembrane tryptase (TMT)/tryptase gamma. *J. Biol. Chem.* **277:** 41906–41915.

Zang, X. and Maizels, R.M. (2001) Serine proteinase inhibitors from nematodes and the arms race between host and pathogen. *Trends Biochem. Sci.* **26:** 191–197.

Zwirner, J., Gotze, O., Sieber, A., Kapp, A., Begemann, G., Zuberbier, T. and Werfel, T. (1998) The human mast cell line HMC-1 binds and responds to C3a but not C3a(desArg). *Scand. J. Immunol.* **47:** 19–24.

4

Animal models for the study of innate immunity: protozoan infections in fish

Maaike Joerink, Jeroen P.J. Saeij, James L. Stafford, Miodrag Belosevic and Geert F. Wiegertjes

1 Introduction

1.1 *Kinetoplastid infections in carp*

Infections with kinetoplastid parasites (e.g. *Trypanosoma brucei*, *T. cruzi*, *Leishmania* spp.) not only are a considerable problem in humans; in fact, African trypanosomiasis has re-emerged as a major health threat to rural Africans with >100 000 new infections per year (Welburn *et al.*, 2001) but also in animal husbandry; the costs of trypanosomiasis in cattle for the whole of Africa exceeds \$1 billion annually (Kristjanson *et al.*, 1999). Studies on these mammalian kinetoplastid parasites and their concomitant immune response have given valuable insight into parasite evasion mechanisms and into the host immune system. However, most of the understanding of trypanosomiasis has been obtained from studies in mice which, with the exception of *T. musculi*, are not a natural host to tsetse-transmitted trypanosomes. Natural parasite–host models such as those in cattle (Naessens *et al.*, 2002) but also natural parasite–host models in fish such as described in this chapter, can be of true importance for providing new insights into the host interaction and phylogeny of the Kinetoplastida. Furthermore, the study of fish kinetoplastid parasites in relation to their host immune response may help to resolve fish diseases caused by these flagellates (Lom, 1979).

Interestingly, the common carp (*Cyprinus carpio* L.) is the natural host of two kinetoplastid parasites (*Trypanoplasma borreli* and *Trypanosoma danilewskyi*) that diverged more than 500 million years ago (Fernandes *et al.*, 1993). Blood-sucking leeches (*Piscicola geometra* or *Hemiclepsis marginata*) act as vectors for transmitting kinetoplastid parasites between cyprinid fish and many carp will in fact carry mixed populations of *Trypanoplasma borreli* (syn. *Cryptobia borreli*) and *Trypanosoma danilewskyi* (syn. *T. carassii*) (Lom and Dyková, 1992). Both protozoan parasites are kineto-

Host–Parasite Interactions, edited by G. F. Wiegertjes & G. Flik.

plastids, with *T. borreli* belonging to the suborder Bodonina, family Cryptobiidae, whereas *T. danilewskyi* is classified in the suborder Trypanosomatina, family Trypanosomatidae. *T. danilewskyi* is a member of the 'aquatic clade' (Figueroa *et al.*, 1999), a group of trypanosomes all transmitted by leeches, that appeared early in the evolution of the genus *Trypanosoma* (Stevens *et al.*, 2001) (*Table 1*).

Table 1 *Classification order Kinetoplastida*

Order	Suborder	Family	Species
Kinetoplastida	Bodonina	Bodonidae	*Bodo saltans*
		Cryptobiidae	*Trypanoplasma borreli*
	Trypanosomatina	Trypanosomatidae	*Trypanosoma danilewskyi*

Infections with *T. borreli* are widespread in farmed populations of common carp. In some European fish farms parasite infestation in juvenile carp may range between 75 and 100%, especially in fish recovering from the first hibernation period (Lom and Dyková 1992; Steinhagen *et al.*, 1989). Likewise, *T. danilewskyi* infects a variety of cyprinid fishes and is commonly found in the blood of fish populations in nature; its prevalence in farmed fish may approach 100% (Lom and Dyková, 1992).

T. borreli and *T. danilewskyi* share the features typical of kinetoplastid flagellate parasites such as a kinetoplastid organelle containing the mitochondrial DNA, a glycosome compartmentalizing glycolysis and a mini-exon; a highly conserved short RNA leader sequence trans-spliced onto every messenger RNA. Trypanosomatids (and thus also *T. danilewskyi* but not *T. borreli*) have a single flagellum and all genera are parasitic, either in vertebrates, invertebrates, ciliates or in flowering plants. The suborder Trypanosomatina is well-studied, because it contains the important mammalian parasites *T. cruzi*, *T. b. rhodesiense*, *T. b. gambiense* and *Leishmania* ssp. *a.o.* Bodonids (and thus also *T. borreli*, but not *T. danilewskyi*) have two flagella and most species are free-living inhabitants of aqueous environments, but some are vertebrate ecto- or endoparasites of fish (e.g. of the skin, gills, alimentary tract, reproductive organs or blood) or parasites of invertebrates (Stevens *et al.*, 2001). Bodonids represent an ecologically and economically important group of organisms, as they are present in all major aquatic ecosystems. They are also crucial components of sewage cleaning units and the causative agents of some fish diseases in aquaculture. Parasitism among Bodonids probably arose by invasion of their hosts via the skin, gills or gut. The evolution of parasitism among Trypanosomatids, especially within the genus *Trypanosoma*, and the acquisition of a digenetic (two host) life cycle have been widely debated. Possibly trypanosomes first appeared as monogenetic (single host) parasites of aquatic (in)vertebrates and subsequently adapted to digenetic transmission cycles involving aquatic vertebrates and leech vectors (Woo, 1970).

Surprisingly, *T. borreli* and *T. danilewskyi* both develop differently in the leech vector. Upon a first blood meal, *T. danilewskyi* transforms in the leech stomach into epimastigotes, which multiply and change into metacyclic trypomastigotes and subsequently invade the proboscis sheath of the leech from where they can be transferred to other fish during a second blood meal. *T. danilewskyi* trypomastigotes can persist in the leech for several months (Lom, 1979). In the case of *T. borreli*, however, the leech is only acting as a vector and not as obligatory intermediate host. In fact, the longest period *T. borreli* was observed to persist in the leech was 11 days only, along with the blood meal in the leech crop (Kruse *et al.*, 1989).

Parasitaemia can easily be established in fish blood with the use of a haemocytometer, but it is not only the possibility to monitor parasitaemia in great detail that allows us to design carefully planned experiments. The possibilities to manipulate, in a controlled manner, parasitaemia through a change in the injection route, dose or temperature (Jones *et al.*, 1993) allows for the optimization of experiments in a cold-blooded vertebrate host natural to these two parasites: the common carp. Moreover, both *T. borreli* and *T. danilewskyi* can be cultured for prolonged periods *in vitro* (Ardelli and Woo, 1998; Bienek and Belosevic, 1997, 1999; Overath *et al.*, 1998; Steinhagen *et al.*, 2000), are highly infectious and cannot be directly transmitted from one carp to another. Genetic differences in resistance to *T. borreli* between different carp lines have been described (Wiegertjes *et al.*, 1995).

1.2 *Comparative host–parasite immunology*

Fish are the first organisms with an immune system showing the same basic aspects of specificity and memory formation as observed in higher vertebrates (Van Muiswinkel, 1995). There are more then 55 000 fish species that, by inhabiting highly different environments, are challenged by all kinds of pathogens. While their immune system evolved under constant pressure from infectious micro-organisms such as viruses, bacteria and larger parasites, it is fascinating to see how the immune system of such an evolutionary successful group of animals evolved under so many different environments. Comparisons (e.g. with the human immune system) may not only give an idea of the history of the immune system, but may also distinguish what has been conserved from what has varied, which will shed light onto what is essential and what is accessory to the human immune system (Paul, 1999).

Studies on kinetoplastid infections of fish enable the evaluation of different modes of adaptation that allow trypanosomes to infect and persist in vertebrates. Antigenic variation observed in the Salivaria (e.g. *T. bruceï*) and intracellular hiding (e.g. *T. cruzi*) are the best-known examples. In analogy to the salivarian trypanosomes both *T. danilewskyi* and *T. borreli* are believed to live exclusively extracellularly in the blood and tissue fluids of their fish hosts but, differently, show no evidence of antigenic variation (Overath *et al.*, 1999). Analogous to the stercorarian trypanosomes, the surface coat of *T. danilewskyi* shows a clear resemblance with the carbohydrate-dominated surface coat of *T. cruzi* (Paulin *et al.*, 1980); highly glycosylated mucin-like surface proteins are abundant, anchored in the plasma membrane by glycosylphophatidylinositol (GPI) residues (Lischke *et al.*, 2000). Although less data are available on the surface coat of *T. borreli*, electron microscopy suggests a much more massive surface coat than found for *T. danilewsky* (Lom and Nohýnková, 1977). Possibly, *T. borreli* and *T. danilewskyi* have different modes of adaptation allowing these parasites to infect and persist in carp. A co-ordinated study of the co-evolution of fish kinetoplastid parasites with the fish immune response will certainly provide new insights in immune evasion mechanisms used by trypanosome species that have not yet been studied in great detail.

Of course, host and parasite are not entities that can be considered separately. Instead, host–parasite interactions need to be considered. It is not of benefit to a parasite to kill the host, at least not until transmission to another host has been ensured. For the fish it is important that the immune response adapts to different developmental stages of the parasite. A highly activated immune system must be down-regulated immediately after control of initial parasite replication. In fact, it is becoming increasingly clear that there is a delicate balance between an effective, protective immune response and one that causes more damage than it prevents. Indeed, in the case of some parasite infections it is not clear whether morbidity should be ascribed to the parasites or is the result of an uncontrolled immune response. For example, although we described a trypanostatic effect of nitric oxide (NO) on *T. borreli in vitro* (Saeij *et al.*, 2000), NO appeared to have an immunosuppressive effect *in vivo* (Saeij *et al.*, 2002). The immunosuppressive (cytostatic) effect was strongest on carp lymphocytes, that were especially susceptible to NO, while high levels of the antioxidant glutathione in neutrophilic granulocytes protected this cell type from NO-mediated cytotoxicity (Saeij *et al.*, 2003b). This is a clear example of a disturbed balance between an immune response (stimulation of macrophages) that should be effective and protective and the same response causing more damage (over-production of NO) than it prevents.

To date, it is well recognized that the innate immune responses against parasites in fish are at least as important for their protection as are the adaptive responses (Jones, 2001). A parasite differs from its host in many thousands of molecules and if fish were to recognize and respond to this large variety of antigens, the resulting massive lymphocyte activation would hinder an effective immune response. Thus, the immune system generally reacts to a limited range of antigenic determinants only, the initiation of the adaptive response being controlled by innate immune recognition. The innate immune system has to differentiate between the large number of pathogen-associated molecular patterns (PAMPs) found in nature, distinguishing these structures from self and discriminating between different pathogens. Protozoan PAMPs that are predominantly recognized as foreign are the GPI anchors (Tachado *et al.*, 1997). It appears that distinct PAMPs from various pathogens are recognized by distinct members of the toll-like receptor (TLR) family for initiation of the immune response (Stafford *et al.*, 2002). TLRs can establish a combinatorial repertoire to discriminate among the large number of PAMPs found in nature (e.g. TLR2 and TLR6 together recognize peptidoglycan, TLR2 alone recognizes lipopeptides) (Ozinsky *et al.*, 2000). Interestingly, recent findings indicate that the recognition system for protozoan GPI has much in common with the recognition system of LPS, while the use of TLR2 or TLR4 knock-out mice indicated an essential role of TLR2, but not TLR4, in the induction of IL-12, TNFα and NO by GPI-activated macrophages (Campos *et al.*, 2001).

For *T. borreli* infections in carp, a heat-labile fraction of *T. borreli* and CpG motifs in its DNA are PAMPs responsible for the induction of NO and most likely also for the induction of expression of TNFα, IL-1β and iNOS by carp head kidney phagocytes *in vitro* (Saeij and Wiegertjes, unpublished data). Probably, in the signal transduction pathway leading to activation of phagocytes, protein tyrosine kinase (PTK) and protein kinase C (PKC) are activated and collaborate in activation of the transcription factor NF-kB. Activation of NF-kB then leads to downstream expression of the NF-kB-dependent loci iNOS, IL-1β and TNFα (Engelsma et al. 2001b; Saeij et al., 2000, 2003a). The heat-labile fraction of *T. borreli* responsible for phagocyte activation could be GPI-anchored proteins.

T. danilewskyi PAMPs did not seem to induce as much NO production as *T. borreli* PAMPs. Probably, divergent GPIs have evolved to the advantage of the various parasitic protozoa to activate or down-regulate the endogenous signalling pathways of the host (Tachado *et al.*, 1999). Most likely, in fish, pattern recognition receptors such as TLRs on cells of the monocyte-macrophage lineage recognize PAMPs on parasites and transduce a signal. In fact, many partial sequences from a variety of fish species have been reported as putative TLR-like receptors. More recently, we described the first full-length TLR cDNA in goldfish, predicted to encode a type II transmembrane protein with an extracellular domain containing leucine-rich repeats and a cytoplasmic tail encoding a toll/interleukin-1 receptor domain (Stafford *et al.*, 2003).

In conclusion, no matter where a parasite enters the body it will encounter resident but highly active macrophages that can recognize PAMPs by their TLRs. These macrophages are maintained by local division and immigration from blood and not only are able to phagocytose large numbers of parasites but also participate in chemotaxis, cytokine-mediated intercellular communication and regulation of coagulation, of the complement cascade and of acute-phase protein production. As such, macrophages perform a key function in the immune response against parasites.

2 Macrophages

2.1 *Natural resistance-associated macrophage protein (NRAMP)*

Macrophages play a critical role in iron metabolism (see also Stafford *et al.*, 2002). Transition metals such as iron participate in many cellular functions such as: (i) regulation of transcription; (ii) the function of hundreds of different enzymes including metalloproteases, superoxide dismutase (SOD) and inducible nitric oxide synthase (iNOS); and (iii) cellular functions such as endosomal fusion. Any dysfunction in the metal-sensing/transporting pathway will have pleiotropic effects and may cause disease. For that reason, iron concentrations in body tissues are tightly regulated; excessive iron can lead to tissue damage as a result of the formation of free radicals. The co-ordinate control of iron uptake and storage is tightly regulated by the feedback system of iron-responsive element-containing gene products and iron-regulatory proteins that modulate the expression levels of the genes involved in iron metabolism (Lieu *et al.*, 2001). Macrophages are responsible for the re-utilization of iron from haemoglobin by phagocytosing senescent erythrocytes. In non-erythroid cells, iron is either stored as ferritin or targeted to iron-containing molecules, such as iron-regulatory proteins. Also, in macrophages, iron is necessary for the production of hydroxyl radicals (OH°) via the Fenton reaction (*Figure 1*), and iron levels control the production of nitric oxide (NO) after activation by antigenic stimuli.

In plasma, iron is transported by transferrin, with high affinity for ferric iron, a 80-kDa glycoprotein consisting of two globular domains, both containing a high-affinity binding site for a single iron molecule. These binding sites are highly conserved through evolution (Ford, 2001). Transferrin exists as a mixture of iron-free (apo), single-iron (monomeric) and two-iron (diferric) forms. Although the relative percentage of each form depends on the concentration of iron and transferrin in plasma, under normal conditions most of the iron molecules are bound to transferrin. Most cells (so not only macrophages) acquire iron from transferrin first by binding of transferrin to transferrin receptors. Following binding, the complex consisting of iron,

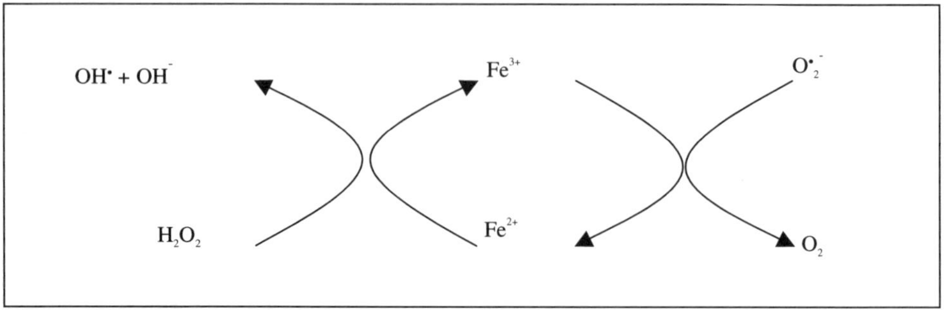

Figure 1 *The Haber–Weiss cycle (left part: Fenton reaction).*

transferrin and transferrin receptor is internalized via a classic receptor-mediated endocytic pathway. Iron is released from transferrin within acidic endosomal compartments and then transported across the endosomal membrane into the cytoplasm by the iron transporter, NRAMP2, or solute carrier family 11 member 2 (Slc11a2). Apotransferrin and the transferrin receptor both return to the cell surface, where they dissociate at neutral pH.

To date, two NRAMP family members have been identified. It is thought that NRAMP1 (Slc11a1), which localizes to *late* endosomes/lysosomes of macrophages, plays a role in the recycling of iron acquired by macrophages through phagocytosis of senescent erythrocytes. NRAMP2, which localizes to *early* recycling endosomes, is thought to influence transferrin-receptor-mediated entry of iron into cells. In the recent past, NRAMP1 polymorphisms in man have been associated with a variety of infectious (HIV, tuberculosis, leprosy, meningococcal meningitis, visceral leishmaniasis) and autoimmune diseases (rheumatoid arthritis, diabetes, sarcoidosis, Crohn's disease), while in mice, a single mutation (Gly→Asp) in Nramp1 renders animals susceptible to intracellular pathogens such as *Mycobacterium* spp., *Salmonella typhimurium*, *Leishmania* spp. and *Toxoplasma gondii* (reviewed by Blackwell *et al.*, 2001). Hence its original name: natural resistance associated macrophage protein (NRAMP).

There are two different views on how exactly NRAMP1 polymorphisms influence disease resistance/susceptibility. One hypothesis is that NRAMP1 functions by increasing intraphagosomal Fe^{2+} in order to provide the catalyst for the Haber–Weiss/Fenton reaction, generating highly toxic hydroxyl radicals for bactericidal activity. The other hypothesis is that NRAMP1 functions by depriving the intraphagosomal pathogen of essential Fe^{2+} (critical for growth) and other divalent cations (Zn^{2+}, Mn^{2+}, for superoxide dismutase), both critical to pathogens for mounting an effective anti-oxidant defence (Wyllie *et al.*, 2002) (*Figure 2*). The broad-specificity divalent cation transporter activities of NRAMP1 provide a rational explanation for the pleiotropic effects of NRAMP1 on macrophage activation and function (e.g. regulation of IL-1β, iNOS, MHC class II, TNFα, NO release, L-arginine flux, oxidative burst and tumouricidal as well as anti-microbial activity) (Blackwell *et al.*, 2001).

As described above, cells acquire iron from transferrin by release of iron within acidic endosomal compartments, using the NRAMP2 protein to transport iron across

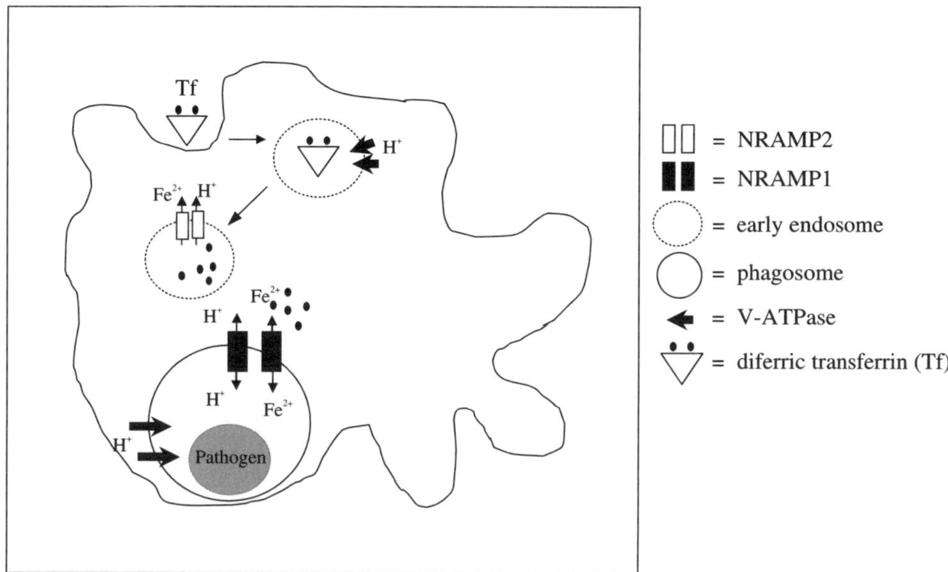

Figure 2 *Current model(s) of NRAMP function in macrophage iron homeostasis: NRAMP2 symport activity delivers* Fe^{2+} *(•) across the early endosomal membrane to the cytosol after recruitment of vesicular ATPase (V-ATPase) and acidification of the vacuole. NRAMP1 antiport activity delivers divalent cations from the cytosol to acidic late endosomes/lysosomes/phagosomes, where the Haber–Weiss cycle generates toxic antimicrobial radicals. Alternatively, it has been proposed that NRAMP1 also delivers* Fe^{2+} *into the cytosol and so depletes microbes from this essential nutrient.*

the endosomal membrane into the cytoplasm. Many cell types show this ability, in contrast to the transport of iron across the membrane of late endosomes/lysosomes by NRAMP1, which is restricted to macrophages. Thus, expression of *NRAMP1* is limited to macrophages while *NRAMP2* expression is ubiquitous. There are, however, indications that no *NRAMP1* exists in fish but that a duplicated *NRAMP2* gene exists with different functions. We described the first *NRAMP* gene in fish as most probably being *NRAMP2*-like (Saeij *et al.*, 1999) and recently identified a second (partial) carp *NRAMP2* gene (Saeij and Wiegertjes, unpublished data). Likewise, in trout (*Oncorhynchus mykiss*), fathead minnow (*Pimephales promelas*), catfish (*Ictalurus punctatis*) and puffer fish (*Fugu rubripes*) *NRAMP2* genes have been found (for trout (Dorschner and Phillips, 1999); for fathead minnow, accession nr. AF190773; for catfish (Chen *et al.*, 2002); for puffer fish (Blackwell *et al.*, 2001)). So far, no fish *NRAMP1* gene has been identified.

However, in fish, in contrast to the situation in mammals, the *NRAMP2* gene is often duplicated. For example, in rainbow trout two NRAMP proteins have been described (α and β), both of them clustering with the mammalian NRAMP2 proteins (as did the carp NRAMP). Also in puffer fish, two NRAMP2 proteins have been found, one of them localizing to late endosomes/lysosomes, consistent with a divergence towards an NRAMP1-like function (Blackwell *et al.*, 2001). In rainbow trout, *NRAMP2*α expression was limited to the head kidney and ovary, while *NRAMP2*β expression was ubiquitous (Dorschner and Phillips, 1999). Three *NRAMP* transcripts were identified in catfish but these were due to alternative splicing in the 3'UTR and alternative polyadenylation and did only yield a single functional protein. However, injection of catfish with LPS increased transcription of *NRAMP2* in the kidney and spleen. Furthermore, expression could be induced in a catfish monocyte/macrophage cell line (Chen *et al.*, 2002). In carp, *NRAMP2* expression in the head kidney was affected during *T. borreli* infection as shown by an early down-regulation (*Figure 3*) that could not easily be explained by a change in cell types populating the head kidney. These data suggest a differential regulation of *NRAMP2* expression supportive of the existence of an NRAMP1-like function in fish. Further research will be needed, however, to verify expression of an NRAMP1-like transcript in fish macrophages.

2.2 *Transferrin*

As described in Section 2.1, cells acquire iron from transferrin first by binding of transferrin to transferrin receptors. Transferrin itself, however, may also exert effects that are not directly linked with maintaining iron levels. These include the induction of neutrophilic end-stage maturation (Evans *et al.*, 1989) and the up-regulation of chemokine synthesis by human proximal tubular epithelial cells (Tang *et al.*, 2002), while transferrin may also be an important component of inflammatory serum that modulates oxygen and nitrogen radical production by chicken macrophages (Xie *et al.*, 2001).

Recently, in goldfish, we showed that transferrin cleavage products act as macrophage-activating factor (MAF) by stimulating macrophages to produce large amounts of nitric oxide (Stafford and Belosevic, 2003; Stafford *et al.*, 2001b). Possibly, neutrophilic

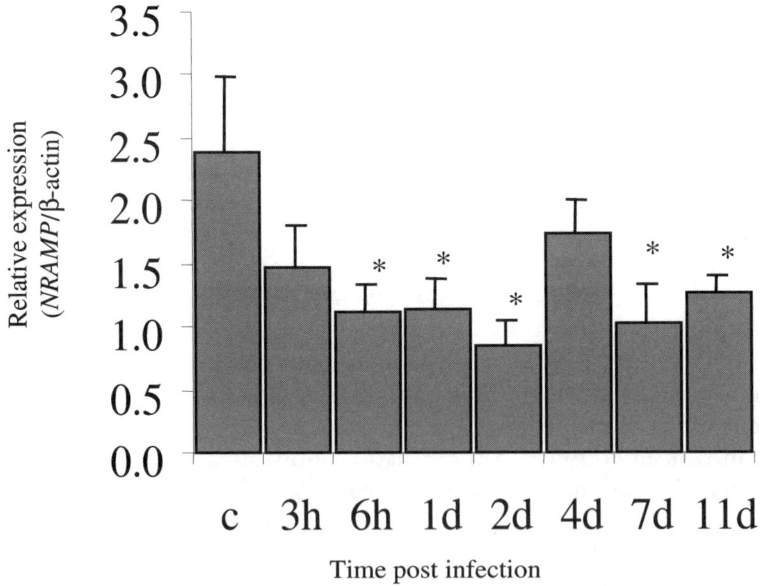

Figure 3 *Relative expression of carp NRAMP in the head kidney after infection with* T. borreli *(C = non-infected control, h = hours, d = days). *, Represents a significant difference as compared with control carp (PBS-injected).*

granulocytes, that are generally the first immune cells recruited to the site of inflammation, initiate the cleavage of transferrin via the production of neutrophil-derived proteases (Miller *et al*., 1996). Likewise, activation by recognition of a proteolytic fragment from a self-protein has also been found in *Drosophila* (a proteolytically processed product of the spätzle gene activates Toll) (Michel *et al*., 2001). The recognition of endogenous 'danger' signals (e.g. molecules produced by stressed cells, or products that are usually found inside a healthy cell) by the immune system is the basis of the 'Danger model' (Matzinger, 1998). At this moment research on a putative goldfish macrophage receptor able to detect these transferrin cleavage products is ongoing.

Carp transferrin is highly polymorph and more than seven different alleles have been identified according to differences in electrophoretic mobility (Valenta *et al*., 1976). So far, no differences in iron-binding capacity have been reported for the transferrin alleles. However, transferrin polymorphisms might give rise to cleavage products with different three-dimensional conformations. Subsequently, these cleavage products might have different abilities to bind to the putative macrophage receptor recognizing transferrin cleavage products. This could result in different capacities to induce nitric oxide production in cyprinid macrophages and would certainly provide a functional rationale for the transferrin polymorphism observed in fish.

Considering host–parasite interactions; pathogenic micro-organisms such as trypanosomes also have evolved a range of mechanisms to acquire iron during their stay in animal hosts. During natural transmission from one host species to another, trypanosomes have to deal with the diversity of host transferrins. For *T. brucei* it has been found that the genome has up to 20 expression sites that encode similar but not identical transferrin receptors, one being active at a given time. By switching between different expression sites, the parasite can express different host-specific transferrin receptors (Steverding, 2000). Transferrin uptake by *T. brucei* involves transferrin binding to a hetero-dimeric transferrin (TF)-binding protein complex (TFBP). The TFBP-TF complex is internalized and transported to lysosomes, where transferrin is proteolytically degraded. Most probably, the receptor is recycled to the membrane of the flagellar pocket (Steverding, 2000). The resulting large peptide fragments are released from the trypanosomes while iron remains cell-associated. The possibility that such parasite-induced fragments of transferrin also might stimulate NO production in activated macrophages is purely speculative but could provide an explanation for the strong induction of NO by *T. borreli* (Saeij *et al*., 2000, 2002).

2.3 *Nitric oxide*

In mammals, macrophages are considered the prime source of nitric oxide (NO), which has been recognized as one of the most versatile players in the immune system. NO is involved in the pathogenesis and control of infectious diseases, tumours, autoimmune processes and chronic degenerative diseases. NO can react with a variety of reaction partners (DNA, proteins, thiols, prosthetic groups, reactive oxygen intermediates (ROIs)) and is widespread produced (by three different NO synthases). Considering the fact that many of the targets of NO are regulatory molecules (e.g. transcription factors) it is not surprising that NO has pleiotropic effects.

As described above (Sections 2.1, 2.2), iron availability is a major factor modulating the immune response. Many of the cytotoxic effects of NO can be explained by its reactivity with iron at the active site of enzymes. Reactive nitrogen intermediates

(RNIs) and ROI such as NO and H_2O_2 can modulate iron availability within the cell by activating iron-responsive element (IRE)-binding by iron-regulatory proteins (IRP)-1 and IRP-2 (*Figure 4*). These IRPs bind to mRNAs of *ferritin*, *NRAMP2* or the *transferrin receptor* (*TfR*) and control the rate of mRNA translation or stability. IRPs are sensitive to changes in iron concentrations. At high iron concentration, a 4Fe-4S cluster is inserted in the IRP, liganded to three cysteine residues, inactivating its IRE-binding. At iron deprivation, IREs in the 5'UTR of *ferritin* (iron storage) inhibit *ferritin* mRNA translation, while IREs in the 3'UTR of the transferrin receptor *TfR* (iron uptake) and *NRAMP2* confer stability to these mRNAs and increase TfR and NRAMP2 protein. Notably, also mRNAs encoding erythroid 5-aminolaevulinic acid synthase (5-ALA-synthase), the first enzyme in haem synthesis for haemoglobin, harbour an IRE in the 5'-UTR. High NO levels, by activating IRP, might inhibit *5'-ALA-synthase* translation. This may be relevant to the role of nitric oxide in parasite-induced anaemia and may explain lowered blood haemoglobin concentrations associated with e.g. parasitic infections (Rafferty *et al.*, 1996). The regulatory cross-talk between iron and NO in macrophages is further highlighted by the transcriptional regulation of iNOS by iron (Hentze and Kuhn, 1996).

What has become clear during the last few years is the importance of the amino acid L-arginine as a factor that can influence immune responses: L-arginine serves not only as substrate for NO production but also as substrate for arginase activity. Arginase converts L-arginine in ornithine that is the precursor for polyamines, important for cell replication (see Section 2.4). In this respect, L-arginine is also of vital importance to the parasite. Parasites need L-arginine for synthesis of polyamines, required for DNA and synthesis of trypanothione (the parasite equivalent of glutathione, the main

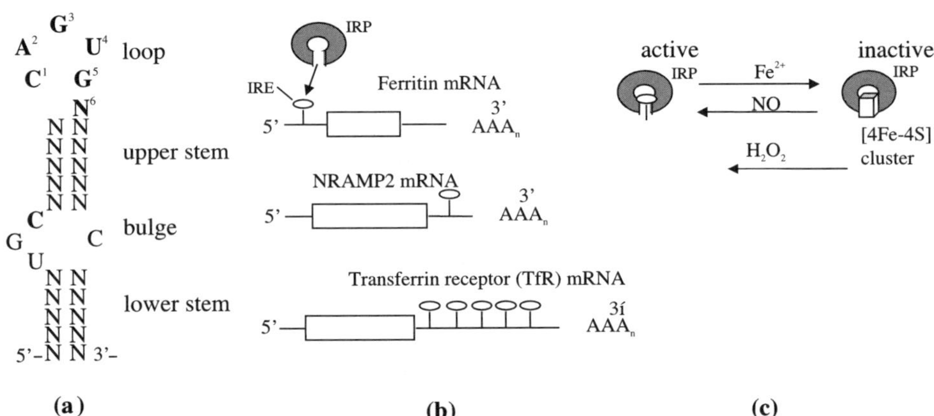

***Figure 4** Effect of iron, NO and H_2O_2 levels on transcriptional regulation.*
(a) Iron-responsive element (IRE) in eucaryotic mRNAs with the consensus sequence (CNNNNNCAGTG) forming a hairpin.
(b) Ferritin *mRNA contains a single IRE in its 5' untranslated region (UTR). Binding of iron-regulatory protein (IRP) blocks the translation of* ferritin *mRNA. NRAMP2 mRNA contains one IRE in its 3' UTR. Binding of IRP stabilizes the NRAMP2 mRNA. Transferrin receptor (*TfR*) mRNA contains five IRE's in its 3' UTR. Binding of IRP stabilizes the* TfR *mRNA.*
(c) In iron-replete cells, IRP assembles a cubane Fe-S cluster that is liganded to IRP cysteines. NO and H_2O_2 can displace the iron cluster and render IRP active.

protectant against nitrosative and oxidative stress, see also Saeij *et al.*, 2003b). In fact, it has been demonstrated that polyamine biosynthesis inhibitors (e.g. DFMO, Berenil) can lead to the destruction of fish trypanosomes (Davies *et al.*, 1999). Trypanosomes actively compete with their hosts for this essential nutrient (Vincendeau *et al.*, 2003). As fish cannot synthesize L-arginine they present an excellent model for studying the role of L-arginine in the immune response.

2.4 *Macrophage polarization*

As mentioned above, arginase converts L-arginine in ornithine that is essential to cells for replication. It is this arginase activity that has provided more insight into the existence of two macrophage subsets; activated macrophages metabolize L-arginine by two pathways involving either the enzyme iNOS or arginase. The balance between the two macrophage subsets is under influence of a competitive regulation by T helper (Th)1 and Th2 cells via their secreted cytokines. In mammals, it is now clear that a wide variety of parasite-induced responses show a dominant Th1 or Th2 cytokine production profile that is associated with either an exacerbative or protective effect on infection. The discovery that cytokines produced by respective Th1/Th2 lymphocyte subsets cross-inhibit each others development and function provides a widely applicable, molecular-based rationale for the understanding of the polarization of immune responses observed in many infectious diseases.

Th2 immune responses are associated with the production of interleukin (IL)-4, IL-10 and IL-13, which are essential for the production of antibodies by B cells, while Th1 immune responses are particularly associated with tumour necrosis factor (TNF)α and interferon (IFN)γ. Immune responses can often be assigned a type I or type II designation based solely on the cytokine production profile without directly implicating the cellular source (Pearce *et al.*, 1999). Classically-activated (or M1) macrophages, occur in a type I cytokine environment and are inhibited by type II cytokines. Alternatively activated (M2) macrophages, develop in a type II cytokine environment and are inhibited by type I cytokines. Classically activated macrophages possess cytotoxic, antimicrobial and antiproliferative functions based on their ability to secrete NO, and play a defensive role in several diseases. However, the same classically activated macrophages also secrete inflammatory mediators (TNFα, IL-1β, IL-6, NO) that are involved in the setting of immunopathologies. In contrast, alternatively activated macrophages secrete anti-inflammatory molecules (IL-10, TGF-β) that down-regulate inflammatory processes and counteract NO synthesis by expressing arginase (Mills, 2001). Accordingly, the type I/type II cytokine balance may influence the development of these different subsets of macrophages that are antagonistically regulated. The level of iNOS or arginase activity in these macrophage subsets may reflect their type I or type II designation. For fish, this allocation would allow for typing immune response as type I or type II, based on the production of NO or arginase activity, respectively (see also Section 2.5).

To date, there is no clear evidence for the existence of Th1/Th2 cells or classically/alternatively activated types of macrophages in fish. However, since IFN-γ, TNFα, IL-1β, IL-6 and NO characterize the type I response, it is worthwhile to consider these molecules in fish. No data on the existence of IL-6-like molecules in fish have been reported. In addition, although several authors have claimed functional evidence for the existence of interferon (IFN)-γ activity as macrophage-activating factor (MAF) (Neumann *et al.*, 1995; Verburg-van Kemenade *et al.*, 1996), so far no IFN-γ but only IFN-induced genes have been cloned and sequenced in fish (Collet and Secombes, 2002). In fact, as discussed above in Section 2.2, we previously suggested that the MAF-like activity of lymphocyte culture supernatants might not be ascribed to IFN-γ activity but might be due to activation by transferrin cleavage products (Stafford and Belosevic, 2003; Stafford *et al.*, 2001b). In contrast, both IL-1β and TNFα have been identified in fish, not only in carp (Engelsma *et al.*, 2001; Saeij *et al.* 2003a), but in several other fish species (Secombes *et al.*, 2001). Also, NO production by, e.g. carp macrophages, is well-recognized in fish (Saeij *et al.*, 2000).

For the type II response, IL-4, IL-10, IL-13 and TGF-β are considered the main players. No data on the existence of IL-4 and IL-13-like molecules in fish have been reported, and database evidence for the existence of a fish IL-10-like molecule is only just emerging. In contrast, several *TGF-β* homologues have been cloned in fish (*O. mykiss*: Hardie *et al.*, 1998; *Morone saxatilis* x *M. chrysops*: Harms *et al.*, 2000; *Pleuronectus platessa*: Laing *et al.*, 2000) and although additional functional studies have to be done with fish TGF-βs, studies with recombinant mammalian TGF-β1 demonstrated inhibition of trout macrophage functions (Jang *et al.*, 1994). In conclusion, an assignation of fish immune responses as type I or type II, based on cytokine production profiles, even without directly implicating the cellular source, is still in its infancy but could tentatively be based on the profiles of those mediators that have been described for fish.

2.5 *Carp macrophages*

Studies on macrophages in common carp (*Cyprinus carpio*) so far have relied on density gradient centrifugation resulting in leucocyte fractions enriched for macrophages. Subsequent adherence to plastic allows for a final enrichment for macrophages up to 60% purity (Verburg-van Kemenade *et al.*, 1994). However, the presence of neutrophilic granulocytes, forming the main impurity, hinders definite conclusions on macrophage functions, for which reason most studies in carp cautiously refer to 'phagocytes'. No macrophage cell line exists for carp.

Previously, in goldfish (*Carassius auratus*), we established and characterized an *in vitro*-derived kidney macrophage (IVDKM) culture system (Barreda and Belosevic, 2001; Barreda *et al.*, 2000; Neumann *et al.*, 1998, 2000). Flow cytometric analysis of this culture system indicated the presence of three distinct macrophage sub-populations, while functional, morphological, cytochemical and developmental characterization of these sub-populations suggested that they represent putative early progenitors (R1), monocytes (R3) and macrophages (R2). Goldfish R2 cells are morphologically similar to mature tissue macrophages of mammals and contain acid phosphatase, localized nuclear myeloperoxidase and non-specific esterase. Both R2 (macrophages) and R3 (monocytes) sub-populations exhibit distinct NO and respiratory burst activities. In addition, the R2 mature macrophage-like cells were capable of self proliferation (Barreda *et al.*, 2000), which capacity makes this sub-population of cells especially interesting for studies on macrophage function in (cyprinid) fish. Here, in addition to the goldfish system, the establishment and characterization of a similar *in vitro*-derived culture system for carp macrophage-like cells is described.

Initial isolation of carp kidney leucocytes and the generation of a first carp IVDKM culture was done as previously described (Neumann *et al.*, 1995, 1998). To establish a primary carp kidney leucocyte culture we used 25% (v/v) cell-conditioned medium (CCM) from established goldfish IVDKM cultures. Control kidney leucocyte cultures without CCM did not develop well. Following four subsequent culture cycles the concentration of the original goldfish CCM was reduced to 0.8% only and CCM could be considered carp-derived. In order to find the optimal seeding density and optimal percentage of CCM for carp IVDKM cultures, different seeding densities and different CCM percentages were tested. Development of carp IVDKM cultures was evaluated by eye and flow cytometry (*Figure 5*).

Typical carp (*C. carpio*) IVDKM cultures were comparable to similar cultures established for goldfish (*C. auratus*) but different from those previously established for rainbow trout (*O. mykiss*). Flow cytometric analysis indicated the presence of three distinct macrophage sub-populations (R1, R2, R3) which were similar for both cyprinid fishes, while for the salmonid rainbow trout three distinct but different populations were observed. In rainbow trout, the R3 population (monocytes) was most abundant (Stafford *et al.*, 2001a) (*Figure 6*).

Having established that carp IVDKM cultures behave similar to goldfish IVDKM cultures and that carp R1, R2 and R3 sub-populations are morphologically indistinguishable from goldfish R1, R2 and R3 sub-populations, carp IVDKM R2 and R3 sub-populations from a 7–10-day-old culture were sorted by flow cytometry and used for cytochemical and functional characterization. Sorted carp R2 cells were morphologically similar to macrophages; they are large irregularly shaped (12–20 μm in diameter) with a low nucleus to cytoplasm ratio and often have extensive vacuolization and membrane ruffling (see *Figure 7a*). Preliminary results suggest that the carp R2 population is positive for esterase (α-naphthyl acetate esterase) and acid phosphatase but negative for myeloperoxidase (data not shown). Sorted carp R3 cells were morpholog-

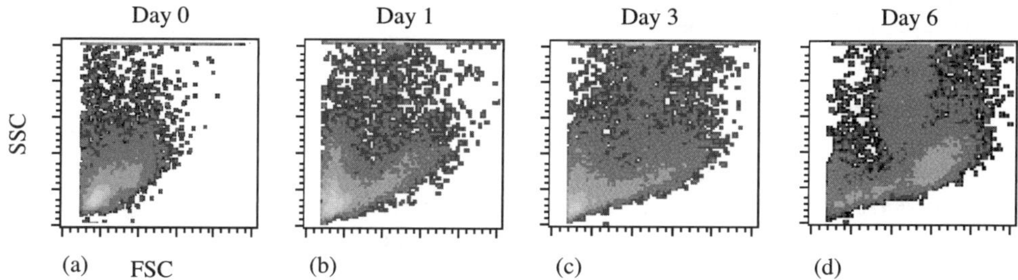

Figure 5 *Flow cytometry (FACS Calibur, Becton Dickinson) profiles showing the development of 3 sub-populations of carp (*Cyprinus carpio*) in-vitro derived kidney macrophages (IVDKM) in time. Profiles (intensity plots) represent forward scatter (FSC) versus side-scatter (SSC) of kidney leucocytes from a single carp. Optimal seeding density was 17.5 x 10^6 leukocytes per 75 cm^2 culture flask, containing 15 ml fresh culture medium (Wang* et al.*, 1995) and 5 ml carp CCM (25%). The data are representative of at least three experiments that were performed. Cells were grown at 20°C.*

(a) Immediately after isolation (day 0)
(b) 1 day culture
(c) 3 days culture
(d) 6 days culture

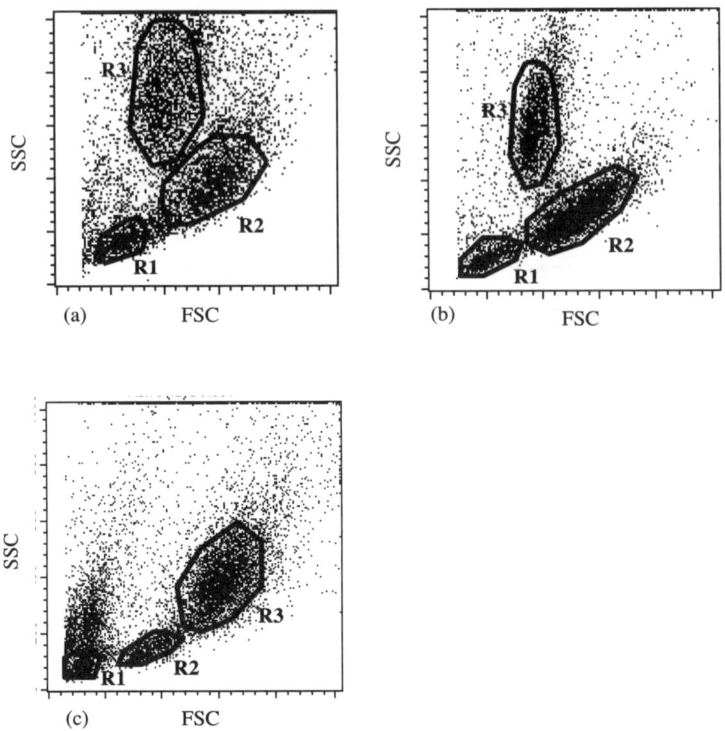

Figure 6 *Flow cytometry profiles showing* in vitro-*derived kidney macrophage (IVDKM) sub-populations of different fish species. Profiles represent forward scatter (FSC) versus side-scatter (SSC) of kidney leucocytes after 8–9 days of culture at 20°C. Typical sub-populations are designated as R1, R2 and R3-type macrophages. The data are representative of at least three experiments that were performed.*
*(a) Carp (*Cyprinus carpio*)*
*(b) Goldfish (*Carassius auratus*)*
*(c) Rainbow trout (*Oncorhynchus mykiss*)*

ically similar to monocytes; they are large round cells (12–15 μm in diameter) with a low nucleus to cytoplasm ratio and often have eccentrically placed kidney-shaped nuclei (*Figure 7b*). The R3 cells seem positive for acid phosphatase but negative for both esterase and myeloperoxidase activity (data not shown). Functional evaluation was done by measuring the nitric oxide production by carp R2 and R3 cells after appropriate stimulation. Both R2 and R3 sub-populations were capable of producing NO; R2 (macrophage-like) cells produced higher amounts than R3 (monocyte-like) cells, which produced only small amounts of NO (*Figure 7c*). In conclusion, cyprinid (i.e. goldfish and carp) IVDKM cultures are comparable in form and function and are useful tools for studying fish macrophage function.

2.6 *Arginase activity in fish*

As described above in Section 2.4, activated macrophages metabolize L-arginine by two alternative pathways involving either iNOS or arginase. Two isoforms of arginase

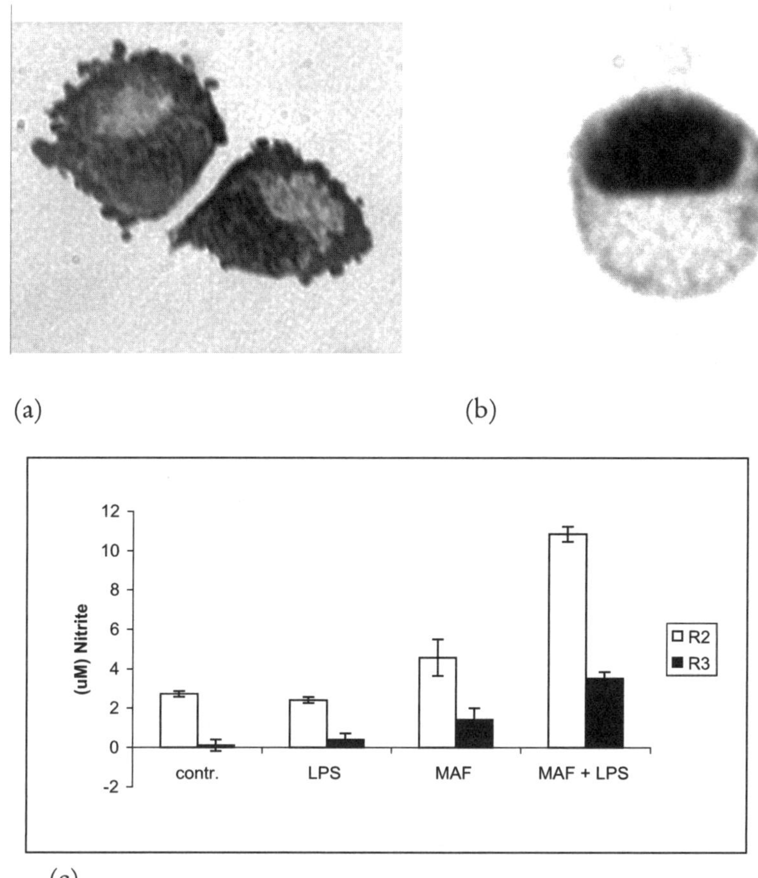

Figure 7 *Characterization of carp (*Cyprinus carpio*) in* vitro*-derived kidney macrophages (IVDKM).*

(a) *Typical R2 (macrophage-like) cell. Carp IVDKM were sorted by flow cytometry (FACS Calibur, Becton Dickinson) after 7–10 days of culture and cytospin slides stained with LeukoStat Solutions (Fisher Scientific Co., Fair Lawn, NJ, USA).*

(b) *Typical R3 (monocyte-like) cell. Carp IVDKM were sorted by flow cytometry after 7–10 days of culture and cytospin slides stained (LeukoStat).*

(c) In vitro *production of nitric oxide (NO) by carp IVDKM. Carp leucocytes were cultured in the presence of 25% (v/v) carp CCM for 7–10 days and sorted by flow cytometry. R2 and R3 cell populations were seeded into half area 96-well plates (5×10^4 cells well $^{-1}$) and stimulated with* E. coli *LPS (10 µg/ml) and/or macrophage activation factor MAF (1:4 dilution; (Neumann* et al., *1995)). NO was measured after 72 h by the detection of nitrite (Griess reaction).*

exist in terrestrial (ureotelic) vertebrates from duplicated genes. One form, essentially present in the liver, is a cytosolic enzyme of the urea cycle and eliminates excess ammonia through the excretion of urea. The extrahepatic (mitochondrial) arginase, more ubiquitous, plays a role in macrophage polarization.

Fish are uricotelic in that they can directly excrete ammonia living in an aqueous environment. Therefore, the synthesis of urea, which is a less efficient nitrogen end product than ammonia, would not only be an unnecessary step in waste conversion but would be a waste of energy. Thus, not surprisingly, although an ancient trait in animals, expression of the ornithine–urea cycle (OUC) does not occur in teleosts, except for some air-breathing fish species. The genes for the OUC are not believed to be suppressed nor deleted but silenced. Indeed, the fifth step in the OUC cycle; the hydrolysis of dietary L-arginine to urea and ornithine by arginase is one step of the OUC that occurs in fish even if the rest of the cycle is incomplete (Wood, 1993). The role of urea formed is still poorly understood in most fish species. Since, like all vertebrates, fish do not have urease for the breakdown of urea, it is difficult to understand the expenditure of irretrievable energy for no logical reason that accompanies the production of urea. What is clear, however, is that fish do express arginase activity in many tissues but mostly in liver and kidney, but also, that the arginase activity in fish without a functional urea cycle may not be centred on ammonia detoxification.

We have found partial carp arginase cDNAs in a head kidney macrophage library and are presently sequencing the remainder of the cDNA sequences. Further, for fish, a number of expressed sequence tags for zebrafish (*Danio rerio*), rainbow trout (*O. mykiss*) and pufferfish (*F. rubripes*) arginase cDNAs are reported in the database. The fish cDNA sequences were used along with several invertebrate *arginase* and mammalian *arginase-1* (liver form) and *arginase-2* (extra-hepatic form) sequences to create a neighbour-joining tree. The fish arginases formed two separate clusters together with the mammalian *arginase-1* and *arginase-2* sequences (data not shown). Thus, the arginase gene duplication that gave rise to two *arginase* genes in the vertebrates likely occurred before the separation of vertebrates and invertebrates (Samson, 2000). Our findings that arginase in carp is expressed in head kidney macrophages suggests the existence of at least an extra-hepatic and functional *arginase-2*-like form in fish. In conclusion, teleost fish can be expected to have two arginase genes that likely encode for functional arginases that convert L-arginine to urea and ornithine but that are not centred on ammonia detoxification.

The above findings on *arginase* cDNAs of fish suggest fish macrophages should express arginase activity. We continued our experiments by using *in vitro* cultures of carp macrophages to study arginase activity in fish. Having established that at 20°C both goldfish and carp IVDKM cultures result in macrophage-like and monocyte-like sub-populations (Section 2.5), we proceeded with characterizing carp leucocytes isolated from the head- rather than the mid-kidney and comparing the effect of *in vitro* culture at a higher temperature. Flow cytometric analysis showed that the use of carp head- rather than mid-kidney leucocytes made no difference to the development of R1–R2–R3 macrophage sub-populations (data not shown). However, preliminary observations suggested that at higher temperature (27°C) carp macrophages primarily develop into macrophage sub-populations (data not shown). This latter cell population (IVDHKM, 27°C) could be readily stimulated to produce nitric oxide and to show arginase activity (*Figure 8*). Thus, carp IVDHKM, the majority being macrophage-like cells, can be induced to express arginase activity after appropriate stimulation. This supports the notion of an extra-hepatic functional arginase gene in fish macrophages. Whether this is the result of a polarized alternatively activated-like macrophage phenotype remains to be confirmed.

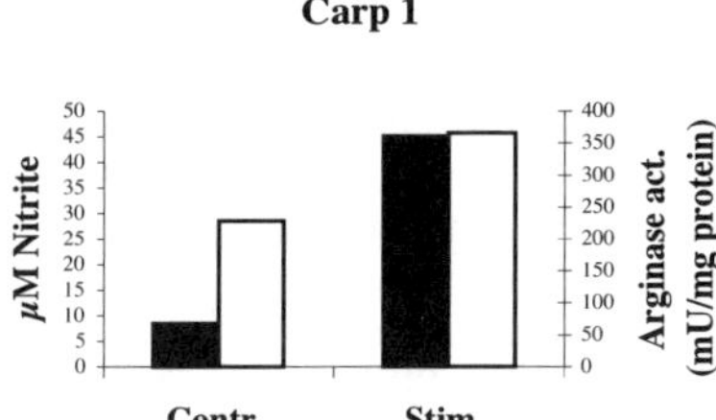

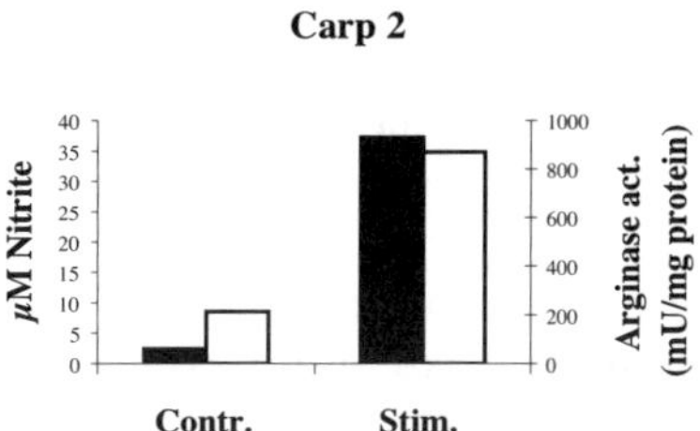

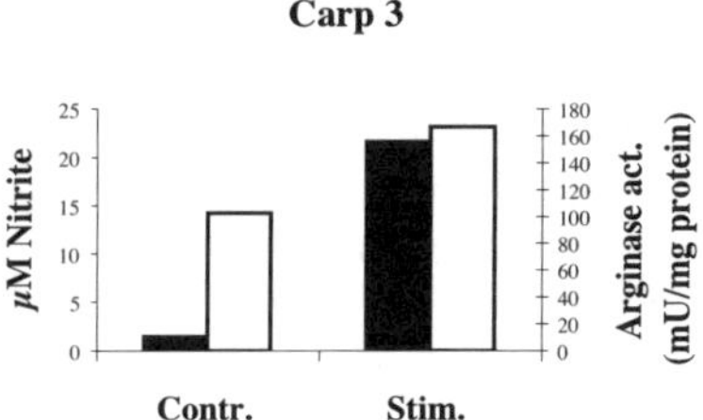

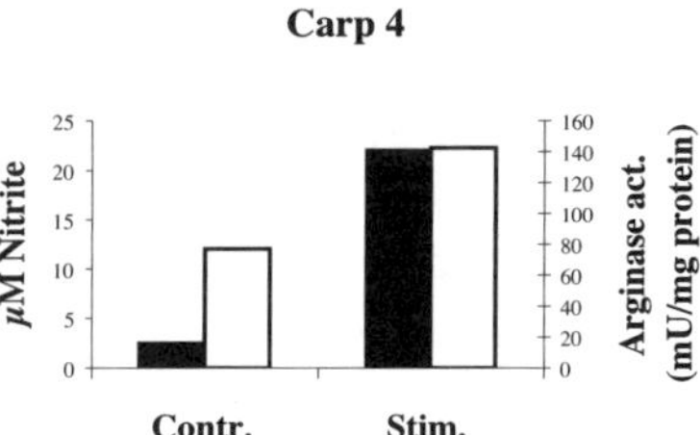

Figure 8 *Nitric oxide (NO) production and arginase activity of carp (*Cyprinus carpio*) in* vitro-*derived head kidney macrophages (IVDHKM).*
Carp (n = 4) head kidney leucocyte cultures were grown in the presence of 25% (v/v) carp CCM for 8 days at 27°C. Subsequently, cells were seeded into 96-well plates (5 x 10^5 cells well $^{-1}$) and NO production (black bars) stimulated with E. coli *LPS (50 µg/ml), arginase activity (white bars) stimulated with dibutyryl cyclic-AMP (0.5 mg/ml). As controls, cells were left untreated. NO was measured after 96 h by the detection of nitrite (Griess reaction), arginase activity was measured after 18 h using the micromethod developed for activated mouse macrophages (Corraliza* et al., *1994).*

3 Macrophage polarization during infection with kinetoplastid parasites

Trypanosomes of the *T. brucei* group are extracellular parasites highly sensitive to NO (Gobert *et al.*, 1998). Peritoneal macrophages from *T. brucei brucei*-infected mice produce NO which is trypanostatic in the presence of L-arginine *in vitro*. Nevertheless, *in vivo*, parasites proliferate in the vicinity of macrophages. We observed that *T. borreli* induces high amounts of trypanostatic NO *in vitro* (Saeij *et al.*, 2000), but immunosuppressive amounts *in vivo* (Saeij *et al.*, 2002), suggesting a similar mechanism may be in place in carp. In mice, an increase in arginase activity was observed in peritoneal macrophages from the first days of infection with *T. brucei brucei* onwards, while later during infection an induction of NO in macrophages and a decrease in plasma L-arginine was observed. Injection of excess L-arginine 'restored' intraperitoneal NO production and NO-dependent killing. The restoration of NO production by local (intraperitoneal) injection of L-arginine is likely to be ineffective systemically, however, due to the excess of NO scavengers in blood. The role of parasite-specific antibodies may then be predominant: it has been suggested that anti-trypanosome antibody mediates the attachment of trypanosomes to activated macrophages and that NO-derived reactive species subsequently affect the juxtaposed parasites (Kaushik *et al.*, 1999). To date, it is suggested that for *T. brucei*, host resistance is associated with the ability to produce IFN-γ and TNFα (type I response) in the early phase of infection, followed by the secretion of IL-4 and IL-10 (type II response) in the late/chronic state of infection. Any perturbations induce tissue damage (exacerbated classically activated macrophage response) or a failure to control early pathogen replication (Baetselier *et al.*, 2001). Similarly, the high production of NO (overactivation of classically activated-like macrophages?) during *T. borreli* infection of carp (Saeij *et al.*, 2002) and related immunosuppression could explain part of the pathology associated with these infections (Bunnajirakul *et al.*, 2000). Interestingly, infections in carp with *T. danilewskyi* do not seem to be associated with high NO levels (Saeij *et al.*, 2002) and do not lead to an excessive inflammatory response (Dyková and Lom, 1979). In theory, it is very well possible that *T. danilewskyi* preferentially induces alternatively activated, rather than classically activated, macrophages in carp.

Recently, more information on the trypanosome molecules responsible for a preferential stimulation of alternatively activated-type macrophages has become available. Cruzipain (syn. cruzain), a highly immunogenic glycoprotein of about 52–58 kDa with a highly mannose glycosylated C-terminal domain is a major *T. cruzi* antigen found in every developmental form of the parasite. Giordanengo and co-authors (2002) found an increase of urea associated with a decrease in nitrate levels after injection of cruzipain, suggesting this cysteine protease preferentially up-regulates the arginase pathway. Macrophages from immune mice cultured with cruzipain showed high urea levels but no increased nitrite levels. Cruzipain was found to directly activate macrophages to increase arginase activity, possibly interacting through the mannose receptor. For *Cryptobia salmositica*, a close relative of *T. borreli*, a cysteine protease with an important role in protein metabolism for the parasite has recently been identified. The chemical properties (e.g. molecular weight, substrate and inhibitor specificity, pH optimum) were found similar to cysteine proteases found in, for example, *T. cruzi* (Zuo and Woo, 1998). This suggests that the presence of cysteine proteases could be a general feature of kinetoplastid parasites and opens possibilities for isolating and testing the supposedly superior ability of 'danilewskyipain' versus 'borrelipain' to induce arginase activity in carp macrophages. We aim to use our *in vitro* culture system for macrophages (IVDHKM, 27°C) to test the ability of *T. borreli* and *T. danilewskyi* to induce either nitric oxide or arginase activity in carp macrophages.

4 Future perspective

Cells belonging to the monocyte–macrophage lineage have long been recognized as heterogeneous, the heterogeneity reflecting the plasticity of the system's response to cytokines and microbial products *a.o.* To date, the existence of polarized (warm-blooded vertebrate) macrophages that differ in terms of receptor expression, effector function and cytokine and chemokine production is rapidly acknowledged. For cold-blooded vertebrates such as teleost fish, the information on T-lymphocytes is sparse and prototypical type I or type II cytokine profiles may not even exist. Macrophages are considered an evolutionary ancient cell type, however, and if macrophage polarization does indeed occur in fish it may for a large part determine the general pattern of innate immune response against foreign stimuli. No doubt, the differentiation of fish macrophages such as seen during our *in vitro* culture system will allow us to perform detailed examinations of this phenomenon in cyprinid fish. If indeed macrophage polarization exists in teleosts, the study of infections with fish kinetoplastid parasites that have evolved with this aspect of the innate immune system in place will certainly bring new insights into the evolution of host–parasite relationships.

Acknowledgements

This work was supported in part by the European Community's Improving Human Potential Programme under contract (HPRN-CT-2001-00214), (PARITY). J.P.J. Saeij acknowledges a grant from NWO-ALW (project No. 806-46.32-P).

References

Ardelli B.F. and Woo, P.T.K. (1998) Improved culture media for piscine hemoflagellates, Cryptobia and Trypanosoma (Kinetoplastida). *J. Parasitol.* **84:** 1267–1271.

Baetselier, P.D., Namangala, B., Noel, W., Brys, L., Pays, E. and Beschin, A. (2001) Alternative versus classical macrophage activation during experimental African trypanosomosis. *Int. J. Parasitol.* **31:** 575–87.

Barreda, D.R., Neumann, N.F. and Belosevic, M. (2000a) Flow cytometric analysis of PKH26-labeled goldfish kidney-derived macrophages. *Dev. Comp. Immunol.* **24:** 395–406.

Barreda, D.R. and Belosevic, M. (2001) Characterisation of growth enhancing factor production in different phases of in vitro fish macrophage development. *Fish Shellfish Immunol.* **11:** 169–185.

Bienek, D.R. and Belosevic, M. (1997) Comparative assessment of growth of *Trypanosoma danilewskyi* in medium containing homologous and heterologous serum. *J. Fish Dis.* **20:** 217–222.

Bienek, D.R. and Belosevic, M. (1999) Macrophage or fibroblast-conditioned medium potentiate growth of *Trypanosoma danilewsky* Laveran & Mesnil, 1904. *J. Fish Dis.* **22:** 359–367.

Blackwell, J.M., Goswami, T., Evans, C.A., *et al.* (2001) SLC11A1 (formerly NRAMP1) and disease resistance. *Cell. Microbiol.* **3:** 773–784.

Bunnajirakul, S., Steinhagen, D., Hetzel, U., Körting, W. and Drommer, W. (2000) A study of sequential histopathology of *Trypanoplasma borreli* (Protozoa: Kinetoplastida) in susceptible common carp *Cyprinus carpio*. *Dis. Aquat. Organ.* **39:** 221–229.

Campos, M.A., Almeida, I.C., Takeuchi, O., *et al.* (2001) Activation of Toll-like receptor-2 by glycosylphosphatidylinositol anchors from a protozoan parasite. *J. Immunol.* **167:** 416–423.

Chen, H., Waldbieser, G.C., Rice, C.D., Elibol, B., Wolters, W.R. and Hanson, L.A. (2002) Isolation and characterization of channel catfish natural resistance associated macrophage protein gene. *Dev. Comp. Immunol.* **26:** 517–531.

Collet, B. and Secombes, C.J. (2002) Type I-interferon signalling in fish. *Fish Shellfish Immunol.* **12:** 389–397.

Corraliza, I.M., Campo, M.L., Soler, G. and Modolell, M. (1994) Determination of arginase activity in macrophages: a micromethod. *J. Immunol. Methods* **174:** 231–235.

Davies, A.J., Thorborn, D.E., Mastri, C. and Daszak, P. (1999) Ultrastructure of culture forms of the eel trypanosome, *Trypanosoma granulosum* Laveran & Mesnil, 1902, exposed to polyamine biosynthesis inhibitors. *J. Fish Dis.* **22:** 285–298.

Dorschner, M.O. and Phillips, R.B. (1999) Comparative analysis of two Nramp loci from rainbow trout. *DNA Cell. Biol.* **18:** 573–583.

Dyková, I. and Lom, J. (1979) Histopathological changes in *Trypanosoma danilewskyi* Laveran & Mesnil, 1904 and *Trypanoplasma borreli* Laveran & Mesnil, 1902 infections of goldfish, *Carassius auratus* (L.). *J. Fish Dis.* **2:** 381–390.

Engelsma, M.Y., Stet, R.J.M., Schipper, H. and Verburg-van Kemenade, B.M.L. (2001) Regulation of interleukin 1 beta RNA expression in the common carp, *Cyprinus carpio* L. *Dev. Comp. Immunol.* **25:** 195–203.

Evans, W.H., Wilson, S.M., Bednarek, J.M., Peterson, E.A., Knight, R.D., Mage, M.G. and McHugh, L. (1989) Evidence for a factor in normal human serum that induces human neutrophilic granulocyte end-stage maturation in vitro. *Leuk. Res.* **13:** 673–682.

Fernandes, A.P., Nelson, K. and Beverley, S.M. (1993) Evolution of nuclear ribosomal RNAs in kinetoplastid protozoa: perspectives on the age and origins of parasitism. *Proc. Natl Acad. Sci. USA* **90:** 11608–11612.

Figueroa, F., Mayer, W.E., Lom, J., Dyková, I., Weller, M., Peckova, H. and Klein, J. (1999) Fish trypanosomes: their position in kinetoplastid phylogeny and variability as determined from 12S rRNA kinetoplast sequences. *J. Eukaryot. Microbiol.* **46:** 473–481.

Ford, M.J. (2001) Molecular evolution of transferrin: evidence for positive selection in salmonids. *Mol. Biol. Evol.* **18:** 639–647.

Giordanengo, L., Guinazu, N., Stempin, C., Fretes, R., Cerban, F. and Gea, S. (2002) Cruzipain, a major *Trypanosoma cruzi* antigen, conditions the host immune response in favor of parasite. *Eur. J. Immunol.* **32:** 1003–1011.

Gobert, A.P., Semballa, S., Daulouede, S., Lesthelle, S., Taxile, M., Veyret, B. and Vincendeau, P. (1998) Murine macrophages use oxygen- and nitric oxide-dependent mechanisms to synthesize S-nitroso-albumin and to kill extracellular trypanosomes. *Infect. Immun.* **66:** 4068–4072.

Hardie, L.J., Laing, K.J., Daniels, G.D., Grabowski, P.S., Cunningham, C. and Secombes, C.J. (1998) Isolation of the first piscine transforming growth factor beta gene: analysis reveals tissue specific expression and a potential regulatory sequence in rainbow trout (*Oncorhynchus mykiss*). *Cytokine* **10:** 555–563.

Harms, C.A., Kennedy-Stoskopf, S., Horne, W.A., Fuller, F.J. and Tompkins, W.A. (2000) Cloning and sequencing hybrid striped bass (*Morone saxatilis* x *M. chrysops*) transforming growth factor-beta (TGF-beta), and development of a reverse transcription quantitative competitive polymerase chain reaction (RT-qcPCR) assay to measure TGF-beta mRNA of teleost fish. *Fish Shellfish Immunol.* **10:** 61–85.

Hentze, M.W. and Kuhn, L.C. (1996) Molecular control of vertebrate iron metabolism: mRNA-based regulatory circuits operated by iron, nitric oxide, and oxidative stress. *Proc. Natl Acad. Sci. USA* **93:** 8175–8182.

Jang, S.I., Hardie, L.J. and Secombes, C.J. (1994) Effects of transforming growth factor beta 1 on rainbow trout *Oncorhynchus mykiss* macrophage respiratory burst activity. *Dev. Comp. Immunol.* **18:** 315–323.

Jones, S.R. (2001) The occurrence and mechanisms of innate immunity against parasites in fish. *Dev. Comp. Immunol.* **25:** 841–852.

Jones, S.R., Palmen, M. and van Muiswinkel, W.B. (1993) Effects of inoculum route and dose on the immune response of common carp, *Cyprinus carpio* to the blood parasite, *Trypanoplasma borreli*. *Vet. Immunol. Immunopathol.* **36:** 369–378.

Kaushik, R.S., Uzonna, J.E., Gordon, J.R. and Tabel, H. (1999) Innate resistance to *Trypanosoma congolense* infections: differential production of nitric oxide by macrophages from susceptible BALB/c and resistant C57Bl/6 mice. *Exp. Parasitol.* **92:** 131–143.

Kristjanson, P.M., Swallow, B.M., Rowlands, G.J. and de Leeuw, P.N. (1999) Measuring the costs of African animal trypanosomiasis: the potential benefits of control and returns to research. *Agri. Systems* **59:** 79–98.

Kruse, P., Steinhagen, D. and Körting, W. (1989) Development of *Trypanoplasma borreli* (Mastigophora: Kinetoplastida) in the leech vector *Piscicola geometra* and its infectivity for the common carp, *Cyprinus carpio*. *J. Parasitol.* **75:** 527–530.

Laing, K.J., Cunningham, C. and Secombes, C.J. (2000) Genes for three different isoforms of transforming growth factor-beta are present in plaice (*Pleuronectes platessa*) DNA. *Fish Shellfish Immunol.* **10:** 261–271.

Lieu, P.T., Heiskala, M., Peterson, P.A. and Yang, Y. (2001) The roles of iron in health and disease. *Mol. Aspects Med.* **22:** 1–87.

Lischke, A., Klein, C., Stierhof, Y.D., Hempel, M., Mehlert, A., Almeida, I.C., Ferguson, M.A. and Overath, P. (2000) Isolation and characterization of glycosylphosphatidylinositol-anchored, mucin-like surface glycoproteins from bloodstream forms of the freshwater-fish parasite *Trypanosoma carassii*. *Biochem. J.* **345 Pt 3:** 693–700.

Lom, J. (1979) Biology of the trypanosomes and trypanoplasms of fish. In: Lumsden, W.H.R. and Evans, D.A. (eds) *Biology of the Kinetoplastida*, pp. 269–337. Academic Press, New York.

Lom, J. and Dyková, I. (1992) *Protozoan Parasites of Fishes*. Elsevier, Amsterdam.

Lom, J. and Nohýnková, E. (1977) Surface coat of the bloodstream phase of *Trypanoplasma borreli*. *J. Protozool.* **24:** 52a.

Matzinger, P. (1998) An innate sense of danger. *Semin. Immunol.* **10:** 399–415.

Michel, T., Reichhart, J.M., Hoffmann, J.A. and Royet, J. (2001) *Drosophila* Toll is activated by Gram-positive bacteria through a circulating peptidoglycan recognition protein. *Nature* **414:** 756–759.

Miller, R.A., Rasmussen, G.T., Cox, C.D. and Britigan, B.E. (1996) Protease cleavage of iron-transferrin augments pyocyanin-mediated endothelial cell injury via promotion of hydroxyl radical formation. *Infect. Immun.* **64:** 182–188.

Mills, C.D. (2001) Macrophage arginine metabolism to ornithine/urea or nitric oxide/citrulline: a life or death issue. *Crit. Rev. Immunol.* **21:** 399–425.

Naessens, J., Teale, A.J. and Sileghem, M. (2002) Identification of mechanisms of natural resistance to African trypanosomiasis in cattle. *Vet. Immunol. Immunopathol.* **87:** 187–194.

Neumann, N.F., Fagan, D. and Belosevic, M. (1995) Macrophage activating factor(s) secreted by mitogen stimulated goldfish kidney leukocytes synergize with bacterial lipopolysaccharide to induce nitric oxide production in teleost macrophages. *Dev. Comp. Immunol.* **19:** 473–482.

Neumann, N.F., Barreda, D.R. and Belosevic, M. (1998) Production of a macrophage growth factor(s) by a goldfish macrophage cell line and macrophages derived from goldfish kidney leukocytes. *Dev. Comp. Immunol.* **22:** 417–432.

Neumann, N.F., Barreda, D.R. and Belosevic, M. (2000) Generation and functional analysis of distinct macrophage sub-populations from goldfish (*Carassius auratus* L.) kidney leukocyte cultures. *Fish Shellfish Immunol.* **10:** 1–20.

Overath, P., Ruoff, J., Stierhof, Y.D., Haag, J., Tichy, H., Dyková, I. and Lom, J. (1998) Cultivation of bloodstream forms of *Trypanosoma carassii*, a common parasite of freshwater fish. *Parasitol. Res.* **84:** 343–347.

Overath, P., Haag, J., Mameza, M.G. and Lischke, A. (1999) Freshwater fish trypanosomes: definition of two types, host control by antibodies and lack of antigenic variation. *Parasitology* **119:** 591–601.

Ozinsky, A., Underhill, D.M., Fontenot, J.D., Hajjar, A.M., Smith, K.D., Wilson, C.B., Schroeder, L. and Aderem, A. (2000) The repertoire for pattern recognition of pathogens by the innate immune system is defined by cooperation between toll-like receptors. *Proc. Natl Acad. Sci. USA* **97:** 13766–13771.

Paul, W.E. (1999) *Fundamental Immunology*. Lippincott-Raven, Philadelphia, USA.

Paulin, J.J., Lom, J. and Nohýnková, E. (1980) The fine structure of *Trypanosoma danilewski*: Structure and cytochemical properties of the cell surface. *Protistologica* **3:** 375–383.

Pearce, E.J., Scott, P.A. and Sher, A. (1999) Immune regulation in parasitic infection and disease. In: Paul, W.E. (ed.) *Fundamental Immunology*, pp. 1271–1294. Lippincott-Raven, New-York.

Rafferty, S.P., Domachowske, J.B. and Malech, H.L. (1996) Inhibition of hemoglobin expression by heterologous production of nitric oxide synthase in the K562 erythroleukemic cell line. *Blood* **88:** 1070–1078.

Saeij, J.P.J., Wiegertjes, G.F. and Stet, R.J.M. (1999) Identification and characterization of a fish natural resistance-associated macrophage protein (NRAMP) cDNA. *Immunogenetics* **50:** 60–66.

Saeij, J.P.J., Stet, R.J.M., Groeneveld, A., Verburg-van Kemenade, B.M.L., van Muiswinkel, W.B. and Wiegertjes, G.F. (2000) Molecular and functional characterization of a fish inducible-type nitric oxide synthase. *Immunogenetics* **51:** 339–346.

Saeij, J.P.J., Van Muiswinkel, W.B., Groeneveld, A. and Wiegertjes, G.F. (2002) Immune modulation by fish kinetoplastid parasites: a role for nitric oxide. *Parasitology* **124:** 77–86.

Saeij, J.P.J., Stet, R.J.M., de Vries, B.J., van Muiswinkel, W.B. and Wiegertjes, G.F. (2003a) Molecular and functional characterization of carp TNF: a link between TNF polymorphism and trypanotolerance? *Dev. Comp. Immunol.* **27:** 29–41.

Saeij, J.P.J., van Muiswinkel, W.B., van de Meent, M., Amaral, C. and Wiegertjes, G.F. (2003b) Different capacities of carp leukocytes to encounter nitric oxide-mediated stress: a role for the intracellular reduced glutathione pool. *Develop. Comp. Immunol.* **27:** 555–568

Samson, M.L. (2000) Drosophila arginase is produced from a nonvital gene that contains the elav locus within its third intron. *J. Biol. Chem.* **275:** 31107–31114.

Secombes, C.J., Wang, T., Hong, S., Peddie, S., Crampe, M., Laing, K.J., Cunningham, C. and Zou, J. (2001) Cytokines and innate immunity of fish. *Dev. Comp. Immunol.* **25:** 713–723.

Stafford, J.L., McLauchlan, P.E., Secombes, C.J., Ellis, A.E., and Belosevic, M. (2001a) Generation of primary monocyte-like cultures from rainbow trout head kidney leukocytes. *Dev. Comp. Immunol.* **25:** 447–459.

Stafford, J.L., Neumann, N.F. and Belosevic, M. (2001b) Products of proteolytic cleavage of transferrin induce nitric oxide response of goldfish macrophages. *Dev. Comp. Immunol.* **25:** 101–115.

Stafford, J.L., Neumann, N.F. and Belosevic, M. (2002) Macrophage-mediated innate host defense against protozoan parasites. *Crit. Rev. Microbiol.* **28:** 187–248.

Stafford, J.L. and Belosevic, M. (2003) Transferrin and the innate immune response of fish: identification of a novel mechanism of macrophage activation. *Develop. Comp. Immunol.* **27:** 539–554

Stafford, J.L., Ellestad, K.K., Magor, K.E., Belosevic, M. and Magor, B.G. (2003b) A toll-like receptor (TLR) gene that is up-regulated in activated goldfish macrophages. *Develop. Comp. Immunol.* **27:** 685–698

Steinhagen, D., Kruse, P. and Körting, W. (1989) The parasitemia of cloned *Trypanoplasma borreli* Laveran and Mesnil, 1901, in laboratory-infected common carp (*Cyprinus carpio* L.). *J. Parasitol.* **75:** 685–689.

Steinhagen, D., Hedderich, W., Skouras, A., Scharsack, J.P., Schuberth, J., Leibold, W. and Körting, W. (2000) *In vitro* cultivation of *Trypanoplasma borreli* (protozoa: kinetoplastida), a parasite from the blood of common carp *Cyprinus carpio*. *Dis. Aquat. Organ.* **41:** 195–201.

Stevens, J.R., Noyes, H.A., Schofield, C.J. and Gibson, W. (2001) The molecular evolution of Trypanosomatidae. *Adv. Parasitol.* **48:** 1–56.

Steverding, D. (2000) The transferrin receptor of *Trypanosoma brucei*. *Parasitol. Int.* **48:** 191–198.

Tachado, S.D., Gerold, P., Schwarz, R., Novakovic, S., McConville, M. and Schofield, L. (1997) Signal transduction in macrophages by glycosylphosphatidylinositols of *Plasmodium*, *Trypanosoma*, and *Leishmania*: activation of protein tyrosine kinases and protein kinase C by inositolglycan and diacylglycerol moieties. *Proc. Natl Acad. Sci. USA* **94:** 4022–4027.

Tachado, S.D., Mazhari-Tabrizi, R. and Schofield, L. (1999) Specificity in signal transduction among glycosylphosphatidylinositols of *Plasmodium falciparum*, *Trypanosoma brucei*, *Trypanosoma cruzi* and *Leishmania* spp. *Parasite Immunol.* **21:** 609–617.

Tang, S., Leung, J.C., Tsang, A.W., Lan, H.Y., Chan, T.M. and Lai, K.N. (2002) Transferrin up-regulates chemokine synthesis by human proximal tubular epithelial cells: implication on mechanism of tubuloglomerular communication in glomerulopathic proteinura. *Kidney Int.* **61:** 1655–1665.

Valenta, M., Stratil, A., Slechtova, V., Kalal, L. and Slechta, V. (1976) Polymorphism of transferrin in carp (*Cyprinus carpio* L.): genetic determination, isolation, and partial characterization. *Biochem. Genet.* **14:** 27–45.

Van Muiswinkel, W.B. (1995) The piscine immune system: innate and acquired immunity. In: Woo, P.T.K. (ed.) *Fish Diseases and Disorders*, pp. 729–750. CAB International, Wallingford.

Verburg-van Kemenade, B.M.L., Groeneveld, A., van Rens, B.T.T.M. and Rombout, J.H.W.M. (1994) Characterization of macrophages and neutrophilic granulocytes from the pronephros of carp (*Cyprinus carpio*). *J. Exp. Biol.* **187:** 143–158.

Verburg-van Kemenade, B.M.L., Daly, J.G., Groeneveld, A. and Wiegertjes, G.F. (1996) Multiple regulation of carp (*Cyprinus carpio* L.) macrophages and neutrophilic granulocytes by serum factors: influence of infection with atypical *Aeromonas salmonicida*. *Vet. Immunol. Immunopathol.* **51:** 189–200.

Vincendeau, P., Gobert, A.P., Daulouede, S., Moynet, D. and Djavad Mossalayi, M. (2003) Arginases in parasitic diseases. *Trends Parasitol.* **19:** 9–12.

Wang, R., Neumann, N.F., Shen, Q. and Belosevic, M. (1995) Establishment and characterization of a macrophage cell line from the goldfish. *Fish Shellfish Immunol.* **5:** 329–346.

Welburn, S.C., Fevre, E.M., Coleman, P.G., Odiit, M. and Maudlin, I. (2001) Sleeping sickness: a tale of two diseases. *Trends Parasitol.* **17:** 19–24.

Wiegertjes, G.F., Groeneveld, A. and van Muiswinkel, W.B. (1995) Genetic variation in susceptibility to *Trypanoplasma borreli* infection in common carp (*Cyprinus carpio* L.). *Vet. Immunol. Immunopathol.* **47:** 153–161.

Woo, P.T.K. (1970) Origin of mammalian trypanosomes which develop in the anterior-station of blood-sucking arthropods. *Nature* **228:** 1059–1062.

Wood, C.M. (1993) Ammonia and urea metabolism and excretion. In: Evans, D.H. (ed.) *The Physiology of Fishes*, pp. 379–425. CRC Press Inc., Baton Rouge, FL.

Wyllie, S., Seu, P. and Goss, J.A. (2002) The natural resistance-associated macrophage protein 1 Slc11a1 (formerly Nramp1) and iron metabolism in macrophages. *Microbes Infect.* **4:** 351–359.

Xie, H., Rath, N.C., Huff, G.R., Balog, J.M. and Huff, W.E. (2001) Inflammation-induced changes in serum modulate chicken macrophage function. *Vet. Immunol. Immunopathol.* **80:** 225–235.

Zuo, X. and Woo, P.T.K. (1998) Characterization of purified metallo- and cysteine proteases from the pathogenic haemoflagellate *Cryptobia salmositica* Katz 1951. *Parasitol. Res.* **84:** 492–498.

5

The pathophysiology of salmonid cryptobiosis and *Glossina*-transmitted mammalian trypanosomiasis in livestock

Patrick T.K. Woo

1 Introduction

The present review is on two groups of haemoflagellates that cause disease in economically important animals, cryptobiosis in salmonids in North America, while mammalian trypanosomiasis is in livestock in Africa. Although the causative agents, *Cryptobia* and *Trypanosoma*, of both diseases are closely related (Order Kinetoplastida) there are some very significant differences between the two diseases they cause. There are both haematozoic and non-haematozoic *Cryptobia*. At least three species of haematozoic *Cryptobia* are known to cause disease in freshwater and marine fishes, and they are *Cryptobia* (*Trypanoplasma*) *salmositica* in salmonids, *C.* (*T.*) *bullocki* in flatfishes, and *C.* (*T.*) *borreli* in cyprinids. These parasites are normally transmitted indirectly by blood-sucking leeches; however *C. salmositica* can also be transmitted directly between fish in the absence of blood sucking leeches (Woo, 1994; Woo and Poynton, 1995). There are also numerous species of *Glossina*-transmitted trypanosomes that infect livestock in tropical Africa, but these pathogens are not just confined to Africa. In areas where tsetse flies are not present, these haemoflagellates (e.g. *Trypanosoma vivax* in South America, *Trypanosoma evansi* in South East Asia) are transmitted mechanically, usually as a result of interrupted feeding, by other blood-feeding flies (Hoare, 1972).

2 Pathogens

Cryptobia salmositica, the causative agent of salmonid cryptobiosis, has been recorded from all species of Pacific salmon (*Oncorhynchus* spp.) along the west coast of North

Host–Parasite Interactions, edited by G. F. Wiegertjes & G. Flik.

America from northern California to southern Alaska (Woo, 1994, 2001). *Glossina*-transmitted *Trypanosoma* occurs in both humans and domestic animals in tropical Africa. The most important pathogenic trypanosomes in domestic animals in sub-Saharan Africa are *Trypanosoma* (*Nanomonas*) *congolense*, *T.* (*Duttonella*) *vivax*, *T.* (*Trypanozoon*) *brucei* and *T.* (*Nanomonas*) *simiae*. These parasites are transmitted indirectly by *Glossina* spp. (tsetse flies) which infest about 10 million km^2 in tropical Africa (Woo, 1977) and this represents about one half of the available arable land. The disease in livestock is considered the most important animal disease in Africa (e.g. Hoare, 1972; Jawara, 1990).

Unless otherwise stated, cryptobiosis in the present review refers to the disease caused by *C. salmositica*, and animal trypanosomiasis is the disease in domestic animals caused by either *T. congolense* and/or *T. vivax*. Both groups of pathogen multiply readily by binary fission in the blood of the vertebrate host and the severity of the disease is usually directly related to the parasitaemias in infected animals (Hoare, 1972; Woo, 1978).

3 Clinical signs

After the pathogen has been injected into the vertebrate host by a blood-sucking invertebrate, the pathogen multiplies. With the onset of parasitaemias, the clinical signs in both diseases include anaemia, anorexia, weight loss, retardation of growth, and loss of conditions (Logan-Henfrey *et al.*, 1992; Woo, 1979, 2001). In salmonid cryptobiosis, other clinical signs are present and they include exophthalmia, general oedema, and abdominal distension with ascites fluid (Woo, 1979). However in mammalian trypanosomiasis, there are also intermittent fever, enlargement of lymph nodes, rapid respiratory heart rates, general weakness, inappetence, scrotal oedema, and abortions especially during the third trimester of pregnancy (e.g. Katunguka-Rwakishaya *et al.*, 1997; Sekoni, 1972; Sekoni *et al.*, 1990a, b; Silva *et al.*, 1998, 1999).

4 Susceptibility, pathogen-tolerance and pathogen-resistance in naive animals

The pathogenic trypanosomes are not host-specific, however, the clinical disease they cause can differ significantly in severity and this in part depends on the susceptibility of the species and/or breed of domestic livestock (e.g. Boran, N'Dama cattle), and on the virulence of the species/strain of the trypanosome (e.g. Katunguka-Rwakishaya *et al.*, 1997; Losos and Ikede, 1972; Soltys and Woo, 1977). Also, there are several West African breeds of cattle (e.g. West African Shorthorn, N'Dama) that are significantly less susceptible to trypanosomiasis than the Boran or Zebu breeds. These trypanotolerant breeds of cattle have evolved mechanisms to survive in tsetse-infested areas (Murray, 1988; Trail *et al.*, 1989). After being infected, the N'Dama trypanotolerant cattle have significantly lower parasitaemias and much milder anaemias than the highly susceptible Zebu cattle (Logan *et al.*, 1988; Paling *et al.*, 1991a, b). However, the mechanism of trypanotolerance is not well understood. One hypothesis is that trypanotolerant cattle have a superior immune system (Vickerman and Barry, 1982). For example, Paling *et al.* (1991b) using N'Dama and Boran cattle raised in a trypanosomiasis-free area in Kenya showed that N'Dama had superior neutralizing anti-mVSG antibody to *T. congolense* compared to Boran cattle. However, a recent study (Naessens *et al.*, 2002) suggests that this natural tolerance to trypanosomes involves at least two mechanisms. One mechanism controls parasite growth and the other involves the haemopoietic system to limit the severity of the anaemia. Hence, this study supports the hypothesis that trypanotolerance in cattle is less reliant on production of antibodies to control the parasite load.

C. salmositica is also not host-specific; it has been reported from all species of Pacific salmon on the west coast of North America (Woo, 1994). In cryptobiosis, there are similar significant variations in mortality in salmon experimentally infected with the pathogen. Briefly, susceptibility to disease and mortality vary between fish species and stocks. Putz (1972) showed that the parasite was much more pathogenic to coho salmon (100% mortality) than to chinook salmon. However, Bowers and Margolis (1984, 1985) found the reverse, i.e. 100% mortality in chinook with 0% mortality in coho. These differences are most likely due to the genetics of the fish. For example, sockeye salmon from the Fulton River stock (in British Columbia, Canada) had high mortality when injected with about 100 parasites per fish while those fish from the Weaver Creek stock (also in British Columbia) suffered light mortality even when injected with about 10^6 parasites per fish. Also, mortality of sockeye was consistent within the same fish stock and to different parasite isolates. Susceptibility to disease even varied significantly between families of Atlantic salmon; this variation was correlated to immune response (Chin and Woo 2002; Chin *et al.*, 2002). Families with detectable circulating antibodies at 3 weeks after infection had significantly lower parasitaemias than those where antibodies were detected at 5 weeks after infection. Besides the genetics of the fish, diet might also be an important factor, e.g. parasitaemias were significantly higher in trout on a high-protein diet (e.g. 3.5 g dl^{-1} plasma protein) than those on a low-protein diet (e.g. 1.0 g dl^{-1} plasma proteins) (Thomas and Woo, 1990). Anorexia, one of the clinical signs of the disease is beneficial to the infected fish as it lowers plasma proteins and subsequently lowers parasitaemias and reduces the severity of the disease (Li and Woo, 1991). Dietary ascorbic acid also has a similar effect; that is, rainbow trout on an ascorbic acid-deficient diet have lower parasitaemias than fish on supplemented diets (Li *et al.*, 1996).

In *Cryptobia*-tolerant brook charr their parasitaemias are as high or higher than those in *Oncorhynchus* spp.; however, they do not suffer from cryptobiosis (Ardelli *et al.*, 1994). A secreted 200-kDa metalloprotease is the important contributing factor to disease in cryptobiosis (e.g. Zuo and Woo, 1997a, d). In infected *Cryptobia*-tolerant brook charr the metalloprotease secreted by the pathogen is neutralized by $\alpha2$ macroglobulin, a natural anti-proteases in the blood. The amount of $\alpha2$ macroglobulin is higher in brook charr than in trout prior to infection and it remains high during the infection, while that in trout drops significantly after infection (Zuo and Woo, 1997b, c). Neutralization of metalloprotease by $\alpha2$ macroglobulin was demonstrated under both *in vivo* and *in vitro* conditions (Zuo and Woo, 1997b, c, d). Since *Cryptobia*-tolerant brook charr do not suffer from clinical disease, their immune system readily controls the infection and they recover much more rapidly than trout from the infection.

Also, some brook charr are also refractive to infection – *Cryptobia*-resistant fish (Ardelli *et al.*, 1994). *Cryptobia*-resistant and *Cryptobia*-tolerant charr have similar growth rates and respond equally well (both humoral and cell-mediated immunity) to a commercial *Aeromonas* vaccine (Ardelli and Woo, 1995). Under *in vitro* conditions the plasma of resistant charr lyse the parasite via the alternative pathway of complement activation (Forward and Woo, 1996). This is innate resistance to infection and is controlled by a dominant Mendelian locus. The protection is inherited, hence it now possible to breed brook charr that are naturally resistant to *Cryptobia* infection (Forward *et al.*, 1995).

5 Anaemia and its mechanism

One of the most consistent clinical sign in both diseases is the anaemia (Losos and Ikede, 1972; Woo, 1979). The main characteristics of the anaemia are low numbers of circulating erythrocytes, low haemoglobin levels, low packed cell volumes and short lives of red cells. Onset of the anaemia in salmonid cryptobiosis (Woo, 1979) and in bovine trypanosomiasis (e.g. Murray, 1979; Sekoni *et al.*, 1990c) is usually about 3 weeks after infection and that the severity of the anaemia is directly related to the parasitaemias. During acute disease the anaemia in cryptobiosis is microcytic and hypochromic (Woo, 1979) while in bovine (and ovine) trypanosomiasis it is characterized as macrocytic and hypochromic (e.g. Katunguka-Rwakishaya *et al.*, 1992; Silva *et al.*, 1999) and with severe leucopenia (e.g. Maxie and Valli, 1979; Silva *et al.*, 1999). At about 4 weeks after *Cryptobia* infection the red blood cells of fish are anti-globulin positive and they are lysed when exposed to complement from a naive fish (Thomas and Woo, 1988). Infected fish show hypocomplementaemia and this explains the presence of Coombs' positive cells in their blood (Thomas and Woo, 1989a). These Coombs' positive cells are lysed under *in vitro* conditions when they are exposed to complement from a naive fish (Thomas and Woo, 1988). Similarly, immune complexes on red blood cells (Facer *et al.*, 1982; Kobayashi *et al.*, 1976; Tabel *et al.*, 1979; Woo and Kobayashi, 1975) and low complement levels in the blood (Authie and Pobel, 1990; Nielsen 1985; Uche and Jones 1993) were demonstrated in animals with mammalian trypanosomiasis.

Haemodilution, splenomegaly (Woo, 1979; Laidly *et al.*, 1988), destruction of erythrocytes (Thomas and Woo 1988, 1989b) and depletion of haemopoietic tissues (Bahmanrokh and Woo, 2001) contribute to the anaemia in cryptobiosis. Thomas and Woo (1988, 1989b) showed that red cells are destroyed by *Cryptobia* secreted 'haemolysin' and the formation of immune complexes on red cells. Red blood cells with immune complexes are Coombs' positive and they may be lysed on reaction with complement or are engulfed by macrophages. Many Coombs' positive cells are not lysed because of the significant depletion of complement in infected fish (Thomas and Woo, 1989b). A 200-kDa metalloprotease has been isolated from *C. salmositica*. The enzyme has been purified and its optimal activity is pH 7.0 (Zuo and Woo, 1997a, d, 1998a). Its high proteolytic activity is inhibited by metal-chelating agents and excess of zinc ions, but is activated by calcium ions (Zuo and Woo, 1997d, 1998a). It lyses fish red cells under *in vivo* and *in vitro* conditions (Zuo and Woo, 1997b, c, 2000) by digesting the proteins in erythrocyte membranes (Zuo and Woo, 1997d) and consequently is an important contributing factor to the anaemia. It is the 'haemolysin' identified earlier as one of the causes of the anaemia in salmonid cryptobiosis (Thomas and Woo, 1988, 1989a). Under *in vitro* conditions, the proteolytic activities of this enzyme can be neutralized by antibodies (Zuo *et al.*, 1997).

In bovine trypanosomiasis plasma volumes in infected animals are higher than normal, while red-cell volumes are significantly reduced, hence there is no change in total blood volume (Dargie, 1979a). It appears the anaemia is mainly haemolytic associated with decreased life span of red cells (Jennings, 1976; Mamo and Holmes, 1975; Valli *et al.*, 1978), and extensive erythrophagocytosis with splenomegaly (e.g. Murray 1979; Valli *et al.*, 1978). Erythrophagocytosis is also an important contributing factor to the anaemia in goats (Witola and Lovelace, 2001). Boran cattle developed a more severe anaemia than N'Dama cattle (trypanotolerant cattle) during *T. congolense* infection while both breeds developed a milder but similar anaemia during *T. vivax* infection (Taiwo and Anosa 2000). However, *T. vivax* infection was more severe in cattle in Bolivia. The clinical signs include fever, anaemia, abortion, progressive weakness, loss of appetite, lethargy, and progressive emaciation (Silva *et al.*, 1998). Disseminated intravascular coagulation (DIC) and haemorrhagic syndrome may occur in cattle with extremely high parasitaemias due to *T. vivax* infections. DIC cause deposition of fibrin thrombi in capillaries and red cells are damaged during passage through these partially blocked capillaries. These damaged red cells are subsequently removed through erythrophagocytosis. Autoantibodies to red cells and thrombocytes have also been detected in calves infected with *T. vivax* (Assoku and Gardiner, 1989). Erythrocyte destruction continues as the disease continues, and the anaemia is likely due to an active and expanded mononuclear phagocytic system, and dyshaemopoiesis (Dargie, 1979b; Murray and Morrison, 1979). However, more recent studies in sheep indicates there is haemodilution but with no evidence of dyshaemopoiesis (Katunguka-Rwakishaya *et al.*, 1992).

When *T. congolense* is allowed to autolyse at 20°C it generates phospholipase A1 which acts on endogenous phosphatidyl choline to generate fatty acids (e.g. linoleic and palmitic). Although these fatty acids can cause haemolysis under *in vitro* conditions they are not likely an important cause of the anaemia under *in vivo* conditions as their activity is blocked by serum albumin (Tizard *et al.*, 1979). Although the phospholipase and its products are not the primary cause of the anaemia, Tizard *et al.* (1979) however suggested that these 'toxins' that are released from dead trypanosomes are capable of causing lesions characteristic of trypanosomiasis and may represent a major pathogenic mechanism in the disease.

6 Pathogenicity

In cryptobiosis, the metalloprotease is also collagenolytic as it readily degrades collagens (types I, IV and V) and laminin (Zuo and Woo, 1997d) under *in vitro* conditions. It is secreted by the pathogen under *in vitro* conditions and its secretion is significantly increased in the presence of either type I or IV collagens and/or their breakdown products (Zuo and Woo, 1998a). Since it is a secreted histolytic enzyme it contributes to the development of histological lesions that include focal haemorrhages, congestion of blood vessels, occlusion of capillaries with parasites and oedomatous changes in kidney glomeruli (Putz, 1972). Bahmanrokh and Woo (2001) characterized the disease as a generalized inflammatory response with severe lesions in connective tissues and in the reticulo-endothelial system. Lesions are in the liver, gills, and spleen soon after infection and these are followed by endovasculitis with mononuclear cells and extravascular parasite infiltrations. The anaemia and necrosis in the liver, kidney, and depletion of haematopoietic tissues are in part responsible for mortality during acute disease. Regeneration and replacement of necrotic tissues, especially haematopoietic and reticular tissues, occur during recovery of the infection.

Acutely infected fish are susceptible to hypoxia partly because of the severe anaemia and high parasitaemias that occludes small blood vessels (Woo and Wehnert, 1986). Also, metabolism and swimming performance of infected trout are significantly reduced (Kumaraguru *et al.*, 1995), and the bioenergetic cost of the disease is very considerable to infected fish. These are contributing factors to the retarded growth as there are significant reductions in food consumption, dry weight and energy gained, energy concentration, and gross conversion efficiency (Beamish *et al.*, 1996). However, the attenuated strain does not cause disease but protects juvenile and adult fish from disease (e.g. Li and Woo, 1995; Sitja-Bobadilla and Woo, 1994; Woo and Li, 1990). It also has no detectable bioenergetic cost to juvenile trout (Beamish *et al.*, 1996).

Tizard *et al.* (1978) in their review suggested that pathogenic mammalian trypanosomes generate toxic catabolites which could contribute to the pathophysiology of the disease. Several cytotoxic enzymes which affect cell membranes have been shown to be present in these pathogenic parasites, and they include neuraminidase (Esievo, 1983), lysophospholipases, phospholipases (Mellors, 1985) and proteases (Authie *et al.*, 1992; Lonsdale-Eccles and Grab, 1986; Lonsdale-Eccles and Mpimbaza, 1986). However, their contributions to the disease need further studies.

T. congolense causes a debilitating disease in calves with pancytopenia persisting throughout the infection (Valli *et al.*, 1979). There is thymic atropy with enlargement of the heart, lungs, liver, kidneys and lymph nodes. There is extensive microvascular dilation with decreased cellularity in all lymphoid tissues and thymic-dependent atropy. Chronic mononuclear inflammatory foci are in the heart, liver and kidneys. Renal glomeruli are diffusely enlarged and hypercellular. Haemosiderin deposits are common in the lungs, liver, spleen, renal epithelium and bone marrow.

7 Thrombocytopenia and hypocomplementaemia

In trypanosomiasis, thrombosis may cause localized ischaemia and subsequently tissue necrosis, and severe thrombocytopenia (Logan-Henfrey *et al.*, 1992). Platelet aggregation is a serious complication of the disease, and proteases are known potent inducers of blood coagulation and thrombocytopenia (e.g. Davis 1982; Nwagwu *et al.*, 1989; Valli *et al.*, 1979).

As indicated earlier, hypocomplementaemia in animals occurs soon after infections with *Cryptobia* (Thomas and Woo, 1989a) and *Trypanosoma* (Authie and Pobel, 1990; Nielsen, 1985) and this could contribute to the immunodepression. There is also evidence that depletion of complement (C3) in rabbits reduces immunological memory (Uche and Jones, 1993). The significant decline in complement levels in infected animals can be caused by at least two obvious processes. Blood complement is depleted soon after the production of complement-fixing antibodies, e.g. to lyse the pathogen (Li and Woo, 1995), and also bind to immune complexes that are subsequently formed in circulation and these are deposited on tissues (Nielsen, 1985). Also, the complement can be depleted as a direct cleavage by the pathogens, for example in cryptobiosis by the histolytic metalloprotease produced by *C. salmositica* (Zuo and Woo, 1997b, c, d).

8 Immunodepression

Morbidity/mortality of an animal infected with an infectious organism may sometimes be due to depression of its immune system by an unrelated disease condition. Also, immunodepression due to an infection may also reduce the efficacy of vaccines against other infectious organisms. Immunodepression to a secondary antigen may be due to factors that would include anorexia, hypocomplementaemia, anaemia, induction of suppressor T cells, defective cytokine production, clonal exhaustion after polyclonal activation of B cells, and antigenic competition. Immunodepression and immunosuppresion are often used synonymously in the literature to describe a decreased response to a second antigenic stimulation in infected animals. I agree with Cox (1986) who indicated that there is a fine distinction between the two phenomena. Since there is usually insufficient evidence to indicate a total dysfunction of the immune system, I prefer to use immunodepression (e.g. Woo and Jones, 1989) to describe decreased response in an infected animal to a second and unrelated antigenic stimulation.

Wehnert and Woo (1981) showed depression of the humoral response to sheep red blood cells in infected fish during acute cryptobiosis. This was confirmed using both sheep red blood cells and a pathogenic bacterium, *Yersinia ruckeri* (Jones *et al.*, 1986). The bacterium is Gram-negative and it causes enteric red mouth disease in salmonids (Horne and Barnes, 1999). Fish with acute cryptobiosis during their initial exposure to *Y. ruckeri* did not produce antibody against the bacterium. Also, the mortality rate (~90%) of these *Yersinia*-exposed fish was similar to *Yersinia*-naive fish when they were later exposed to a lethal dose of the bacterium. Immunodepression was not in fish that had recovered from cryptobiosis at the time they were first exposed to *Yersinia*; these fish were protected from yersiniosis and mortality when they were later challenged with the bacterium (Jones *et al.*, 1986). The anorexia also contributes to the immunodepression in cryptobiosis (Thomas and Woo, 1992).

Immunodepression in laboratory animals infected with pathogenic trypanosomes is well documented (e.g. Goodwin *et al.*, 1972; Vickerman and Barry, 1982); however, this is less well studied in livestock which often die with concurrent microbial infections (Morrison *et al.*, 1985). Holmes *et al.* (1974) were the first to provide sufficient evidence of immunodepression in cattle infected with *T. congolense*. Subsequent studies using microbial antigens confirmed immunodepression in cattle suffering from trypanosomiasis (Scott *et al.*, 1977; Whitelaw *et al.*, 1979). Recent *in vitro* studies showed suppressor macrophages in lymph nodes but not in the blood of cattle infected with *T. congolense* (Flynn *et al.*, 1990; Sileghem *et al.*, 1990). Also, nitric oxide production was not involved in the loss of T-cell proliferation associated with trypanosomiasis in cattle although this was seen in mice. The production of nitric oxide by monocytes and macrophages was actually down-regulated in infected cattle (Taylor *et al.*, 1996).

9 Endocrine system and reproduction

Little is know about the effects *C. salmositica* has on the endocrine system in fish. Plasma cortisol levels were not significantly different after infection, however the plasma levels of T3, T4 (thyroxine), protein and glucose concentration, and liver glycogen were much lower in infected trout. There were no recoveries even during the chronic phase of the disease. These hormonal and metabolic changes clearly show that infected fish were under considerable physiological duress (Laidley *et al.*, 1988). However, in sheep infected with *T. congolense*, there were significant increases in cortisol in rams after onset of parasitaemias (9–16 days after infection), followed by transient non-significant decreases, and variable and parasitaemia-dependent levels between 44–79 days after infection (Mutayoba *et al.*, 1995). Testosterone secretions decreased within 4 weeks of infection and the decreases were likely due to testicular rather than pituitary effects (Mutayoba *et al.*, 1994, 1995). In infected goats, the T4 levels declined significantly within a week of infection and remained low until chemotherapy, after which they returned to pre-infection values (Mutayoba *et al.*, 1988a). Normocyclic female goats infected with *T. congolense* developed irregular and shorter oestrous cycles before complete cessation at the fourth cycle. This was followed by declines in plasma progesterone and oestradiol-17 beta levels. The rapidity of ovarian dysfunction was associated with the degree of susceptibility of goats to trypanosomiasis (Mutayoba *et al.*, 1988b). Irregular oestrous cycles and significant declines in progesterone and oestradiol-17 beta were also in infected trypanotolerant East African goats. Declines were less in more resistant individuals at least in the first 2 months after infection. Resumption of ovarian cycles in resistant goats occurred at about 5 months after infection (Mutayoba *et al.*, 1988c).

Reproduction is often lower in animals with trypanosomiasis but this has not been studied in cryptobiosis. In trypanosomiasis there are lesions in the ovaries and testes of infected goats, sheep and cattle. The disease affects the oestrous cycle in female animals and this causes infertility (e.g. Anosa and Isoun, 1980; Ikede *et al.*, 1988; Kaaya and Oduor-Okelo, 1980; Llewelyn *et al.*, 1987, 1988; Mutayoba *et al.*, 1988b, d; Ogwu *et al.*, 1984). Weight loss and anaemia are expected to contribute to reproductive impairment in infected animals. Susceptible bulls had poor-quality semen, low volume of ejaculates and low sperm count, and an increased percentage of abnormal sperm soon after infection. By the sixth week after infection, experimentally infected bulls lacked libido (Sekoni *et al.*, 1988). Histopathological lesions were obvious in the reproductive organs, especially in the testes and epididymides (Sekoni *et al.*, 1990a). Some improvements were observed in semen characteristics of some bulls at 10 weeks after chemotherapy. However, all infected bulls (either with or without chemotherapy) still had poor semen characteristics in that they had decreased volume of semen, oligospermia, azoospermia and high numbers of morphologically abnormal sperms (Sekoni *et al.*, 1990b).

Similarly, there was progressive deterioration in semen quality in rams after infection and this included a decrease in volume or cessation of semen production, decrease in sperm motility with high numbers of dead sperm, and in most animals all their sperm showed morphological abnormalities (Sekoni, 1992). Ewes in their third trimester of pregnancy were more susceptible to trypanosome infections than those in the second trimester – this was evidenced by the number of abortions and the number of deaths of ewes. During the study, no abortions were observed in uninfected ewes. Also, fetuses from infected ewes had lower body weights than those from infected ewes that had undergone chemotherapy (Bawa *et al.*, 2000). In general, these pathogenic trypanosomes cause either serious infertility or even sterility in cattle and sheep.

10 Conclusions

The pathophysiology of the two diseases is not only intriguing but also very challenging. It is hoped that this review brings up gaps in our knowledge and that it will stimulate further studies, especially at the molecular level.

References

Anosa, V.O. and Isoun, T.T. (1980) Further observations on the testicular pathology in Trypanosoma vivax infection of sheep and goats. *Res. Vet. Sci.* **28:** 151–160.

Ardelli, B.F. and Woo, P.T.K. (1995) Immune response of *Cryptobia*-resistant and *Cryptobia*-susceptible *Salvelinus fontinalis* to an *Aeromonas salmonicida* vaccine. *Dis. Aquat. Organ.* **23:** 33–38.

Ardelli, B.F., Forward, G.M. and Woo, P.T.K. (1994) Brook charr (*Salvelinus fontinalis*) and cryptobiosis: A potential salmonid reservoir host for *Cryptobia salmositica* Katz 1951. *J. Fish Dis.* **17:** 567–577.

Assouku, R.K.G. and Gardiner, P.R. (1989) Detection of antibodies to platelets and erythrocytes during infections with haemorrhage-causing *Trypanosoma vivax* in Ayrshire cattle. *Vet. Parasitol.* **31:** 199–216.

Authie, E. and Pobel, T. (1990) Serum haemolytic complement activity and C3 levels in bovine trypanosomiasis under natural conditions of challenge – early indications of individual susceptibility to disease. *Vet. Parasitol.* **35:** 43–59.

Authie, E., Muteti, D.K., Mbawa, Z.R., Lonsdale-Eccles, J.D., Webster, P. and Wells, C.W. (1992) Identification of a 33-kilodalton immuno-dominant antigen of *Trypanosoma congolense* as a cysteine protease. *Mol. Biochem. Parasitol.* **56:** 103–116.

Bahmanrokh, M. and Woo, P.T.K. (2001) Relationships between histopathology and parasitaemias in *Oncorhynchus mykiss* infected with *Cryptobia salmositica*, a pathogenic haemoflagellate. *Dis. Aquat. Organ.* **46:** 41–45.

Bawa, E.K., Ogwu, D., Sekoni, V.O., Oyedipe, E.O, Esievo, K.A. and Kambai, J.E. (2000) Effects of *Trypanosoma vivax* on pregnancy of Yankasa sheep and the results of homidum chloride chemotherapy. *Theriogenology* **54:** 1033–1040.

Beamish, F.W.H., Sitja-Bobadilla, A., Jebbink, J.A. and Woo, P.T.K. (1996) Bioenergetic cost of cryptobiosis in fish: rainbow trout (*Oncorhynchus mykiss*) infected with *Cryptobia salmositica* and with an attenuated live vaccine. *Dis. Aquat. Organ.* **25:** 1–8.

Bower, S.M. and Margolis, L. (1984) Distribution of *Cryptobia salmositica*, a haemoflagellate of fishes in British Columbia and the seasonal pattern of infection in a coastal river. *Can. J. Zool.* **62:** 2512–2518.

Bower, S.M. and Margolis, L. (1985) Effects of temperature and salinity on the course of infection with the haemoflagellate *Cryptobia salmositica* in juvenile Pacific salmon (*Oncorhynchus* spp.). *J. Fish Dis.* **8:** 25–33.

Chin, A. and Woo, P.T.K. (2002) Susceptibility and immunological response in families of Atlantic salmon (*Salmo salar*) against *Cryptobia salmositica*. 10th International Congress of Parasitology, Vancouver, Canada.

Chin, A., Eldridge, M., Glebe, B.D. and Woo, P.T.K. (2002.) *Salmo salar* and *Cryptobia salmositica*: variations in susceptibility and humoral response in families of Atlantic salmon to the pathogen. AquaNet II, Annual General Meeting, Moncton, Canada.

Cox, F.E.G. (1986) Interactions in protozoan infections. In: Howell, M.J. (ed.) *Parasitology – Quo Vardit?*, pp. 569–575. Proceedings of the Sixth International Congress of Parasitology, Brisbane, Australia.

Dargie, J.D. (1979a) Effects of *T. congolense* and *T. brucei* on the circulatory volumes of cattle. In: Losos, G. and Chouinard, A. (eds) *Pathogenicity of Trypanosomes*, pp. 140–144, International Development Research Centre Publication 132e, Ottawa, Canada.

Dargie, J.D. (1979b) Erythropoietic response in bovine trypanosomisis. In: Losos, G. and Chouinard, A. (eds) *Pathogenicity of Trypanosomes*, pp. 128–134. International Development Research Centre Publication 132e, Ottawa, Canada.

Davis, C.E. (1982) Throbocytopenia: a uniform complication of African trypanosomiasis. *Acta Tropica* **39:** 123–133.

Esievo, K.A.N. (1983) Trypanosoma vivax, stock V953: Inhibitory effect of type A influenza virus anti-HAV8 serum on in vitro neuraminadase (sialidase) activity. *J. Parasitol.* **69:** 491–495.

Facer, C.A., Crosskey, J.M., Clarkson, M.J. and Jenkins, G.C. (1982) Immune haemolyticanaemia in bovine trypanosomiasis. *J. Comp. Pathol.* **92:** 393–401.

Flynn, J.N., Sileghem, M.R. and Williams, D.J.L. (1990) Induction of immunosuppression in cattle infected with *Trypanosoma congolense*. European Federation of Immunological Societies, 10th Meeting, Edinburgh, UK [Abstract].

Forward, G.M. and Woo, P.T.K. (1996) An *in vitro* study on the mechanism of innate immunity in *Cryptobia*-resistant brook charr (*Salvelinus fontinalis*) against *Cryptobia salmositica*. *Parasitol. Res.* **82:** 238–241.

Forward, G.M., Ferguson, M.M. and Woo, P.T.K. (1995) Susceptibility of brook charr, *Salvelinus fontinalis* to the pathogenic haemoflagellate, *Cryptobia salm*ositica, and the inheritance of innate resistance by progenies of resistant fish. *Parasitology* **111:** 337–345.

Goodwin, L.G., Green, D.G., Guy, M.W. and Voller, A. (1972) Immunosuppression during trypanosomiasis. *Br. J. Exper. Pathol.* **53:** 40–43.

Hoare, C.A. (1972) *The Trypanosomes of Mammals: A Zoological Monograph*, Blackwell Scientific Publications, Oxford and Edinburgh, UK.

Holmes, P.H., Mammo, E., Thomson, A., Knight, P.A., Lucken, R., Murray, P.K., Murray, M., Jennings, F.W. and Urquhardt, G.M. (1974) Immunosuppression in bovine trypanosomisis. *Vet. Rec.* **95:** 86–87.

Horne, M.T. and Barnes, A.C. (1999) Enteric redmouth disease (*Yersinia ruckeri*). In: Woo, P.T.K. and Bruno, D.W. (eds) *Fish Diseases and Disorders, Volume 3: Viral, Bacterial and Fungal Infections*, pp. 455–477. CABI Publishing, Oxford, UK.

Ikede, B.O., Elhassan, E. and Alpavie, S.O. (1988) Reproductive disorders in African trypanosomiasis: a review. *Acta Tropica* **45:** 5–10.

Jawara, A.S.D.K. (1990) Animal disease as a factor limiting economic development in Africa. *Cornell Vet.* **80:** 17–25.

Jennings, F.W. (1976) The anaemias of parasitic infections. In: Soulsby, E.J.L. (ed.) *Pathophysiology of Parasitic Infection*, pp. 41–67. Academic Press, New York, USA.

Jones, S.R.M., Woo, P.T.K. and Stevenson, R.M.W. (1986) Immunosuppression in *Salmo gairdneri* caused by the haemoflagellate, *Cryptobia salmositica*. *J. Fish Dis.* **9:** 931–938.

Kaaya, G. and Oduor-Okelo A. (1980) The effects of *Trypanosoma congolense* infection on the testis and epididymis of the goat. *Bulletin Anim. Health Prod. Africa* **28:** 1–5.

Katunguka-Rwakishaya, E., Murray, M. and Holmes, P.H. (1992) Pathophysiology of ovine trypanosomiasis: ferrokinetics and serthrocyte survival studies. *Res. Vet. Sci.* **53:** 80–86.

Katunguka-Rwakishaya, E., Murray, M. and Holmes, P.H. (1997) Susceptibility of three breeds of Ungandan goats to experimental infection with *Trypanosoma congolense*. *Trop. Anim. Health Prod.* **29:** 7–14.

Kobayashi, A., Tizard, I.R. and Woo, P.T.K. (1976) Studies on the anaemia in experimental African trypanosomiasis. II The pathogenesis of the anaemia in calves infected with *Trypanosoma congolense*. *Am. J. Trop. Med. Hygiene* 25: 401–406.

Kumaraguru, A.K., Beamish, F.W.H. and Woo, P.T.K. (1995) Impact of a pathogenic haemoflagellate, *Cryptobia salmositica* on the metabolism and swimming performance of rainbow trout, *Oncorhynchus mykiss* (Walbaum). *J. Fish Dis.* **18:** 297–305.

Laidley, C.W., Woo, P.T.K. and Leatherland, J.F. (1988) The stress response of rainbow trout to experimental infection with the blood parasite, *Cryptobia salmositica* Katz 1951. *J. Fish Biol.* **32:** 253–261.

Llewelyn, C.A., Luckins, A.G., Munro, C.D. and Perrie, J. (1987) The effect of *Trypanosoma congolense* infection on the oestrous cycle of the goat. *Br. Vet. J.* **143:** 423–431.

Llewelyn, C.A., Munro, C.D., Luckins, A.G., Jordt, T., Murray, M. and Lorenzini, E. (1988) The effect of *Trypanosoma congolense* infection on the oestrus cycle of the Boran cow. *Br. Vet. J.* **144:** 379–387.

Li, S. and Woo, P.T.K. (1991) Anorexia reduces the severity of cryptobiosis in *Oncorhynchus mykiss*. *J. Parasitol.* **77:** 467–471.

Li, S. and Woo, P.T.K. (1995) Efficacy of a live *Cryptobia salmositica* vaccine, and the mechanism of protection in vaccinated *Oncorhynchus mykiss* (Walbaum) against cryptobiosis. *Vet. Immunol. Immunopathol.* **48:** 343–353.

Li, S., Cowey, C.B. and Woo, P.T.K. (1996) The effects of dietary ascorbic acid on *Cryptobia salmositica* infection and on vaccination against cryptobiosis in *Oncorhynchus mykiss*. *Dis. Aquat. Organ.* **24:** 11–16.

Logan, L.L., Paling, R.W., Moloo, S.K. and Scott, J.R. (1988) Comparative responses of N'Dama and Boran cattle to experimental challenge with tsetse transmitted *T. congolense*. In: *Livestock Production in Tsetse Affected Areas in Africa*, pp. 152–160. International Livestock Centre for Africa, International Laboratory for Research on Animal Diseases, Nairobi, Kenya.

Logan-Henfrey, L.L., Gardiner, P.R. and Mahmoud, M.M. (1992) Animal trypanosomiasis in Sub-Saharan Africa. In: Kreier, J.P. and Baker, J.R. (eds) *Parasitic Protozoa,* Volume 2, 2nd edn., pp. 157–276. Academic Press, San Diego, USA.

Lonsdale-Eccles, J.D. and Grab, D. (1986) Proteases in African trypanosomes. In: Turk V. (ed.) *Cysteine Proteinases and Their Inhibitors*, pp. 189–197. Walter de Gruyter, Berlin, Germany.

Lonsdale-Eccles, J.D. and Mpimbaza, G.W.N. (1986) Thiol-dependent proteases of African trypanosomes. Analysis by electrophoresis in sodium dodecyl sulphate/polyacrylamide gels co-polymerized with fibrinogen. *Euro. J. Biochem.* **155:** 469–473.

Losos, G.J. and Ikede, B.O. (1972) Review of the pathology of diseases in domestic and laboratory animals caused by *Trypanosoma congolense, T. vivax, T. brucei, T. rhodesiense* and *T. gambiense*. *Vet. Pathol.* **9:** 1–71.

Mamo, E. and Holmes, P.K. (1975) The erythrokinetics of zebu cattle chronically infected with *Trypanosoma congolense*. *Res. Vet. Sci.* **18:** 105–106.

Maxie, M.G. and Valli, V.E.O. (1979) Pancytopenia in bovine trypanosomiasis. In: Losos, G. and Chouinard, A. (eds) *Pathogenicity of Trypanosomes*, pp. 135–136. International Development Research Centre Publication 132e, Ottawa, Canada.

Mellors, A. (1985) Phospholipases in trypanosomes. In: Tizard I.R. (ed.) *The Immunology and Pathogenesis of Trypanosomiasis*, pp. 67–74. CRC Press Inc., Boca Raton, USA.

Morrison, W.L., Murray, M. and Akol, G.W.O. (1985) Immune response of cattle to African trypanosomes. In: Tizard I.R. (ed.) *The Immunology and Pathogenesis of Trypanosomiasis*, pp. 103–132. CRC Press Inc., Boca Raton, USA.

Murray, M. (1979) Anaemia of bovine African trypanosomiasis: an overview. In: Losos G. & Chouinard A. (eds) *Pathogenicity of Trypanosomes*, pp. 121–127. International Development Research Centre Publication 132e, Ottawa, Canada.

Murray, M. (1988) Trypanotolerance, its criteria and genetic and environmental influences. In: *Livestock Production in Tsetse Affected Areas in Africa*, pp. 133–151. International Livestock Centre for Africa, International Laboratory for Research on Animal Diseases, Nairobi, Kenya.

Murray, M. and Morrison W.I. (1979) Parasitaemia and host susceptibility to African trypanosomiasis. In: Losos, G. and Chouinard, A. (eds) *Pathogenicity of Trypanosomes*, pp. 71–86. International Development Research Centre Publication 132e, Ottawa, Canada.

Mutayoba, B.M., O'Hara-Ireri, H.B. and Gombe, S. (1988a) Trypanosome-induced depression of plasma throxine. Levels in prepubertal and adult female goats. *Acta Endocrinol.* **119:** 21–26.

Mutayoba, B.M., Gombe, S., Waindi, E.N. and Kaaya G.P. (1988b) Depression of ovarian function and plasma progesterone and estradiol-17B in female goats chronically infected with *Trypanosoma congolense*. *Acta Endocrinol.* **117:** 477–484.

Mutayoba, B.M., Gombe, S., Kaaya, G.P. and Waindi, E.N. (1988c) Trypanosome-induced ovarian dysfunction. Evidence of higher residual fertility in trypanotolerant small East African goats. *Acta Tropica* **45:** 225–237.

Mutayoba, B.M., Gombe, S., Kaaya, G.P. and Waindi, E.N. (1988d) Effects of chronic experimental *Trypanosoma congolense* infection on the ovaries, pituitary, thyroid and adrenal glands in female goats. *Res. Vet. Sci.* **44:** 140–146.

Mutayoba, B.M., Eckersall, P.D., Jeffcoate, I.A., Cestnik, V. and Holmes, P.H. (1994) Effects of *Trypanosoma congolense* infection in rams on the pulsatile secretion of L.H. and testerone and responses to injection of GnRH. *Reproduction Fertility* **102:** 425–431.

Mutayoba, B.M., Eckersall, P.D., Cestnik, V., Jeffcoate, I.A., Gray, C.E. and Holmes, P.H. (1995) Effects of *Trypanosoma congolense* on pituitary and adrenocortical function in sheep: changes in the adrenal gland and cortisol secretion. *Res. Vet. Sci.* **58:** 174–179.

Naessens, J., Teale, A.J. and Sileghem, M. (2002) Identification of mechanisms of natural resistance to African trypanosomiasis in cattle. *Vet. Immunol. Immunopathol.* **87:** 187–194.

Nielson, K.H. (1985) Complement in trypanosomiasis. In: Tizard, I.R. (ed.) *The Immunology and Pathogenesis of Trypanosomiasis*, pp. 134–144. CRC Press Inc., Boca Raton, USA.

Nwagwu, M., Inyang, A.L., Molokwu, R.I. and Essien, E.M. (1989) Platelet aggregating activity of released factors from *Trypanosoma brucei*. *African J. Med. Sci.* **18:** 283–287.

Ogwu, D., Njoku, C.O., Osori, D.I.K., Ezeokoli, C.D. and Kumi-Diaka, J. (1984) Effects of experimental *Trypanosoma vivax* on fertility of heifers. *Theriogenology* **40:** 625–633.

Paling, R.W., Moloo, S.K., Scott, J.R., Gettinby, G., McOdimba, F.A. and Murray, M. (1991a) Susceptibility of N'Dama and Boran cattle to sequential challenges with tsetse-transmitted clones of *Trypanosoma congolense*. *Parasite Immunol.* **13:** 427–445.

Paling, R.W., Moloo, S.K., Scott, J.R., McOdimba, F.A., Logan-Henfrey, L., Murray, M. and Williams, D.J.L. (1991b) *Trypanosoma congolense*: Comparison of immune responses in N'Dama and Boran cattle exposed to tsetse fly-transmitted primary and rechallenge infections with a homologous serodeme. *Parasite Immunol.* **13:** 413–425.

Putz, R.E. (1972) Biological studies on the haemoflagellates *Cryptobia cataractae* and *Cryptobia salmositica*. Technical Paper, Bureau of Sport Fisheries and Wildlife (US) **63:** 3–25.

Scott, J.M., Pegram, R.G., Holmes, P.H., Pay, T.W.F., Knight, P.A., Jennings, F.W. and Urquhardt, G.M. (1977) Immunosuppression in bovine trypanosomisis: Field studies using foot-and-mouth disease vaccine and clostridial vaccine. *Trop. Animal Health Prod.* **9:** 159–165.

Sekoni, V.O. (1992) Effect of *Trypanosoma vivax* infection on semen characteristics of Yankasa rams. *Br. Vet. J.* **148:** 501–506.

Sekoni, V.O., Kumi-Diaka, J., Saror D. and Njoku, C. (1988) The effect of *Trypanosoma vivax* and *Trypanosoma congolense* infections on the reaction time and semen characteristics in the Zebu bull. *Br. Vet. J.* **144:** 388–394.

Sekoni, V.O., Njoku, C.O., Kumi-Diaka, J. and Saror, D.I. (1990a) Pathological changes in male genitalia of cattle infected with *Trypanosoma vivax* and *Trypanosoma congolense*. *Br. Vet. J.* **146:** 175–180.

Sekoni, V.O., Njoku, C., Saror, D., Sannusi, A., Oyejola, B. and Kumi-Diaka, J. (1990b) Effect of chemotherapy on elevated ejaculation time and deteriorated semen characteristics consequent to bovine trypanosomiasis. *Br. Vet. J.* **146:** 368–373.

Sekoni, V.O., Saror, D.I., Njoku, C.O., Kumi-Diaka, J. and Opaluwa, G.I. (1990c) Comparative haematological changes following *Trypanosoma vivax* and *T. congolense* infections in Zebu bulls. *Vet. Parasitol.* **35:** 11–19.

Sileghem, M., Flynn, J.N. and Williams, D.J.L. (1990) *Suppression of IL2 production and IL2 receptor expression on bovine trypanosomiasis*. European Federation of Immunological Societies, 10th Meeting, Edinburgh, UK [Abstract].

Sitja-Bobadilla, A. and Woo, P.T.K. (1994) An enzyme-linked immunosorbent assay (ELISA) for the detection of antibodies against the pathogenic haemoflagellate, *Cryptobia salmositica* Katz and protection against cryptobiosis in juvenile *Oncorhynchus mykiss* (Walbaum) inoculated with a live *Cryptobia* vaccine. *J. Fish Dis.* **17:** 399–408.

Silva, R.A., Morales, G., Eulert, E., Montenegro, A. and Ybanez, R. (1998) Outbreaks of trypanosomiasis due to *Trypanosoma vivax* in cattle in Bolivia. *Vet. Parasitol.* **76:** 153–157.

Silva, R.A.M.S., Ramirez, L., Souza, S.S., Ortiz, A.G., Pereira, S.R. and Davila, A.M.R. (1999) Hematology in natural bovine trypanosomiasis in the Brazilian Pantanal and Bolivian wetlands. *Vet. Parasitol.* **85:** 87–93.

Soltys, M.A. and Woo, P.T.K. (1977) Trypanosomes producing disease in livestock in Africa. In: Krier, J.P. (ed.) *Parasitic Protozoa*, Volume 1, pp. 239–268. Academic Press, New York, USA.

Tabel, H., Rurangirwa, F.W. and Losos, G.J. (1979) Is the anaemia in bovine trypanosomiasis caused by immunologic mechanisms? In: Losos, G. and Chouinard, A. (eds) *Pathogenicity of Trypanosomes*, pp. 91–93. International Development Research Centre Publication 132e, Ottawa, Canada.

Taiwo, V.O. and Anosa, V.O. (2000) *In vitro* erythrophagocytosis by cultured macrophages stimulated with extraneous substances and those isolated from the blood, spleen and bone marrow of Boran and N'Dama cattle infected with *Trypanosoma congolense* and *Trypanosoma vivax*. *Onderstepoort J. Vet. Res.* **67:** 273–287.

Taylor, K., Lutje, V. and Mertens, B. (1996) Nitric oxide synthesis is depressed in *Bos indicus* cattle infected with *Trypanosoma congolense* and *Trypanosoma vivax* and does not mediate T-cell suppression. *Infection Immunity* **64:** 4115–4122.

Thomas, P.T. and Woo, P.T.K. (1988) *Cryptobia salmositica*: *in vitro* and *in vivo* study on the mechanism of anaemia in infected rainbow trout, *Salmo gairdneri*. *J. Fish Dis.* **11:** 425–431.

Thomas, P.T. and Woo, P.T.K. (1989a) An *in vitro* study on the haemolytic components of *Cryptobia salmositica*. *J. Fish Dis.* **12:** 389–393.

Thomas, P.T. and Woo, P.T.K. (1989b) Complement activity in *Salmo gairdneri* Richardson infected with *Cryptobia salmositica* and its relationship to the anaemia in cryptobiosis. *J. Fish Dis.* **12:** 395–397.

Thomas, P.T. and Woo, P.T.K. (1990) Dietary modulation of humoral immune response and anaemia in *Oncorhynchus mykiss* (Walbaum) infected with *Cryptobia salmositica* Katz 1951. *J. Fish Dis.* **13:** 435–446.

Thomas, P.T. and Woo, P.T.K. (1992) Anorexia in *Oncorhynchus mykiss* infected with *Cryptobia salmositica* (Sarcomastigophora: Kinetoplastida): its onset and contribution to the immunodepression. *J. Fish Dis.* **15:** 443–447.

Tizard, I.R., Nielsen, K.H., Seed, J.R. and Hall, J.E. (1978) Biologically active products from African trypanosomes. *Microbiol. Rev.* **42:** 661–681.

Tizard, I.R., Nielsen, K.H., Mellors, A. and Assoku, R.K.G. (1979) Biologically active lipids generated by autolysis of *T. congolense*. In: Losos, G. and Chouinard, A. (eds) *Pathogenicity of Trypanosomes*, pp. 103–110. International Development Research Centre Publication 132e, Ottawa, Canada.

Trail, J.C.M., D'Ieteren, G.D.M. and Teale, A.J. (1989) Trypanotolerance and the value of conserving livestock genetic resources. *Genome* **31:** 805–812.

Uche, U.E. and Jones, T.W. (1993) Effect of complement (C3) depletion on the generation of memory in rabbits primed with antigens of *Trypanosoma evansi*. *Vet. Parasitol.* **47:** 205–213.

Valli, V.E.O., Forsberg, C.M. and McSherry, B.J. (1978) The pathogenesis of *Trypanosoma congolense* in calves. II. Anaemia and erythroid response. *Vet. Pathol.* **15:** 732–745.

Valli, V.E.O., Forsberg, C.M. and Mills, J.N. (1979) Pathology of *T. congolense* in calves. In: Losos, G. and Chouinard, A. (eds) *Pathogenicity of Trypanosomes*, pp. 179–183, International Development Research Centre Publication 132e, Ottawa, Canada.

Vickerman, K. and Barry J.D. (1982) African trypanosomiasis. In: Cohen, S. and Warren, J.S. (eds) *Immunology of Parasitic Infections*, pp. 204–260. Blackwell Scientific Publications, Oxford, UK.

Wehnert, S.D. and Woo, P.T.K. (1981) The immune response of *Salmo gairdneri* during *Trypanoplasma salmositica* infection. *Can. Soc. Zool. Bull.* **11:** 100.

Whitelaw, D.D., Scott, J.M., Reid, H.W., Holmes, P.H., Jennings, F.W. and Urquhardt, G.M. (1979) Immunosuppression in bovine trypanosomiasis: studies with louping-ill vaccine. *Res. Vet. Sci.* **26:** 102–107.

Witola, W.H. and Lovelace, C.E. (2001) Demonstration of erythrophagocytosis in *Trypanosoma congolense*-infected goats. *Vet. Parasitol.* **96:** 115–126.

Woo, P.T.K. (1977) Salivarian trypanosomes producing disease in livestock outside of Sub-Saharan Africa. In: Kreier J.P. (ed.) *Parasitic Protozoa: Vol. 1*, pp. 270–296. Academic Press, New York, USA.

Woo, P.T.K. (1978) The division process of *Cryptobia salmositica* in experimentally infected rainbow trout (*Salmo gairdneri*). *Can. J. Zool.* **56:** 1514–1518.

Woo, P.T.K. (1979) *Trypanoplasma salmositica*: experimental infections in rainbow trout, *Salmo gairdneri*. *Exper. Parasitol.* **47:** 36–48.

Woo, P.T.K. (1994) Flagellate parasites of fish. In: Kreier, J.P. (ed.) *Parasitic Protozoa: Vol. 8, 2nd edition*, pp. 1–80. Academic Press, New York, USA.

Woo, P.T.K. (1998) Protection against *Cryptobia* (*Trypanoplasma*) *salmositica* and salmonid cryptobiosis. *Parasitol. Today* **14:** 272–277.

Woo, P.T.K. (2001) Cryptobiosis and its control in North American fishes. *Intern. J. Parasitol.* **31:** 566–574.

Woo, P.T.K. and Jones, S.R.M. (1989) The piscine immune system and the effects of parasitic protozoans on the immune system. In: Ko, R.C. (ed.) *Current Concepts in Parasitology*, pp. 47–64, Hong Kong University Press, Hong Kong.

Woo, P.T.K. and Kobayshi, A. (1975) Studies on the anaemia in experimental African trypanosomiasis. I. A preliminary communication on the mechanisms of the anaemia. *Annales Soc. Belge Med. Trop.* **55:** 37–45.

Woo, P.T.K. and Li, S. (1990) *In vitro* attenuation of *Cryptobia salmositica* and its use as a live vaccine against cryptobiosis in *Oncorhynchus mykiss*. *J. Parasitol.* **76:** 752–755.

Woo, P.T.K. and Poynton, S. (1995) Diplomonadida, Kinetoplastida and Amoebida. In: Woo, P.T.K. (ed.) *Fish Diseases and Disorders, Volume 1: Protozoan and Metazoan Infections*, pp. 27–96. CAB International, Oxford, UK.

Woo, P.T.K. and Wehnert, S.D. (1986) *Cryptobia salmositica*: susceptibility of infected trout, *Salmo gairdneri*, to environmental hypoxia. *J. Parasitol.* **72:** 392–396.

Zuo, X. and Woo, P.T.K. (1997a) Proteases in pathogenic and nonpathogenic hemoflagellates, *Cryptobia* spp. (Sarcomastigophora: Kinetoplastida) of fishes. *Dis. Aquat. Organ.* **29:** 57–65.

Zuo, X. and Woo, P.T.K. (1997b) Natural antiproteases in rainbow trout, *Oncorhynchus mykiss*, and brook charr, *Salvelinus fontinalis*, and the *in vitro* neutralization of fish alpha2-macroglobulin by the metalloprotease from the pathogenic haemoflagellate, *Cryptobia salmositica*. *Parasitology* **114:** 375–382.

Zuo, X. and Woo, P.T.K. (1997c) The *in vivo* neutralization of proteases from *Cryptobia salmositica* by α2-macroglobulin in the blood of rainbow trout, *Oncorhynchus mykiss* and brook charr, *Salvelinus fontinalis*. *Dis. Aquat. Organ.* **29:** 67–72.

Zuo, X. and Woo, P.T.K. (1997d) Purified metallo-protease from the pathogenic haemoflagellate, *Cryptobia salmositica*, and its *in vitro* proteolytic activities. *Dis. Aquat. Organ.* **30:** 177–185.

Zuo, X. and Woo, P.T.K. (1998a) Characterization of purified metallo- and cysteine proteases from the pathogenic haemoflagellate, *Cryptobia salmositica* Katz 1951. *Parasitol. Res.* **84:** 492–498.

Zuo, X. and Woo, P.T.K. (1998b) *In vitro* secretion of metallo-protease (200 kDa) by the pathogenic piscine haemoflagellate, *Cryptobia salmositica* Katz, and stimulation of protease production by collagen. J. Fish Dis. **21:** 249–255.

Zuo, X. and Woo, P.T.K. (2000) *In vitro* haemolysis of piscine erythrocytes by purified metallo-protease from the pathogenic haemoflagellate, *Cryptobia salmositica* Katz. *J. Fish Dis.* **23**: 227–230.

Zuo, X., Feng, S. and Woo, P.T.K. (1997) The *in vitro* inhibition of proteases from *Cryptobia salmositica* Katz by a monoclonal antibody (MAb-001) against a glycoprotein on the pathogenic haemoflagellate. *J. Fish Dis.* **20:** 419–426.

6

The biology of parasites from the genus *Argulus* and a review of the interactions with its host

Peter D. Walker, Gert Flik and Sjoerd E. Wendelaar Bonga

1 Introduction

During the 20th century crustacean ectoparasites of fish did not receive the same research focus that other piscine pathogens experienced. This has resulted in a delay in our understanding of these economically important organisms. However, it is now recognized by fin fish producers and researchers alike that parasitic lice can indeed play a significant role in the economic success of aquaculture organizations. Around the globe there are continuous reports on the deleterious effects of these pathogens on fish farm stock (e.g. Bauer, 1959; Costello, 1993; Menezes *et al.*, 1990) and more recently on wild fish populations (e.g. Johnson *et al.*, 1996; Poulin and Fitzgerald, 1987, 1988; Whelan and Poole, 1996). Even parasitic infections in recreational fisheries are under current investigation to evaluate the effects of such infections on fishery economics (Taylor *et al.*, 2003). These reports have led to an increase in the number of researchers concentrating on the study of ectoparasitic crustaceans found on fish but there is still a considerable lack of knowledge in certain areas. This chapter aims to give a review on the current ecological knowledge available for the Branchiuran genus *Argulus* with a focus upon the general biology and parasite–host interactions.

All argulids are described as obligate ectoparasites of fish but they are also frequently encountered swimming freely in the water column as they seek out new hosts, mates or when females detach from their hosts to deposit eggs (Bower-Shore, 1940; Mikheev *et al.*, 1998). Morphologically, they bear a close resemblance to several parasitic copepod species and this similarity has led to some conflict over their classification during the 20th century (Martin, 1932). To date the majority of researchers are in agreement that similarities exist as a result of convergent evolution

Host–Parasite Interactions, edited by G. F. Wiegertjes & G. Flik.

rather than the sharing of a common ancestor (Walker, 2002). *Table 1* details the taxonomic classification of the genus *Argulus* according to descriptions by Bowman and Abele in 1982.

Table 1 *Taxonomic classification of the genus Argulus. Reprinted from* The Biology of Crustacea, v.1, Systematics, the Fossil Record, and Biogeography, *Bowman, T. E. and Abele, C. G. Classification of the Recent Crustacea. pages 1–27. © 1982, with permission from Elsevier.*

Taxonomic level	Taxa
Phylum	Arthropoda
Subphylum	Crustacea (Pennant, 1977)
Class	Maxillopoda (Dahl, 1959)
Subclass	Branchiura (Thorell, 1864)
Order	Arguloida (Raffinesque, 1815)
Family	Argulidae (Leach, 1819)
Genus	*Argulus* (Muller, 1785)

When factors such as parasite distribution and relative lack of host specificity are considered, the genus *Argulus* can be regarded as one of the most widespread and economically important groups of crustacean ectoparasites affecting freshwater fish around the globe (Bower-Shore, 1940; Menezes *et al.*, 1990; Shafir and Oldewage, 1992). Whilst morbidity is not always linked to infections of *Argulus* spp. the direct and indirect results of louse infestations can still be significantly costly to aquaculture and sport fishing operations (Menezes *et al.*, 1990; Northcott *et al.*, 1997). In addition to the deleterious impacts resulting from parasitic feeding and attachment, secondary infections from bacteria and fungi (Shimura, 1983; Singhal *et al.*, 1990; Stammer, 1959) are very common and argulids have also been shown to act as vectors for other pathogens including nematodes (Moravec, 1978) and viruses (Ahne, 1985; Cusack and Cone, 1986; Dombrowski, 1952).

There are several reviews available for argulids concentrating on factors such as distribution (Rushton-Mellor, 1992; Gurney, 1948; Poly, 1997, 1998), development (Rushton-Mellor and Boxshall, 1994) and morphology (Benz and Otting, 1996; Martin, 1932). Many of these reviews however, focus on just a single species from the *Argulus* genus. This chapter aims to give a review on the current knowledge available for this group of parasites with a focus on those ecological and biological factors common throughout all species from this genus. In addition we will address some of those host–parasite interactions which are considered by us to be of significance in terms of recent advances in the knowledge of this louse–fish association.

2 Life cycle, growth and development

The life cycle of argulids has been described by several authors in the past (e.g. Bower-Shore, 1940; Kollatsch, 1959; Mclaughlin, 1980) and evidence shows that the cycle is very similar for all members of the *Argulus* genus. The only differences appear to be in the number of developmental stages between hatching and maturity. *Figure 1* illustrates the key stages involved in the life cycle of a generalized argulid.

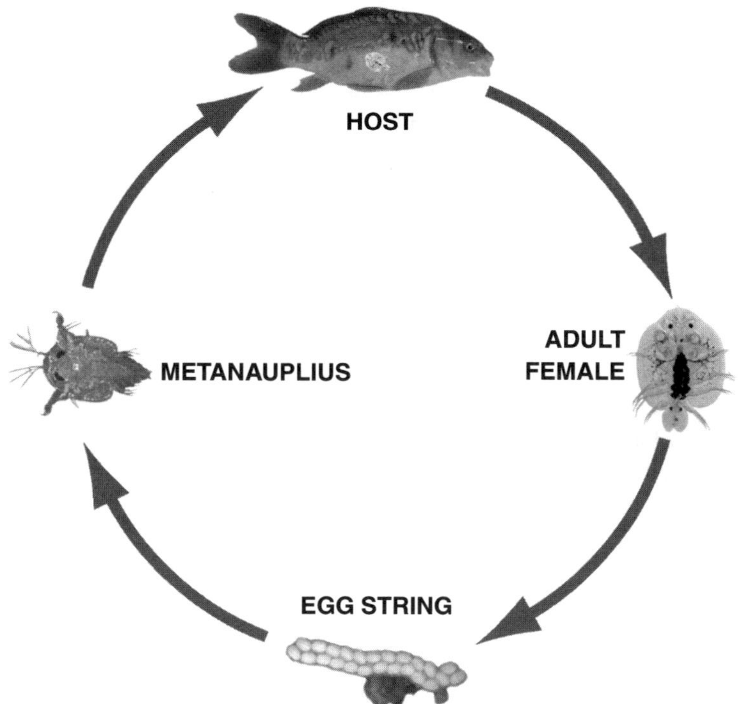

Figure 1 *Generalized life cycle for the genus* Argulus. *Copulation occurs on the host after which the female detaches to deposit eggs in rows or clumps on a suitable substrate (e.g. a rock, plant stem or even glass sides of aquaria). The tiny larvae are immediately parasitic and upon locating a suitable host they undertake several moults until the adult stage is achieved. Adult lice may continue to moult throughout their life.*

Eggs are deposited on a suitable surface (e.g. plant stems, stones or glass sides of aquaria) in clumps or more typically parallel rows (*Figure 1*), a process reported to be unusual for aquatic ectoparasites (Williams, 1997). In parasitic copepods (e.g. *Caligus elongatus, Lepeophtheirus salmonis*) the females generally possess egg sacs in which eggs are maintained until hatching. The number of eggs deposited by any female argulid varies considerably between individuals (from less than ten up to several hundred) and to date there is no hard evidence on factors affecting clutch size. Even between different species from this genus there are no obvious differences in clutch sizes. We speculate that gravidity is dependent upon such factors as meal quality (i.e. health status of host fish), parasite size (or age) and parasite species. Development time of egg stages is heavily dependent upon the ambient temperature of the surrounding water with development being more rapid at higher temperatures. *Argulus foliaceus* eggs, for example, have been shown to hatch after just 8 days at 26°C (Fryer, 1982) or after several months for eggs deposited at temperatures below 10°C (e.g. Lester and Roubal, 1995; Mikheev *et al.*, 2001). This could be a strategy that evolved to enable parasites in temperate regions to survive harsh winter conditions as an egg stage and to take maximum advantage of favourable summer conditions to increase and/or maintain population sizes. Certainly overwintering egg stages have been reported for several species of this genus (e.g. Bower-Shore, 1940; Mikheev *et al.*, 2001).

Upon hatching, the late nauplius larva (*Figure 1*) closely resembles the larger adult form. Although the general body form of the different life history stages is similar there are distinctive morphological features that enable differentiation of the various developmental stages. The copepodid-like metanauplius larvae are immediately parasitic (Lester and Roubal, 1995) and perish within a few days of hatching if a suitable host is not located (Kollatsch, 1959). After just a few days of feeding these hatchlings moult into a second stage which possesses most of the features of an adult and can therefore be referred to as juveniles (Rushton-Mellor and Boxshall, 1994). A key morphological difference between juvenile lice and adults is a lack of the prominent maxillary suckers, which only begin to appear in later developmental stages. A succession of moults takes place approximately every 5 days depending on the individual species and the ambient temperature (Fryer, 1982). The total number of developmental stages involved is species dependent but examples include seven stages for *A. japonicus* (Tokioka, 1936), nine stages for *A. coregoni* (Shimura, 1980) and ten stages for *A. foliaceus* (Rushton-Mellor and Boxshall, 1994). After approximately 4–6 weeks a mature adult louse is recognizable. This development time is generalized here for ease of explanation but the reader should be aware that this period can be significantly shorter or longer depending upon temperatures and individual *Argulus* species examined (see Fryer, 1982; Hindle, 1949; Shimura, 1980).

Although all argulid species are dioecious, i.e. have separate sexes (Benz and Otting, 1996), the sexes do not differ enormously in terms of their morphology. Separate sexes are distinguishable by examination of the abdominal lobes located at the posterior end of the parasite's body. Females possess small spermathecae whereas males possess large testes which are clearly visible in live specimens due to the transparent properties of these animals' exoskeletons. In addition, white eggs can often be seen within the pigmented ovaries located along the midline in adult, female lice. Copulation typically occurs on the host although our observations have revealed that lice will also copulate whilst detached from their host fish, i.e. whilst swimming freely in the water column. Copulation involves the transfer of sperm from the male directly to the female. Sperm cells are then stored in the female's spermathecae until she fertilizes her eggs during the deposition process (Kollatsch, 1959). Eggs are typically protected by a mucus-like coating which presumably protects them from some smaller predators or opportunistic bacteria and fungi or possibly plays a role in maintaining the hydro–mineral balance of the fertilized eggs.

3 Morphology

Argulid morphology has been the topic of attention for several authors in the past (e.g. Bauer, 1959; Kabata, 1985; Martin, 1932) and in light of this only a superficial summary of those structures considered important to the parasite's life style and ecology are described here along with some discussion of their possible function. Similarly, a detailed description of the morphology of larval and intermediate juvenile stages is also omitted here. *Figure 2* shows the key morphological features of a typical argulid. Much of an argulid's morphological design can be linked to its ectoparasitic life style. The dorso-ventrally flattened body covered by a large, rounded carapace presents a streamlined surface offering little resistance to water currents that may otherwise dislodge a parasite when a host fish moves through the water column. This general form can also be witnessed in other ectoparasitic organisms of host animals such as fish and birds that need to maintain a streamlined form, e.g. *Lepeophtheirus salmonis* (salmon louse: see also Chapter 7) *Caligus elongatus* (sea louse) and *Crataerina melbae* (swift louse/flat fly). Such a large number of parasites demonstrate this ectoparasitic mode of existence and exhibit this flattened, streamlined form, that we must consider it to be an evolutionary successful trait.

The cuticle of all argulids, like most crustaceans, is chitinous forming a rigid exoskeleton that provides the support needed for the animal to maintain its form in a similar way to the internal skeleton of vertebrates. The body can be divided into three distinct regions: (i) cephalothorax; (ii) thorax; and (iii) abdomen (*Figure 2*). When live specimens are observed one can often see numerous pigment cells (chromatophores) which are typically associated with the gut and ovaries. Respiratory areas can be viewed as those areas of the carapace lacking in the small spines and scales that adorn the ventral surface of the carapace. These areas possess a much thinner cuticle and are located adjacent to a large blood sinus, which facilitates the diffusion of oxygen into the bloodstream of the animal. The shape and position of these areas is also useful for taxonomic purposes (Benz and Otting, 1996).

3.1 *Attachment*

Streamlined bodies, however, do not provide all the necessary tools to keep an ectoparasite attached to its host and argulids typically show an array of structures that assist in

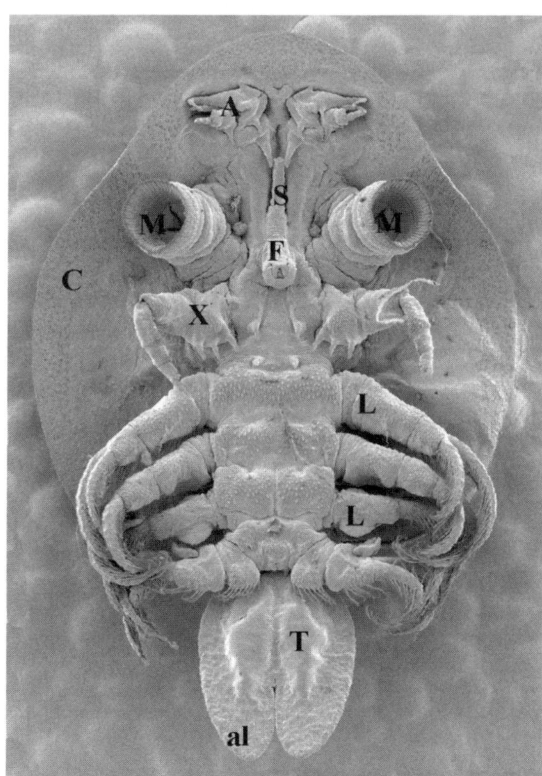

Figure 2** Scanning electron micrograph of the ventral surface of an adult male* Argulus japonicus *(actual size = 6 mm). The key morphological features are labelled here and discussed in* **Section 6.3.** *Carapace,* **C; Antennae,* **A***; Sheath containing pre-oral stylet,* **S***; Maxillary suckers,* **M***; Feeding proboscis,* **F***; Second maxilla,* **X***; Thoracopods (legs),* **L***; Testes,* **T***; Abdominal lobe,* **al**.

keeping the parasite connected to its food source. Not surprisingly the attachment structures are all located on the ventral surface of the animal which is the surface that is found to be in contact with the animal's host organism. The most conspicuous of these structures are the large maxillary suckers. These suckers are actually modified first maxillae (commonly referred to as 'maxillules'). Their chitinous support structures (formed by rods composed of sclerites stacked on top of each other, (Benz and Otting, 1996), and associated musculature provide a powerful suctorial action that keeps these animals 'stuck' to their hosts. These suction cups are positioned upon a moveable stalk, allowing the parasite to move the suckers independently across the host's surface, and this means that the louse can travel over the body of its host with relative ease and surprising speed! In addition to these highly specialized structures, argulids also possess modified first antennae that appear as hooks. Numerous small setae, spines and bristles are also believed to play a role in attachment and may also have a defence purpose. These various spines and scales can be observed on the majority of the ventral surface.

3.2 *Locomotion*

Section 2 discussed the life cycle of these parasites and drew attention to the fact that these animals can and frequently do leave their hosts for a variety of reasons (e.g. accidental dislodging, mate location, egg deposition, new host location). During these 'off-host' periods argulids must propel themselves through the water column with a great deal of efficiency, especially if they are aiming to successfully locate a new host. Indeed, argulids are quite proficient swimmers. Propulsion is provided in the main by the four pairs of thoracopods located on the posterior portion of the animal's ventral surface. These appendages exhibit the primitive crustacean form (Martin, 1932) in that they are cirriform and biramous with two segmented sympods and rami each with lateral rows of pinnate setae (Benz and Otting, 1996). The long setae found on these swimming appendages form a paddle-like surface as they beat backwards, propelling the animal forwards. These limbs are also frequently moved backwards and forwards whilst the louse is attached to its host, presumably to provide a continual flow of fresh water across the respiratory areas. In addition argulids can also 'catapult' themselves very quickly over a short distance using a method similar to that employed by lobsters and shrimps. The lice rapidly flick their abdominal lobes ventrally and towards the anterior of their bodies at the same time as flexing the whole carapace. This action is typically used in predator avoidance although it is possible that lice will also use the same technique to 'jump' onto a passing fish.

3.3 *Feeding*

The mechanisms employed by argulids to access a meal from their hosts have been described by many authors in the past and yet still remain a topic of debate. Here we use evidence provided by past studies combined with our own observations of *A. japonicus* to try and form solid conclusions on the feeding mechanisms of argulids (*Figures 3* and *4*). In the simplest description the feeding apparatus of an argulid is composed of a pre-oral stylet (also referred to as a sting or stiletto) (*Figure 4*) and a feeding proboscis or mouth tube (*Figure 3*). The components are however somewhat more complex than this commonly used description portrays. Here we begin with a brief description of the pre-oral stylet and some discussion of its possible use and following this we will describe the feeding proboscis and feeding mechanisms in some detail.

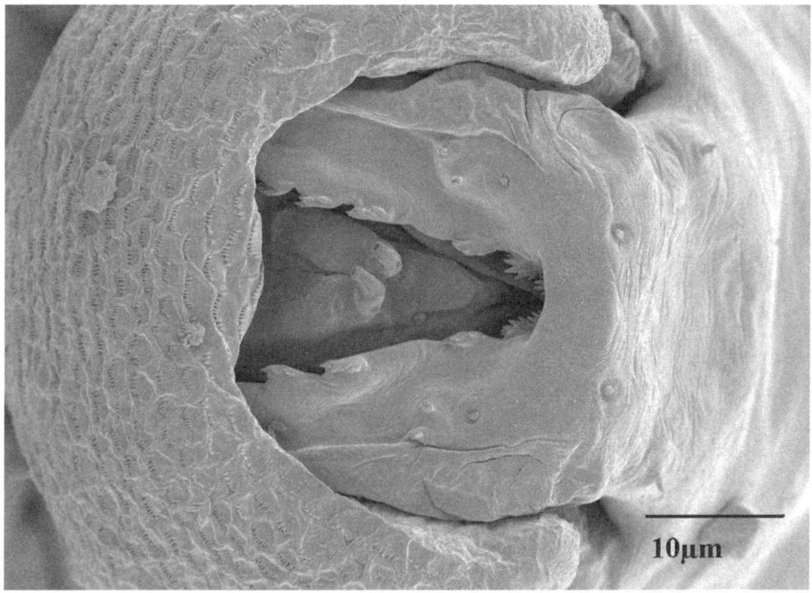

Figure 3 *Scanning electron micrograph of the feeding proboscis of an adult* Argulus japonicus.

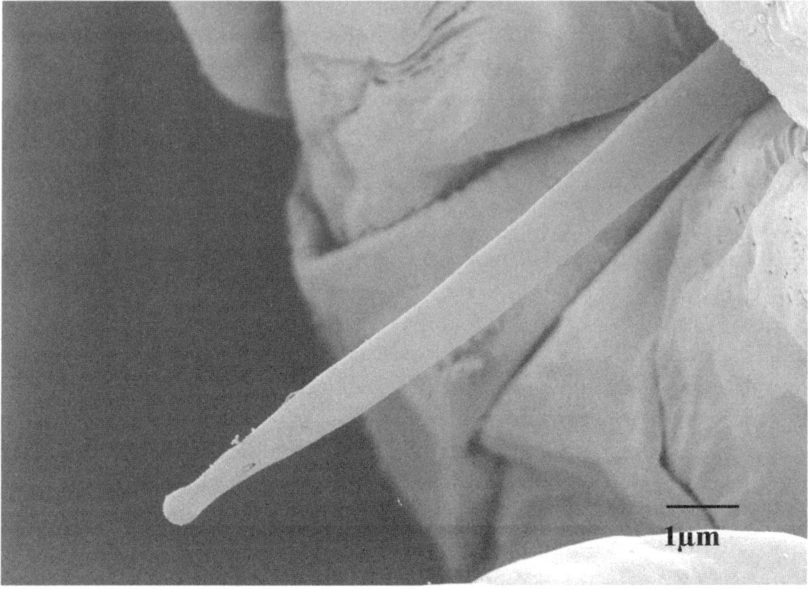

Figure 4 *Scanning electron micrograph of the pre-oral stylet of an adult* Argulus japonicus.

The pre-oral stylet is located on the midline of the lice just anterior to the feeding proboscis and in specimens prepared for scanning electron microscopy the stylet may be completely or partially extended or may even be retracted completely within its associated sheath. According to Martin (1932) the relative size of this stylet varies considerably between species and although often considered as part of an argulid's mouthparts it bears no direct relation to the other mouthparts associated with the feeding proboscis. The stylet itself takes the form of a hollow spine possessing two openings near to its tip (*Figure 4*). Shimura (1983) claimed that one of these openings appears to have a secretory function whilst the other has a sensory function. Most authors agree that secretions/poisons transmitted into the host's integument by this spine play a key role in the feeding process of argulids. However, Martin claimed that the fragile nature of this structure suggested that its importance was debateable (Martin, 1932). Many authors state that these secretions actually digest host skin cells, whilst others suggest that secretions from the stylet merely serve to cause subcutaneous haemorrhaging (e.g. Shimura and Inoue, 1984). Labial spines (referred to as siphons by some authors) have also been suggested as structures used to secrete digestive enzymes.

The feeding proboscis itself also lies along the midline of the animal, just posterior to the aforementioned stylet. When an argulid is not feeding this proboscis rests in a medial groove but during feeding activities it is extended away from the parasite's body so that it meets the host's integument at right angles. The tube terminates with an anterior labrum and a posterior labium. The labrum is commonly covered in serrated scales, which are employed to scrape mucus and other debris away from feeding sites. Serrated mandibles located just inside the buccal cavity (the true mouth is found beyond this cavity) can be everted and it is believed these structures are responsible for lacerating host tissue so that the animal can access the blood of its host. Accessing a blood meal is understood to be facilitated also by the haemorrhagic response to the pre-oral stylet secretions and some extracellular digestion caused by secretions from the labial spines. Ultimately blood is ingested and this can be observed in live specimens as their mid gut (or crop) and associated diverticula turn red from the haemoglobin containing, red blood cells (Walker, 2003).

3.4 *Sensory organs*

The structures/organs utilized by an organism to sense its environment are typically diverse and can take the form of light-sensitive organs (including eyes), mechanoreceptors that detect pressure waves (e.g. the lateral line in fishes) and chemosensory/olfactory organs (including antennae and noses).

Argulids certainly possess well-developed compound eyes and it is the belief of several authors (e.g. Mikheev *et al.*, 1998), including the authors here, that these function as the main organs utilized by argulids to locate their hosts. A median, naupliar eye is also present but this relatively primitive structure can probably detect nothing more than simple changes in light intensity but this would provide information for the parasite regarding its orientation. In addition antennae, setae and sensillae can be observed on various parts of the animal's surface and these structures may have a sensory function. Several authors have speculated on the role of mechanosensation and chemosensation in locating hosts and possibly mates

(Galarowicz and Cochran; 1991; Russon, personal communication). To date evidence is inconclusive but future research may show that these sensory mechanisms are indeed very important to the success of argulids.

4 Distribution and seasonality

One of the factors that make *Argulus* spp. so successful as a parasite and so threatening as a pathogen is its lack of specificity. Argulids are often described as being able to infect any freshwater fish and several species have even been recorded from amphibians, especially tadpoles. This lack of specificity appears to be one of the key factors that have resulted in the cosmopolitan distribution of many species. In addition many species of *Argulus* can tolerate temperature ranges from as low as 3–4°C up to 30°C. However, the disease argulosis (the causative agent being *Argulus* spp.) is often referred to as being a seasonal disease, particularly in temperate regions. Certainly *A. foliaceus, A. japonicus* and *A. coregoni* are all known to reach their maximum abundance in Europe during the late summer/early autumn months. Relatively few lice are found during the winter and for many years researchers believed that lice populations only survived winters due to overwintering egg stages deposited in the autumn. Over the last few decades however, there have been many reports of adult argulids being found on fish sampled during winter months (Bower-Shore, 1940; Kimura, 1970; Shafir and Van As, 1986). Some believe this could, again, be an unnatural phenomenon facilitated by global warming.

5 Parasite–host interactions

In general, parasites are significantly smaller than their host organisms (think of a single flea on a dog). Despite their relatively small size many parasites have the potential to elicit strong effects on their host's biology, the consequences of which are often detrimental to the individual host organism. Many parasites are even capable of causing mortality of their hosts although this does not make evolutionary sense because many parasites are in fact obligate parasites meaning they depend on their host organisms for nutrition and survival. This 'harm-causing' ability is one of the criteria proposed by Begon *et al.* (1990) as being a defining characteristic of a parasite.

The relationships between host animals and the organisms that parasitize them are very intimate and have typically been subject to generations of co-evolution, which has resulted in the characteristics of infection that we observe. In this section we will discuss several key aspects of the host–parasite relationship between argulids and their fish hosts. We will begin with some of the more ecological subjects of this relationship, e.g. host choice, and then expand to introduce topics such as the immune response to argulids and the stress response of fish to ectoparasites. Due to the embryonic research status of this topic we will also use examples from other host–parasite relationships to illustrate key topics.

Whilst argulids are typically described as obligate ectoparasites of fish this statement carries some flaws. Argulids are indeed ectoparasitic organisms because they attach to external surfaces of their hosts and feed from this attachment site. It is also fair to state that they are obligate parasites because they must locate a host to feed. However, they are not necessarily obligate parasites of fish. Several species of the genus *Argulus* have been documented from amphibian hosts and we have witnessed *Argulus foliaceus* successfully attaching to *Xenopus* toads under laboratory conditions. It is important to note here that argulids can and frequently do detach from their hosts for periods of time making them temporary or intermittent parasites compared to permanent parasites which spend their entire adult lives within or on their hosts (Roberts and Janovy, 1996). For this reason argulids can even be assigned the term 'micropredators' which is a term also used to describe some other temporary ectoparasites such as mosquitoes and ticks.

5.1 *Host choice and specificity*

Many species of lice from the genus *Argulus* are described as being non-host-specific. Poulin (1998) defines parasite specificity as 'the extent to which a parasite taxon is restricted in the number of host species used at any given stage in the life cycle' and we will adopt this definition here. In Western Europe there are three known freshwater species from this genus; *A. foliaceus*, *A. coregoni*, and the introduced *A. japonicus*. When examining lists of fish species that these louse species have been found on, then the genus quickly appears to show very low specificity. However many researchers, ourselves included, have undertaken studies which show that under certain conditions these parasites may show some preference for certain fish over others. For the purpose of this discussion we will concentrate on the common European species because host preference studies are most common for these lice species and they are also some of the more common species known from around the globe.

In 2002 (Walker, 2002), the *A. foliaceus* population of a small mixed coarse fishery in south-west England underwent an intensive 5-month study to examine population dynamics and infection characteristics. A total of eight fish species were recorded and all eight species yielded individual fish harbouring lice at some point during the study. However, the prevalence (proportion of infected fish) and infection intensity (number of lice per fish) differed markedly between individual fish and also species and in turn the infection characteristics varied over the sampling period. During the warm summer months the infection levels (both prevalence and intensity) increased significantly on adult carp (*Cyprinus carpio*). These fish spend a large portion of their time basking in the warmer shallow regions of the lake and as a result may prove to be easier hosts for the lice to locate. Adult bream (*Abramis brama*) showed similar infection characteristics.

A comparative study of argulosis from two sites in SW England revealed differences in preferred host species for *A. foliaceus* (Walker, 2002). The two sites differed somewhat in their fish community structure and species present with one being a densely populated commercial coarse fishery and the other a sparsely populated natural lake with a much greater volume and depth. The results of this study suggested that host choice by argulids cannot be characterized definitively. Host choice instead depends on the combination of several factors, such as host species present, population densities and prevailing environmental conditions both in terms of the physical and chemical parameters (e.g. depth, temperature, dissolved oxygen).

Other biological factors may also play a significant role in host selection by argulids. Poulin and Fitzgerald (1988) noted that during daylight hours *A. canadensis* would inhabit the lower half of the water column. Certain fish species (e.g. rudd, *Scardinius erythropthalmus*) spend much of their time in the upper few centimetres of the water column feeding on small zooplankton and other invertebrates that may drop onto the water surface. This would reduce the likelihood of these fish encountering lice, at least during daylight hours. At the other end of the scale eels spend much of their time hidden amongst submerged objects or half buried in the bottom silt and again this would reduce the chances of them encountering lice. In fact this behaviour may account for the fact that very few eels have ever been found harbouring lice (Evans and Mathews, 2000).

Mikheev *et al.* (1998) noted that light intensity significantly affected the host choice of *A. foliaceus*. In darkened conditions perch (*Perca fluviatilis*) were favoured over roach (*Rutilus rutilus*) and when light intensity was increased roach became increasingly more preferable to the lice. Reduced activity and vertical positioning of the fish probably accounts for some of this variation in host choice, however, the host location strategies employed by the lice suggest something more interesting.

During light periods the free-swimming lice spent most of their time hovering almost stationary in the water column whilst during dark periods they appeared to undertake a much more active searching strategy (Mikheev *et al.*, 1998). The 'sit and wait' strategy probably serves to conserve energy stores whilst host fish are more active during daylight hours. The active searching strategy would then be more successful during darker periods when many fish lie dormant. This also suggests that argulids can either see extremely well in dark conditions or that they have some kind of chemosensory capability. If these lice do indeed have chemosensory capabilities then it is plausible that certain fish species will have a more appealing set of chemical discharges that argulids could home in on. In the case of the experiments conducted by Mikheev *et al.* (1998), perch would have emitted the more appealing 'smell'. Galarowicz and Cochran (1991) showed that *A. japonicus* would indeed respond to host chemical cues, however Y-maze style investigations undertaken by ourselves have failed to yield results to substantiate this theory.

5.2 *The effects of argulids on fish*

The effects of *Argulus* parasites on their fish hosts are quite diverse and range from physical damage caused by attachment and feeding activities of the lice to behavioural changes associated with stress. The following sections will highlight the current knowledge on the effects of the parasites on fish with a particular emphasis on stress and behaviour. Other topics such as physical damage and immune responses related to infection will also be introduced.

Effects on host behaviour

Many parasites are known to affect their host's behaviour during the course of an infection. There are numerous parasite species that are known to have taken advantage of these host behavioural changes. For example, studies involving the cestode *Schistocephalus solidus* and its intermediate host the three-spine stickleback (*Gasterosteus aculeatus*) have shown that this parasite alters its host's behaviour in a way that facilitates transmission of the tapeworm to its definitive host, piscivorous birds. Behavioural changes included increased time spent foraging away from cover, reduced swimming performance and suppression of the host's anti-predator response (Barber *et al.*, in press).

Argulid parasite species are also known to cause behavioural changes in their fish hosts. Some of these appear to facilitate parasite transmission whereas others are clearly behaviours that have evolved as mechanisms employed by the host fish to try and dislodge lice from their external surfaces. Here we will discuss some of those changes documented by other researchers along with lice-induced behaviours witnessed by the authors here.

One of the first behavioural changes people observe in fish harbouring an *Argulus* infection is the so-called flashing or scratching behaviour. Fish will repeatedly 'flash' their flanks against submerged objects such as plants, stones or even gravel. This behaviour is very common in fish infected with a wide range of ectoparasites and is believed to be an attempt at dislodging the culprits (e.g. *Gyrodactalus* spp., *Ichthyoptheirus multifilis* (white-spot), *Lernaea* spp.). Similarly some fish are known to leap clear of the water and certainly argulids will often detach from their hosts when exposed to the air. Communication between the authors and a commercial trout fishery in The Netherlands revealed that trout leaping clear of the water increases in frequency as water temperature increases and this coincides with an increase in the abundance of *A. foliaceus* within the lake. Sticklebacks have even been witnessed leaping onto the concrete surrounding of a garden pond in an attempt to rid themselves of a heavy lice infection. Sadly the efforts of these particular fish were thwarted by some resident frogs, which viewed the floundering sticklebacks as an easy meal (Van der Veld, personal communication)! The determined effort by some fish to rid themselves of attached lice shows that lice are in fact extremely irritating to a fish's skin. Personal observations also showed that the African clawed toad (*Xenopus laevis*) skin is also extremely sensitive to attaching parasites.

Other general behavioural changes associated with argulid infections include loss of appetite, denser shoaling behaviour, lethargy and changes in vertical positioning within a water column. Our broodstock carp (*Cyprinus carpio*) harbouring *A. japonicus* were also observed to spend significant amounts of time crowding over air stones in their aquaria. This behaviour probably offers some relief to the irritation caused by lice living on the fishes' bodies or it is even possible that the continual bombardment of air bubbles actually persuades the lice to dislodge from their hosts.

Research undertaken in Canada on the argulid species *Argulus canadensis* has also revealed some interesting behaviour changes associated with *Argulus* infections. Dugatkin *et al.* (1994) showed that under experimental conditions juvenile sticklebacks will actually avoid schools of parasitized conspecifics even though the parasite itself did not elicit an avoidance response. Poulin (1999) showed that juvenile trout respond to the release of alarm substances emitted by conspecifics infected by *Diplostomum* parasites and it is possible that a similar event was occurring in the sticklebacks of Dugatkin's 1994 experiments.

Poulin and colleagues have also shown some other behavioural changes associated with fish infected with argulids (e.g. Poulin and Fitzgerald, 1988). These researchers have shown that parasitic infections can in fact have significant consequences on the community and population structures of fish. They demonstrated that in the presence of parasites fish will form larger, denser shoals. This strategy will decrease the chance that an individual will be targeted by parasites seeking a new host. However, the strategy also benefits the parasite by giving it easy access to a wide range of potentially suitable hosts. Similar changes in shoaling behaviour were observed by Northcott

(1997) during an *Argulus* epizootic in a Scottish stillwater trout fishery. Sticklebacks were also noted to adjust their vertical positioning in the water column when infected with lice. Heavily infected fish would often just lie motionless for long periods on the bottom of ponds or experimental tanks.

Damage to the host's integument

Argulus spp. cause direct damage to their host's integument through their attachment and feeding mechanisms. This damage can result from either mechanical actions (i.e. from the sharp mouth parts) or from chemical secretions.

As mentioned in Section 3.1, the main attachment organs in adult lice are the maxillary suckers. In addition various appendages are modified to form hooks or spines and microscopic examination of these structures clearly reveals how they may be damaging to the host's integument. Whilst attached to their hosts argulids continuously beat their thoracopods back and forth to maintain a flow of fresh water over their bodies for respiratory purposes. This results in pressure atrophy and small 'bruised' areas can often be observed when lice are removed.

The majority of the damage caused to the fishes' skin results from the feeding activities of these parasites. The pre-oral stylet and labial spines have all been suggested as capable of secreting various toxins or digestive enzymes that facilitate the parasites' feeding. The next sections will deal with these substances in more detail and here we state simply that these reported substances probably degenerate cells making the mechanical feeding processes less strenuous for the lice.

Section 3.3 described the feeding mechanisms of argulids and also detailed the apparatus involved. This feeding action creates significant damage to the host's integument. In fact, Lester and Roubal (1995) demonstrated that small craters can even be formed in the host skin as a result of the feeding activities of these lice. Epidermal hyperplasia at the wound margins is also visible. Typically, these craters do not penetrate much deeper than the epidermis but examples of wounds penetrating as deep as the stratum compactum have been recorded (Lester and Roubal, 1995). Mucus cells are generally absent within the craters themselves but a proliferation of these cells is frequently evident at wound margins (Lester and Roubal, 1995).

Immune response

The immune response in mammals has been a topic addressed by multiple research organizations throughout the 20th century. Much of this focus has been on those pathogens that may ultimately have a significant implication for humans. This impact can take the form of disease symptoms in human populations directly or in the form of infections affecting livestock or domestic animals. With the increasing importance of fish for supplying protein to human populations researchers have begun to investigate those factors which may impact upon fish production, i.e. fish diseases. In order to combat these pathogens it is vital that we gain an understanding of the natural defence mechanisms employed by fish. This knowledge requirement has led to a new wave of research focusing upon the piscine immune system. In this section we will discuss some of those immune responses exhibited by fish that harbour parasites from the genus *Argulus*.

The immune system in fish can be described as having two parts: (i) the innate/non-specific immune response; and (ii) the acquired/specific immune response. Recently there has been a renewed interest in the innate immune response of fish to pathogens because this consists of the first defensive 'barriers' that any foreign invader will encounter (e.g. epithelial barriers, acidic conditions in the gut, IgM, etc.). These mechanisms are typically not pathogen specific and similar infection types will illicit similar responses (i.e. localized inflammation, increase in mucus cell numbers, macrophage activation, etc.). The adaptive or acquired immune response is so called due to its 'memory' properties. It normally takes several hours or days for the animal to mount a response against a first infection from a pathogen but subsequent infections from the same pathogen will be met with an increasingly faster response. Typically this involves the production of antibodies specific to certain antigen-binding sites.

The immune response of fish to ectoparasitic organisms and particularly crustacean ectoparasites has only begun to receive attention in the last decade or so and even then the research is somewhat unfocused. It is however recognized now that this knowledge is of paramount importance if we ever hope to develop commercially viable and environmentally sound methods of treatment and control, i.e. vaccines.

Typically the most notable immune response to argulid infestations is observed as localized inflammation (appears as small red spots on the fish's skin). The causative factor is often reported to be secretions from the pre-oral stylet of *Argulus* individuals but attachment of the parasite is also likely to have an impact on this response and certainly pressure atrophy has been documented previously (Lester and Roubal, 1995).

Lester and Roubal (1995) suggested that secretions of the parasite have low antigenicity due to inflammation at feeding sites not being a major component of the histological changes. Ruane *et al.* (1995) demonstrated a humoral antibody response in rainbow trout (*Oncorhynchus mykiss*) after they were immunized with an antigen extract from *A. foliaceus*. A similar type of response was seen in rainbow trout and Atlantic salmon (*Salmo salar*) that had been immunized with sea lice antigens (Grayson *et al.*, 1991; Reilly and Mulcahy, 1993). This work provides information that may prove invaluable for the future development of vaccines against ectoparasitic crustaceans.

Inflammation is commonly associated with argulid infections and indeed large red spots can frequently be observed even within just a few hours post infection. The mechanisms involved in the inflammatory response have been well documented for mammals but considerably less information is available for fish. In a preliminary investigation of the early inflammatory immune response of carp to *A. japonicus* Haond and Wiegertjes (unpublished data) measured the blood leucocyte redistribution over a period of 40 days post infection. The relative percentage of granulocytes and monocytes, identified on the basis of typical forward/sideward scatter profiles in a flow cytometer, increased dramatically over time. The relative percentage of these phagocytic cell types was highest 30 days post infection, when numbers of parasites on the skin also peaked, and declined thereafter with declining numbers of parasites. Skin samples from these infected fish were analysed for early (1 hour post infection) gene expression with RT-PCR, indicating increased expression of the pro-inflammatory cytokines interleukin-1 and tumour necrosis factor alpha. These preliminary data support the existence of an inflammatory immune response to *Argulus*. Furthermore, Huising *et al.* (2003), analysing the same samples, demonstrated increased expression of particular CXC chemokines at the inflammation site, which could possibly explain the increase in granulocytes in the blood. Future work on the inflammatory response and other innate immune responses of fish are likely to show several comparisons with the mammalian systems and as a result we may be able to combine knowledge to develop successful control methods.

Stress response

Cannon was one of the first researchers to introduce the concept of stress (Cannon, 1935). He claimed that when the normal homeostasis of an organism was threatened by one or more stimuli that the organism could be considered as stressed. In 1992, Chrousos and Gold provided a more comprehensive definition of stress: 'stress is a condition in which the dynamic equilibrium of animal organisms, called homeostasis, is threatened or disturbed as a result of the actions of intrinsic or extrinsic stimuli, commonly defined as stressors'. In this section we will focus upon the stress response of fish.

The physiological mechanisms of the teleost-integrated stress response have been shown to share many similarities with that of terrestrial vertebrates (Wendelaar Bonga, 1997). The responses exhibited by a fish subjected to a stressor will involve physiological and behavioural responses that are induced as mechanisms to try and protect homeostasis or maintain the dynamic equilibrium of the stressed organism (Wendelaar Bonga, 1997). These responses have been categorized by previous authors as primary, secondary and tertiary responses.

The primary response involves dramatic increases in the blood levels of catecholamines and glucocorticoids. These hormones are the dominant hormones involved in the stress response and are the primary messengers of the two main routes through which the brain co-ordinates the stress response. These two routes include the hypothalamic–autonomic nervous system and the hypothalamic–pituitary–inter-renal axis (Wendelaar Bonga, 1997). The secondary response comprises metabolic changes (i.e. plasma glucose and lactate levels), hydromineral disturbance (fluctuations in plasma chloride and sodium levels), haematological changes (e.g. in haematocrit and haemoglobin content), and changes in the host's immune system (i.e. immunosuppression). Finally, the tertiary response consists of changes in the organism as a whole including reduced growth, impaired swimming performance, lower reproductive success and reduced disease resistance, all of which can impact negatively upon the survival of the organism.

When examining the effects of a stressor, researchers typically use one or a combination of several parameters. Cortisol is the most widely used hormone involved in the stress response and because it is usually obtained from a fish's blood many researchers will also examine blood glucose levels in conjunction with cortisol and to a lesser extent serum sodium and chloride levels may be measured.

Other factors that can be indicative of stress include changes in the skin and gill epithelia. For example, skin from stressed fish typically exhibits abnormally high levels of apoptosis and necrosis, which, if not fully compensated by cell proliferation, may result in epithelial disruption. In addition mucus cell discharge is typically stimulated and infiltration of the epithelia by leucocytes can be observed and is indicative of the immune response being activated. Several effects of cortisol on the skin of rainbow trout (*Oncorhynchus mykiss*) *in vivo* have also been documented by Iger *et al.* (1995). Gill lamellae of stressed fish can appear irregular and swollen due to increased blood flow (and blood pressure).

Parasites as stressors

Several studies in previous years have explored the hypothesis that sea lice infestations elicit a stress response in fish (e.g. Nolan *et al.*, 1999; Poole *et al.*, 2000; Ruane *et al.*, 2000). In contrast to this, very few studies have investigated the physiological effects of freshwater lice (*Argulus* spp.) on their hosts, including whether they induce or affect the stress response of their hosts. For this reason we will include the effects of sea lice as stressors in this discussion.

Several authors have previously used challenge experiments to examine the stress response of fish infected with ectoparasites. These studies have included a range of different host species and sizes but in general they have not been conducted using standard methods. What is clearly evident from many of these studies is that the level of stress caused by the infection is influenced by the intensity of the infection, host size, host condition and possibly species of host (van Ham, 2003).

To enable some comparison between studies we have reported intensity of infection as the number of *L. salmonis* per gram of host body weight. In a study of sea lice, we examined the effects of ectoparasites in post-smolt Atlantic salmon, *Salmo salar* (Nolan *et al.*, 1999). The direct effects of the parasite were the damage caused by parasite attachment and feeding on the body surface. The indirect stress effects included the effects on the overall integrity of the skin and gill epithelia such as increased apoptosis and necrosis of the superficially located epithelial cells and decreased numbers of mucus cells in the skin. Reduced mucus cell numbers as a result of ectoparasitic infestations have also been reported in brown trout (*Salmo trutta*) epidermis (Pottinger *et al.*, 1984). In the gills, where no lice were found, uplifting of the epithelium, intercellular swelling and infiltration by leucocytes is commonly observed in filaments and lamellae. High cell turnover of chloride cells was associated with significantly elevated gill Na^+/K^+-ATPase activities. These indirect stress effects are predominantly hormone mediated as a consequence of the parasite being perceived by the host, causing a stress response in the fish, and likely resulting in increased levels of blood cortisol and catecholamines.

Some of the differences between the stress response of terrestrial vertebrates and the stress response of teleosts can be attributed to the aquatic life style of fishes. The most important of these differences is the disturbance of the hydromineral balance in fishes, expressed by the changes in plasma sodium and chloride levels. Fish are directly exposed to the water over a large area via the epithelia covering the skin and gills. These epithelia are a complex assembly of many types of living cells (Whitear, 1986) and maintenance of epithelial integrity, particularly that of the gills, is essential for maintaining hydromineral balance, protection against waterborne pathogens, and thus ultimately the fish's health. High levels of catecholamines can influence the integrity of the branchial epithelium, probably by increasing the blood flow (and blood pressure) through the gills, and the permeability of the epithelium to water and ions (Wendelaar Bonga, 1997).

Nolan *et al.* (2000) examined the effects of low numbers of *Argulus foliaceus* on the epidermis of the rainbow trout. No effects were noticed on the number of mucocytes, but electron microscopic analysis of the upper cell layers revealed stimulated mucus cell discharge, and increased production of the small secretory vesicles of the pavement cells. These have been attributed antimicrobial activity, and the presence of peroxidase activity in these vesicles has been demonstrated. In addition to these signs of higher cellular activity, increased rates of apoptosis and necrosis were noticed. The intracellular spaces in the skin epithelium of the parasitized fish were enlarged and contained many leucocytes, most likely cells that had permeated the epithelium after leaving the blood system, since the number of circulating leucocytes was reduced 48 hours post infection.

In an attempt to distinguish between the direct effects of the parasites on the skin and the indirect effects mediated by hormonal messengers connected with the stress response, the effects of administration of cortisol (via the diet, to prevent the stress associated with injections) were studied (van der Salm *et al.*, 2000). This showed that cortisol stimulates mucus discharge, secretion of small vesicles by the pavement cells, apoptosis (but not necrosis) of epidermal cells, and leucocyte infiltration. Nolan *et al.* (2000) also noticed that low numbers of *A. foliaceus* (six lice per fish) did not significantly elevate plasma cortisol levels 48 hours post infection. They attributed this to the fact that ectoparasites have co-evolved with their host organisms and therefore the host fish have evolved a tolerance to low numbers of lice (Nolan *et al.*, 2000). Ruane *et al.* (1998) did however demonstrate elevated plasma cortisol levels in rainbow trout infected with *Argulus* 48 hours after a 4-h confinement stress. This suggested that whilst the effect of lice may not be seen immediately after infection the effects become apparent when the response to a second stressor is examined (Nolan *et al.*, 2000).

Infection with low numbers of the salmon louse *L. salmonis* generally does not result in significant increases in plasma cortisol levels of the host (Bjorn and Finstad, 1997; Johnson and Albright, 1992; Ross *et al.*, 2000). However, heavy infestations may cause substantial elevation, far beyond those which cause immunosuppression (Johnson and Albright, 1992; Mustafa *et al.*, 2000). Exposure of *O. mykiss* to juvenile stages of *L. salmonis* increased blood cortisol levels after 4-h net confinement to levels that were significantly higher than those in confined, but unparasitized fish (Ruane *et al.*, 2000). Similar results were obtained with *O. mykiss* confined after 21 days of infestation with adult *A. foliaceus* (Ruane *et al.*, 1998).

The conclusions we can draw from these few studies include the fact that ectoparasitic lice can induce similar responses in fish skin to those stress responses observed for toxic stressors (Nolan *et al.*, 2000). In addition it was observed that certain fish species can tolerate low numbers of lice providing additional stressors are not encountered. This stress effect has implications for other host systems including the immune response and it is known that fish harbouring lice are frequently subjected to secondary infections possibly due to immunosuppression influenced by stress-related responses. The next section highlights some of the more common secondary infections that fish infected with argulids may encounter.

Secondary infections

In addition to the damage and stress caused by *Argulus* itself, one of the main concerns for fin fish producers is the associated secondary infections that can result from infections with parasites. Several studies have examined the role of parasites as vectors for other diseases (e.g. Cusack and Cone, 1985, 1986; Jones and Hine, 1983; Nigrelli, 1950;) and *Argulus* spp. have been the topic of some of these reviews (e.g. Ahne, 1985; Dombrowski, 1952). The wounds created by this parasite's feeding action are an obvious site for infection and bacterial, e.g. *Aeromonas salmonicida* (Shimura, 1983) and fungal, e.g. *Saprolegnia spp.* (Bower-Shore, 1940; Stammer, 1959) infections are often concurrent with *Argulus* spp. infections (Lester and Roubal, 1995). Some nematodes (e.g. Anguillicolidae and Skrjabillanidae) also use argulids as intermediate hosts (Moravec, 1978). The most worrying of these transmitted pathogens, however, is spring viraemia of carp (also known as infectious dropsy). This acute viral disease of carp wipes out hundreds of wild and captive fish every year. Argulids have long been suspected as vectors for this virus (Dombrowski, 1952) but it was not until the 1980s that this was proved satisfactorily through controlled laboratory experiments (Ahne, 1985).

6 Control and prevention

There are numerous options available for the control, prevention and treatment of *Argulus* infections. Much of the literature recognizes that argulosis, like many other diseases, is best defended against by good fish husbandry and stock management. Quarantining fish is vital for the aquarist but not always feasible for large-scale fish production (e.g. trout farming). Individual fish can then be examined for the presence of parasites and any encountered can be carefully removed using forceps (Benz *et al.*, 2001). Care must be taken to examine buccal and gill cavities also because on occasion *Argulus* individuals have been found there. Large-scale fish production requires other control methods suitable for treating large numbers of fish.

There are numerous options available for the control, prevention and treatment of *Argulus* infections. Many traditional treatment methods rely on the use of toxic chemicals such as malachite green or other insecticide-type chemicals. The damaging effects of many of these chemicals on wildlife are now widely recognized and as a result many countries now prohibit their use. There is also an increasing pressure from consumers for food fish that have not been subjected to chemical treatments. The wide range of chemical treatments available for infections of *Argulus* spp. is well documented. Due to the extensive literature available on this topic no effort is made here to discuss them in detail. The author therefore refers the reader to selected texts (e.g. Kabata, 1970, 1985; Lester and Roubal, 1995; Van Duijn Jnr, 1973); Williams, 1997).

In addition to the chemical treatments available scientists are constantly examining new methods that may prove cost-effective and more importantly today, environmentally friendly. The use of invertebrate developmental inhibitors (IDIs) is now under review for the treatment of fish ectoparasites. For example, Williams (1997) recently examined the effectiveness of a chitin-inhibiting treatment (Diflubenzuron) for the treatment of *Argulus* infestations and particularly the success from oral administration of these chemicals. Whilst the results showed some effect against the parasite the study was not conclusive. Combining the drug with the fish feed in a way that proved palatable for the fish and viable as a treatment proved to be the main difficulty.

During the last couple of decades, researchers seeking cost-effective, environmentally friendly methods to control crustacean ectoparasites have come up with some interesting biological control methods. In wild situations many fish utilize cleaner fish (commonly various species of wrasse) to help rid themselves of parasites. In Norway, fish farmers have employed wrasse in salmon cages to help control ectoparasitic copepod numbers (Bjordal, 1991). This method has shown some promise but has not been universally successful. In 2002, Gualt *et al.* published a paper detailing the use of novel egg-laying boards to control argulid numbers in a commercial, stillwater trout fishery. Boards were positioned in the water column with the idea that lice would use them as sites for egg deposition. The results were very promising and as a result further investigations are being undertaken by this research group.

7 Conclusions

Argulus spp. rarely have significant impacts upon natural fish populations. Epizootics are typically observed when the natural equilibrium is perturbed by one or more factors and in many cases anthropogenic actions have been implicated. For example, increased population densities in fish farms and even commercial sport fisheries facilitate parasite transmission and stress resulting from crowding, capture, handling and confinement can have a deleterious effect upon the fish's immune response. Menezes *et al.*, (1990) provided evidence that stocking water bodies with non-native fish species can also provide easy targets for lice and as we have seen with *Argulus japonicus*, anthropogenic transfer of fish can also facilitate parasite dispersal (Rushton-Mellor, 1992).

Much of the research with argulids in the past has focused on morphological and ecological aspects of these organisms. Whilst this information is useful in understanding certain aspects of the parasite's life cycle and mode of existence the more recent research focusing on the host–parasite interactions shows considerable promise in helping to develop economically viable and environmentally sound methods of controlling these parasites. If the fish and lice have co-evolved over many generations it seems logical that we should consider both organisms when trying to solve the problems we face.

With continuing advances in the field of fish immunology future research is likely to increase our knowledge of the intimate relationship between argulids and their fish hosts. Many of the advances in sea lice research may also assist in our understanding of the relationship between freshwater lice and their hosts. In addition to increasing our knowledge of fish defence mechanisms and the host–parasite interactions we must also consider the parasite itself. Much of the basic knowledge regarding these organisms' biology is still not fully understood and in many cases may prove vital to our understanding of the complex relationships between pathogens and their hosts.

Acknowledgements

This work was supported in part by the European Community's Improving Human Potential Programme under contract (HPRN-CT-2001-00214), (PARITY). In addition, the authors would like to thank Raymond Duijf for the donation of his scanning electron microscope images to this chapter and Geert Wiegertjes for useful comments on early versions of the manuscript.

References

Ahne, W. (1985) *Argulus foliaceus* L. and *Pisicola geometra* L. as mechanical vectors of spring viraemia of carp virus (SVCV). *J. Fish Dis.* **8:** 241–242.

Barber, I. Walker, P. and Svensson, P.A. (2004) Behavioural responses to simulated avian predation in three-spined sticklebacks: the effect of experimental *Schistocephalus* infections. *Behaviour* (in press).

Bauer, O.N., Musselius, V.A. and Strelkov, Y.A. (1973) *Diseases of Pond Fishes.* Israel Program For Scientific Translations. Keter Press, Jerusalem. Translated from Russian.

Begon M., Harper J.L. and Townsend C.R. (1990) *Ecology: Individuals, Populations and Communities.* Blackwell Scientific Publications, London, UK.

Benz, G.W., Bullard, S.A. and Dove A.D.M. (2001) Metazoan parasites of fishes: synoptic information and portal to the literature for aquarists. pp 1–15. In Regional Conference Proceedings 2001, Anonymous (ed). American Zoo and Aquarium Association, Silver Spring, MO. 160p.

Benz, G.W. and Otting, R.L. (1996) Morphology of the Fish Louse (*Argulus*: Branchiura). *Drum Croaker* **27:** 15–22.

Bjordal, Å. (1991) Wrasse as cleaner-fish for farmed salmon. *Prog. Underwater Sci.* **16:** 17–29.

Bjorn, P.A. and Finstad, B. (1997) The physiological effects of salmon lice infection on sea trout smolts. *Nordic Journal of Freshwater Research* **73:** 60–72.

Bower-Shore, C. (1940) An investigation of the common Fish Louse, *Argulus foliaceus* (Linn.). *Parasitology* **32:** 361–371.

Bowman, T.E. and Abele, L.G. (1982) Classification of the Recent Crustacea. Pages 1-27 In: Abele, L.G. (ed.) *The Biology of Crustacea, v. 1. Systematics, The Fossil Record, and Biogeography*, pp. 1–27. Academic Press, New York.

Cannon W.B. (1935) Stresses and strains of homeostasis. *Am. J. Med. Sci.* **189:** 1–14.

Chrousos, G.P. and Gold, P.W. (1992) The concepts of stress and stress system disorders. Overview of physical and behavioral homeostasis. *JAMA* **267:** 1244–1252.

Costello, M.J. (1993) Review of methods to control sea lice (Caligidae: Crustacea) infestations on salmon (*Salmo salar*) farms. In: Boxshall, G.A. and DeFaye, D. (eds) *Pathogens of Wild and Farmed Fish: Sea Lice*, pp. 219–252. Ellis Horwood, London.

Cusack, R. and Cone, D.K. (1985) A report on the presence of bacterial microcolonies on the surface of *Gyrodactylus* (Monogenea). *Journal of Fish Diseases.* **8:** 125–127.

Cusack, R. and Cone, D.K. (1986) A review of parasites as vectors of viral and bacterial diseases of fish. *J. Fish Dis.* **9:** 169–171.

Dombrowski, H. (1952) Die Karpenlaus, *Argulus foliaceus. Fishbauer* **2:** 145–147.

Dugatkin, L.A., FitzGerald, G.J. and Lavoie, J. (1994) Juvenile three-spined sticklebacks avoid parasitized conspecifics. *Environ. Biology Fishes* **39:** 215–218.

Evans, D.W. and Matthews, M.A. (2000) First record of *Argulus foliaceus* on the European eel in the British Isles. *J. Fish. Biol.* **57:** 529–530.

Fryer, G. (1982) *The parasitic Copepoda and Branchiura of British freshwater fishes: A handbook and key.* F.B.A. Scientific Publications of the Freshwater Biological Association, Ambleside. **46:** pp 87.

Galarowicz, T. and Cochran, P.A. (1991) Response by the parasitic crustacean *Argulus japonicus* to host chemical cues. *J. Freshwater Ecol.* **6:** 455–456.

Gault, N.F.S., Kilpatrick, D.J., and Stewart, M.T. (2002) Biological control of the fish louse in a rainbow trout fishery. *J. Fish. Biol.* **60:** 226–237.

Grayson, T.H., Jenkins, P.G., Wrathmell, A.B. and Harris, J.E. (1991) Serum responses to the salmon louse, *Lepeophtheirus salmonis* (Kroyer, 1838) in naturally infected salmonids and immunised rainbow trout *Oncorhynchus mykiss* (Walbaum) and rabbits. *Fish Shellfish Immunol.* **1:** 141–156.

Gurney, R. (1948) British species of the fish louse of the genus *Argulus. Proc. Zool. Soc. London* **118:** 553–558.

Hindle, E. (1949) Notes on the treatment of fish infected with Argulus. *Proc. Zool. Soc.* **119(1):** 79–81.

Huising, M.O., Stolte, E., Flik, G., Savelkoul, H.F., and Verburg-van Kemenade, B.M. (2003) CXC chemokines and leukocyte chemotaxis in common carp (Cyprinus carpio L.). *Dev. Comp. Immunol.* **27(10):** 875–888.

Iger, Y., Balm, P.H.M., Jenner, H.A. and Wendelaar Bonga, S.E. (1995) Cortisol induces stress-related changes in the skin of rainbow trout (*Oncorhynchus mykiss*). *Gen. Comp. Endocrinol.* **97:** 188–198.

Johnson, S.S. and Albright, L.J. (1992) Comparative susceptibility and histopathology of the response of naïve Atlantic, chinook and coho salmon to experimental infection with *Lepeophtheirus salmonis* (Copepoda: Caligidae). *Dis. Aquat. Organ.* **14:** 179–193.

Johnson, S., Blaylock, R.B., Elphick, J. and Hyatt, K.D. (1996) Disease caused by the sea louse (*Lepeophtheirus salmonis*) (Copepoda:Caligidae) in wild sockeye salmon (*Oncorhynchus nerka*) stocks of Alberni inlet, British Columbia. *Can. J. Fish. Aquat. Sci.* **53:** 2888–2897.

Jones, J.B. and Hine, P.M. (1983) *Ergasilus rotundicorpus* n.sp. (Copepoda: Ergasilidae) from *Siganus guttatus* (Bloch) in the philipines. *System. Parasitol.* **5:** 241–244.

Kabata, Z. (1985) *Parasites and Diseases of Fish Cultured in the Tropics*. Taylor and Francis, London.

Kimura, S. (1970) Notes on the reproduction of water lice (*Argulus japonicus* Thiele). *Bull. Freshwater Fish. Res. Lab. Tokyo* **20:** 109–126.

Kollatsch, D. (1959) Untersuchungen uber die biologie und okologie der karpenlause (*Argulus foliaceus* L.). *Zool. Beitr.* **5:** 1–36.

Lester, R.J.G. and Roubal, F. (1995)Phylum Arthropoda, 475–598. In: Woo, P. (ed.) *Fish Diseases And Disorders: Volume 1: Protozoan and Metazoan Infections*, pp. 475–598. CAB International, Wallingford, UK.

Martin, F.M. (1932) On the morphology and classification of *Argulus* (Crustacea). *Proc. Zool. Soc. London* 771–806.

Mclaughlin P.A. (1980). *Comparative Morphology of Recent Crustacea*. W.H. Freeman and Co. San Francisco, USA.

Menezes, J., Ramos, M.A., Pereira, T.G. and Moreira da Silva, A. (1990) Rainbow trout culture failure in a small lake as a result of massive parasitosis related to careless introductions. *Aquaculture* **89:** 123–126.

Mikheev, V.N., Valtonen, E.T. and Bintamäki-Kinnunen, P. (1998) Host searching in *Argulus foliaceus* L. (Crustacea: Branchiura): the role of vision and selectivity. *Parasitology* **116:** 425–430.

Mikheev, V.N. Mikheev, A.V., Pasternak, A.F. and Valtonen, E.T. (2000) Light-mediated host searching strategies in a fish ectoparasite, *Argulus foliaceus* L. (Crustacea: Branchiura). *Parasitology* **120:** 409–416.

Mikheev, V.N., Pasternak, A.F., Valtonen, E.T. and Lankinen, Y. (2001) Spatial distribution and hatching of overwintered eggs of a fish ectoparasite, *Argulus coregoni* (Crustacea: Branchiura). *Dis. Aquat. Organ.* **46:** 123–128.

Moravec, F. (1978) Redescription of the nematode *Philometra obturans* (Prenant, 1886) with a key to the philometrid nematodes parasitic in European freshwater fishes. *Folia Parasitol.* **25:** 115–124.

Mustafa, A., MacWilliams, C., Fernandez, N., Matchett, K., Conboy, G.A. and Burka, J.F. (2000) Effects of sea lice, *Lepeophtheirus salmonis* infestation on non-specific defence mechanisms in Atlantic salmon, *Salmo salar*. *Fish Shellfish Immunol.* **10:** 47–59.

Nigrelli, R.F. (1950) Lymphocystis disease and ergasilid parasites in fishes. *J. Parasitol.* **36:** 36.

Nolan, D.T., Reilly, P. and Wendelaar Bonga, S.E. (1999) Infection with low numbers of the sea louse *Lepeophtheirus salmonis* (Krøyer) induces stress-related effects in the Atlantic salmon (*Salmo salar* L.). *Can. J. Fish. Aquat. Sci.* **56:** 947–959.

Nolan, D.T., van der Salm, A.L. and Wendalaar Bonga, S.E. (2000) The host–parasite relationship between the rainbow trout (*Oncorhynchus mykiss*) and the ectoparasite *Argulus foliaceus* (Crustacea: Branchiura): epithelial mucous cell response, cortisol and factors which may influence parasite establishment. *Contrib. Zool.* **69:** 57–63.

Northcott, S.J. (1997) An outbreak of freshwater fish lice, *Argulus foliaceus* L., seriously affecting a Scottish stillwater fishery. *Fish. Manage. Ecol.* **4:** 73–75.

Poly, W.J. (1997) Host and locality records of the fish ectoparasite, *Argulus* (Branchiura), from Ohio (U.S.A.). *Crustaceana* **70(8):** 867–874.

Poly, W.J. (1998) New state, host, and distribution records of the fish ectoparasite, *Argulus* (Branchiura), from Illinois (U.S.A.). *Crustaceana* **71(1):** 1–8.

Poole, W.R., Nolan, D.T. and Tully, O. (2000) Modelling the effects of capture and sea lice (*Lepeophtheirus salmonis* (Krøyer) infestation on the cortisol stress response in trout. *Aquacult. Res.* **31:** 835–841.

Pottinger T.G., Pickering A.D. and Blackstock N. (1984) Ectoparasite induced changes in epidermal mucification of the brown trout, *Salmo trutta* L. *J. Fish Biol.* **25:** 123–128.

Poulin, R. (1998) **Evolutionary Ecology of Parasites.** Chapman and Hall, New York.

Poulin, R. and FitzGerald, G.J. (1987) The potential of parasitism in the structuring of a salt marsh stickleback community. *Can. J. Zool.* **65:** 2793–2798.

Poulin, R. and FitzGerald, G.J. (1988) Water temperature, vertical distribution, and risk of ectoparasitism in juvenile sticklebacks. *Can. J. Zool.* **66:** 2002–2005.

Poulin R., Marcogliese, D.J. and McLaughlin, J.D. (1999) Skin-penetrating parasites and the release of alarm substances in juvenile rainbow trout. *J. Fish. Biol.* **55:** 47–53.

Reilly, P. and Mulcahy, M.F. (1993) Humoral antibody response in Atlantic salmon (*Salmo salar* L.) immunised with extracts derived from the ectoparasitic caligid copepods, *Caligus elongatus* (Nordmann, 1832) and *Lepeophtheirus salmonis* (Krøyer, 1838). *Fish Shellfish Immunol.* **3:** 59–70.

Roberts, L.S. and Janovy, J. (1996) *Foundations of Parasitology*, 5th edn. WCB Publishers New York, USA.

Ross, N.W., Firth, K.J., Wang, A., Burka, J.F. and Johnson, S.C. (2000) Changes in hydrolytic enzyme activities of naive Atlantic salmon (*Salmo salar*) skin mucus due to infection with the salmon louse (*Lepeophtheirus salmonis*) and cortisol implantation. *Dis. Aquat. Org.* **41:** 43–51.

Ruane, N., McCarthy, T.K. and Reilly, P. (1995) Antibody response to crustacean ectoparasites immunized with *Argulus foliaceus* L. antigen extract. *J. Fish Dis.* **18:** 529–537.

Ruane, N.M., Nolan, D.T., Rotllant, J. Tort, L., Balm, P.H.M. and Wendelaar Bonga, S.E. (1998) Modulation of the response of rainbow trout (*Oncorhynchus mykiss* Walbaum) to confinement, by an ectoparasitic (*Argulus foliaceus* L.) infestation and cortisol feeding. *Fish Physiol. Biochem.* **20:** 43–51.

Ruane, N.M., Nolan, D.T., Rotllant, J. and Wendelaar Bonga, S.E. (2000) Experimental exposure of rainbow trout *Oncorhynchus mykiss* (Walbaum) to the juvenile stages of the sea louse *Lepeophtheirus salmonis* (Krøyer) modifies the response to a second stressor. *Fish Shellfish Immunol.* **10:** 451–463.

Rushton-Mellor, S.K. (1992) Discovery of the fish louse, *Argulus japonicus* Theile (Crustacea: Branchiura), in Britain. *Aquacult. Fish. Manage.* **23:** 269–271.

Rushton-Mellor, S. and Boxshall, G.A. (1994) The developmental sequence of *Argulus foliaceus*. *J. Natl Hist.* **28:** 763–785.

Shafir, A. and Oldewage, W.H. (1992) Dynamics of a fish ectoparasite population: opportunistic parasitism in *Argulus japonicus* (Branchiura). *Crustaceana* **62(1):** 50–64.

Shafir, A. and van As, J.E. (1986) The opportunistic nature of *Argulus japonicus* in maintaining host-parasite relationships. *South Afr. J. Sci.* **81:** 635.

Shimura, S. (1980) The larval development of *Argulus coregoni* Thorell (Crustacea, Branchiura). *J. Natl Hist.* **15:** 331–348.

Shimura, S. (1983) SEM observation on the mouth tube and preoral sting of *Argulus coregoni* Thorell and *Argulus japonicus* Thiele (Crustacea: Branchiura). *Fish Pathol.* **18:** 151–156.

Shimura, S. and Inoue, K. (1984) Toxic effects of extract from the mouth-parts of Argulus coregoni Thorell (Crustacea: Branchiura). *Bull. Jap. Soc. Sci. Fish.* **50:** 729.

Singhal, R.N., Jeet, S. and Davies, R.W. (1990) The effect of argulosis-saprolegniaisis on the growth and production of *Cyprinus carpio*. *Hydrobiologia* **202:** 27–32.

Stammer, H.J. (1959) Beitrage zur Morphologie, Biologie und Bekampfung der karpfenlause. *Zeits. Parasiten.* **19:** 135–208.

Taylor, N., Sommerville, C. and Wooten, R. (2003) *A lousy day's fishing! The epidemiology of the fish louse (Argulus) in UK Stillwater trout fisheries. Book of Abstracts, 11th International Conference of the EAFP-Diseases of Fish and Shellfish.*

Tokioka, T. (1936) Preliminary report on Argulidae in Japan. *Annot. Zool. Japon.* **15:** 334–343.

van Ham E.H. (2003) The physiology of juvenile turbot (*Scophthalmus maximus* Rafinesque) and Atlantic halibut (*Hippoglossus hippoglossus*, L.) under different stress conditions. Katholieke Universiteit Nijmegen diss.

van der Salm, A.L., Nolan, D.T., Spannings, F.A.T. and Wendelaar Bonga, S.E. (2000) Effects of infection with the ectoparasite *Argulus japonicus* (Thiele) and administration of cortisol on cellular proliferation and apoptosis in the epidermis of common carp, *Cyprinus carpio* L., skin. *J. Fish Dis.* **23:** 173–184.

Walker, P.D. (2002) An investigation into the population structure, distribution, host selection and effects of abiotic factors on the off-host longevity of, the crustacean ectoparasite of fish, *Argulus foliaceus* L. (Crustacea: Branchiura). MRes thesis University of Plymouth.

Walker, P.D., Haond, C., Russon, I.J. and Wendelaar Bonga, S.E. (2003) Evidence of blood feeding by the crustacean ectoparasite of fish, *Argulus japonicus* Thiele (Crustacea: Branchiura). *Book of Abstracts, 11th International Conference of the EAFP-Diseases of Fish and Shellfish.*

Wendelaar Bonga, S.E. (1997) The stress response in fish. *Physiol. Rev.* **77:** 591–625.

Whelan, K.F. and Poole, W.R. (1996) The sea trout stock collapse, 1989–1992. In: *The Conservation of Aquatic Systems*, Reynolds, J.D. (ed.) Dublin, Royal Irish Academy.

Whitear, M. (1986) The skin of fishes including cyclostomes – Epidermis. In: Bereiter-Hahn, J., Matoltsy, A.G. and Richards, K.S. (eds) *Biology of the Integument*, Vol 2., pp. 8–38. Springer-Verlag, Berlin.

Williams, C. (1997) Investigation of Diflubenzuron in the control of *Argulus foliaceus* L. in carp, *Cyprinus carpio* L. in relation to management strategies. MSc thesis, University of Plymouth.

Woo, P. (1995) *Fish Diseases And Disorders: Volume 1: Protozoan and Metazoan Infections*. CAB International, Wallingford.

Van Duijn Jnr, C. (1973) *Diseases of Fishes.* Cox & Wyman Ltd, London.

7

Interactions between sea lice and their hosts

Stewart C. Johnson and Mark D. Fast

1 Introduction

Although most people think of copepods as primarily free-living animals this is not the case as over half of all the known species of copepods live in association with other organisms (Huys and Boxshall, 1991). Of these species the nature of their associations ranges from being symbiotic to truly parasitic during some or all of their life history stages. For the purpose of this chapter we are defining sea lice as those species that are in the genera *Lepeophtheirus* (Nordmann, 1832) and *Caligus* (Müller, 1785). These genera belong to the order Siphonostomatoida and family Caligidae. Species within these genera are ectoparasites of a wide variety of species of fish in marine and brackish waters.

Members of these genera are the most intensively studied of all parasitic copepods due to their economic importance as parasites of fish cultured in marine and brackish waters (Johnson *et al.*, in press). Of all of the sea species, the salmon louse, *Lepeophtheirus salmonis* (Krøyer, 1837) has been studied in the most detail due to its economic impact on salmonid aquaculture in the northern hemisphere. Economic losses due to sea lice are related to: the costs of treatments, the costs of the management strategies used to control their abundance, the costs associated with reduced growth rates that are a direct result of infection and/or treatment and the costs of carcass downgrading at harvest. Indirect and direct losses due to *L. salmonis* are estimated to be in excess of US$100 million annually (Johnson *et al.*, in press). Estimations of costs associated with *L. salmonis* infection for the different salmon growing regions in the northern hemisphere are available (Mustafa *et al.*, 2001a; Pike and Wadsworth, 1999; Rae, 2002; Sinnott, 1999).

Greater than 28 other species of sea lice have been identified from fish cultured in marine and brackish waters (Johnson *et al.*, in press). With respect to causing disease in aquaculture the most important species include *Caligus elongatus*, *Caligus epidemicus*, *Caligus orientalis*, *Caligus punctatus* and *Caligus rogercresseyi* (cf. Ho, 2000; Johnson *et al.*, in press). Estimates of economic losses due to these species of sea lice are not available, however they are likely to be as large or larger then those reported for *L. salmonis*.

Host–Parasite Interactions, edited by G. F. Wiegertjes & G. Flik.

In this chapter we provide a brief review of aspects of the morphology and biology of sea lice that are important with respect to understanding the nature of their relationships with their hosts. We also critically examine literature on their host–parasite interactions and separate those studies that have studied these interactions from those that have examined the process of disease. We will also review other host–parasite relationships to provide insight into other, as yet not described, mechanisms that may allow sea lice to establish and maintain themselves on their hosts.

2 The life cycle

Sea lice have direct life cycles that consist of two free-living non-feeding planktonic naupliar stages, one free-swimming infectious copepodid stage, four to six attached chalimus stages, up to two pre-adult stages, and one adult stage (Johnson and Albright, 1991; Lin *et al.*, 1996a, 1997; Ogawa, 1992; Piasecki, 1996; Piasecki and Mackinnon, 1995; Pike and Wadsworth, 1999). Reported differences in the lifecycles between different species of sea lice may represent differences in the interpretation of what constitutes a particular developmental stage, or the failure to recognize developmental stages that are highly morphologically similar. Life cycle descriptions are available for a number of sea lice species including the economically important species: *L. salmonis* (cf. Johnson and Albright, 1991; Schram, 1993), *C. elongatus* (cf. Piasecki, 1996) and *C. epidemicus* (cf. Lin *et al.*, 1996a).

Several features of their lifecycles are important in our discussion of the host–parasite relationship. The infectious copepodid stage locates and establishes itself on suitable hosts using the second antennae and maxillipeds to maintain its position on the host (Bron *et al.*, 1991, 1993a). The mechanism by which this stage locates its host is not understood. Copepodids have been shown under laboratory conditions to display a variety of behaviours that may be important to position them within an environment where the probability of encountering a host is increased (Bron *et al.*, 1993a; Heuch 1995, Heuch *et al.*, 1995). Once within this environment it is thought that visual and/or mechanical and/or chemical clues may direct them to potential hosts (Bron *et al.*, 1993a; Heuch and Karlsen, 1997; Ingvarsdóttir, 2002). Upon attachment to a potential host chemoreception is thought to be important in determining whether it is a suitable host (Bron *et al.*, 1993a; Fast *et al.*, 2003; Ingvarsdóttir, 2002).

Once attached to a suitable host the copepodid stage begins feeding using mouthparts that are modified to form a structure referred to as the oral cone. The oral cone is a characteristic feature of all caligid copepods and it is maintained with little modification to its structure throughout the other developmental stages (see Section 3). After a period of feeding, the duration of which is primarily dependent upon water temperature, the copepodid moults into the first chalimus stage. The chalimus stages are characterized by the presence of a frontal filament that physically attaches these stages to the host. The frontal filament is described in more detail in Section 3. In *L. salmonis* the frontal filament is replaced at each moult during the chalimus stages (Gonzalez-Alanis *et al.*, 2001). In *C. elongatus* the frontal filament is not replaced at each chalimus moult, however additional filament material is added at the moult (Piasecki and MacKinnon, 1993). The ability of *L. salmonis* and possibly other species of sea lice to move during the chalimus stages may be advantageous due to the soft nature of the tissues to which they attach or possibly to avoid host immune responses. The pre-adult stages are similar in morphology to the adult stages. These stages are free-moving on the surface of the host with the exception of a short period at the moult when they also attach to the host by means of a temporary frontal filament. As in the copepodid stages the second antennae and maxilliped are important in the maintenance of these stages on the host, however these stages also rely upon the use of suction, which they create beneath their cephalothorax (Kabata, 1979). Immediately after the moult to the adult stage females are inseminated and will continue to produce egg strings throughout their life. Under laboratory conditions the lifespan of female *L. salmonis* may be up to 7 months and the lifespan of *C. elongatus* up to 260 days (Mustafa *et al.*, 2001b; Piasecki and MacKinnon, 1995).

3 Morphological features of sea lice

For the purposes of this review there are several morphological features of sea lice that are important with respect to their host–parasite interactions. These features include the oral cone, the frontal filament and the alimentary tract and its associated glands.

The oral cone or mouth tube of sea lice is formed by the labium and labrum and the mandibles (Kabata, 1974, 1979). At the tip of the labium there is a structure called the strigil that is armed with numerous fine sharp teeth. The teeth of the strigil along with the teeth on the tips of the manibles serve to abrade the surface of the host and to pass the resulting debris into the oral cone. Once this material is within the oral cone it is passed to the oesophagus by the action of muscles within the labrum (Kabata, 1974, 1979). Using the immunohistochemical stain 3',3-diaminobenzidine tetrahydrochloride (DAB) numerous exocrine glands have been identified in the different developmental stages of *L. salmonis* and *C. elongatus* (cf. Bell *et al.*, 2000). Positive reactions with this stain occur in the presence of exogenous peroxidases and catalase, enzymes that are involved in a variety of processes including free radical neutralization and prostaglandin production. In their study DAB-positive glands ('circum-oral glands') were identified in association with the oral cone. These authors postulate that these glands may be involved in the production of substances that protect sea lice from host reactive oxygen species and/or are involved with host immunomodulation through the production of prostaglandins (see Section 9). *Lepeophtheirus salmonis* is known to secrete proteases onto the surface of its hosts to aid in feeding and/or to avoid host immune responses (Section 9; Fast *et al.*, 2003; Firth *et al.*, 2000; Ross *et al.*, 2000). It is likely that other compounds are present in these secretions and that salivary-type glands may produce some of these compounds.

As mentioned previously the chalimus stages of sea lice are physically attached to their hosts by means of a frontal filament. During the moult the pre-adult stages also form temporary filaments that allow them to maintain their position on the host until the exoskeleton has hardened sufficiently to allow them to once again attach using their appendages and cephalothorax. The structure of the frontal filament, the timing of its production and the tissues associated with its production have been studied in detail for *L. salmonis* and *C. elongatus* (cf. Bron *et al.*, 1991; Gonzalez-Alanis *et al.*, 2001; Johnson and Albright, 1992a; Piasecki and Mackinnon, 1993; Pike *et al.*, 1993). There are differences in the timing, the method of production and the physical structure of the frontal filament between chalimus and pre-adult stages as well as between species. In *L. salmonis* and *C. elongatus* copepodids an internal frontal filament begins to develop shortly after establishment and initiation of feeding on the host (Gonzalez-Alanis, 2001; Piasecki and MacKinnon, 1993). At the moult to the first chalimus stage the filament is attached to the host with an adhesive-like substance. The filament and the area surrounding it are then pulled out to form the anterior margin of the chalimus larva. In *L. salmonis* the frontal filament is replaced at each of the moults between the chalimus stages and the production of the new filaments is related to the moult cycle (Gonzalez-Alanis, 2001; Johnson and Albright, 1992a). In contrast *C. elongatus* has been reported to retain the original frontal filament throughout the chalimus stages with additional material being added to the filament at each of the moults (Piasecki and MacKinnon 1993; Pike *et al.*, 1993). The ability of *L. salmonis* to reposition itself on the host's body at each moult may be advantageous in avoidance of host immune responses (Section 7).

In *L. salmonis* the frontal filament is continuous with the anterior margin of the chalimus larvae. It consists of a fibrous shaft and basal plate, the latter of which is made up partially by an adhesive-like compound which serves to adhere the filament to the host tissues (Bron *et al.*, 1991; Johnson and Albright, 1992a; Pike *et al.*, 1993). In *C. elongatus* the frontal filament is distinct from the anterior margin of the chalimus stages. Its shaft consists of fibres made up of bundles of extremely fine fibrils. These fibres are arranged in helicoids at the proximal end and in parallel at the distal end of the filament. There is no evidence of an adhesive-like secretion as seen in *L. salmonis* (cf. Pike *et al.*, 1993). Bron *et al.* (1991) and Gonzalez-Alanis (2001) described three groups of cells (A, B and C) that are responsible for the production of frontal filaments in *L. salmonis*. Bell *et al.* (2000) report that the 'A cells' stain positively with DAB but the significance of this result is unclear. The 'A cells' are thought to be involved with the production of the adhesive-like secretion (Bron *et al.*, 1991; Gonzalez-Alanis, 2001).

The gross anatomy and morphology of the cells of the alimentary tract has been described for *L. salmonis* (cf. Bron *et al.*, 1993b; Johnson, 1991; Nylund *et al.*, 1992) (*Figure 1*). The alimentary tract consists of four parts: the oral cone, a cuticle-lined oesophagus, a narrow tubular midgut, and a short cuticle-lined hindgut. The midgut of *L. salmonis* lacks distinct cell zones as seen in free-living copepods and appears to have similar cell types present throughout its length (Johnson, 1991). Epithelial cell division and growth results in cells of the midgut being forced apically, forming a bulbous topography of the midgut lumen. Cells of the midgut appear to undergo distinct cell cycles with undifferentiated cells being present near the base of these ingrowths and mature cells being present at their apex. The cell cycle of the midgut cells of *L. salmonis* is similar to that reported for the other parasitic copepods *Peroderma cylindricum* (cf. Monterosso, 1930) and *Ergasilus sieboldi* (cf. Einszporn, 1965). Five cell types have been identified from the midgut of *L. salmonis*. These include: cell type I (similar to F-cells of calanoid copepods) that are thought to have a secretory function; cell type II (similar to R-cells of calanoid copepods) that are thought to have enzyme secretion and digestive functions; and cell type III (similar to B-cells of calanoids and decapod crustaceans) that are thought to have absorptive and digestive functions similar to the hepatopancreas (Bron *et al.*, 1993b; Johnson, 1991; Nylund *et al.*, 1992). Johnson (1991) described two additional cell types from the midgut, including; 'type E' cells that are relatively undifferentiated cells and are believed to be responsible for replacing mature cells that are lost by degeneration or holocrine secretion, and 'type D' cells which by their ultrastructure were thought to be responsible for absorption. Recently, Johnson *et al.* (2002) demonstrated a high level of trypsinogen expression in all cell types of the anterior midgut. Their results supported the view of Nylund *et al.* (1992) that all cells of the midgut are responsible for producing digestive enzymes either for intracellular digestion in the case of type III cells or enzymes for secretion into the lumen of the gut (type I and II cells). The presence of trypsin and other proteases within the secretions of *L. salmonis* supports the view that cells of the midgut produce some of the secreted compounds (Section 7; Fast *et al.*, 2003; Firth *et al.*, 2000; Johnson *et al.*, 2002).

The peritrophic matrix, also known as the peritrophic membrane, is a non-cellular sheath that separates ingested food from the absorptive/secretory surface of the alimentary tract. This sheath forms a size-selective barrier to macromolecules that have a variety of cut-off sizes depending on the species. This structure is present in most Arthropods including many species of free-living copepods where it serves to improve efficiency of digestion, provide mechanical protection of the gut epithelium and act as a barrier to infection from ingested pathogens (Lehane, 1997; Shao *et al.*, 2001; Terra, 2001; Yoshikoshi and Kô, 1988). There are conflicting reports as to whether a peritrophic matrix is present in sea lice. No mention was made of the presence of a peritrophic membrane in early studies of the gut structure of *L. salmonis* copepodids, chalimus larvae and adults (Bron *et al.*, 1993b; Johnson, 1991; Nylund *et al.*, 1992). More

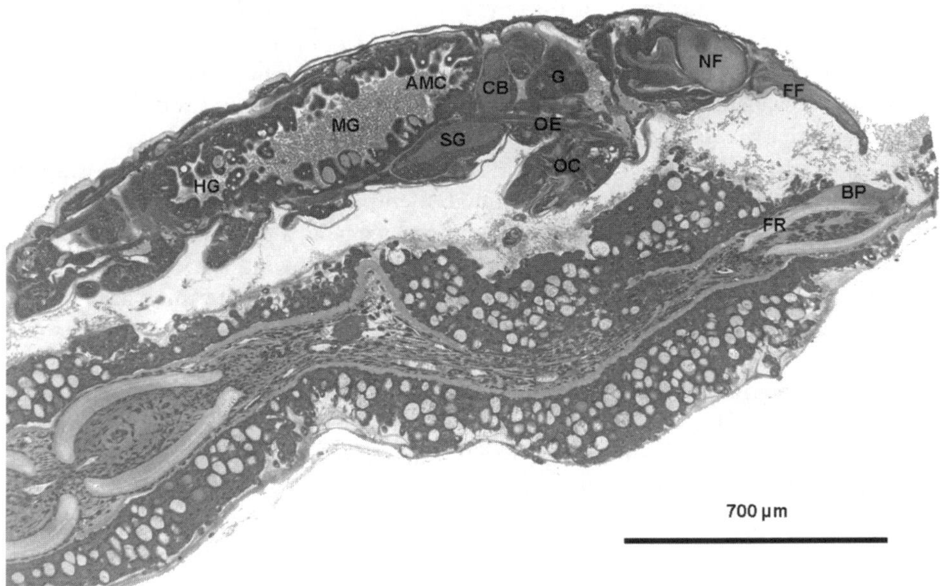

Figure 1 *Sagittal section through a third-stage* Lepeophtheirus salmonis *chalimus larva attached to a fin of a naive Atlantic salmon. Note the lack of a host response to the presence of the copepod and the presence of a new frontal filament that will be extruded at the moult. AMC, anterior midgut caecum; BP, basal plate; CB, cerebrum; FF, frontal filament; FR, fin ray; G, glandular structure; HG, hind gut; MG, midgut; NF, new frontal filament; OC, oral cone; OE, oesophagus; SG, suboesophageal ganglion.*

recently peritrophic matrix-like structures have been reported in the mid and posterior midgut of adult *L. salmonis* (Raynard *et al.*, 2002). The presence or absence of this structure in *L. salmonis* is of interest due to its potential, if present, to prevent ingested antibody from binding to the surfaces of the gut. This would result in reducing or eliminating the potential effectiveness of vaccines produced against gut cell antigens (see Section 8). Whether this is important or not is unknown, especially as it has been reported for free-living copepods that the permeability of their peritrophic matrixes is considerably greater than that of Diptera (Hansen and Peters, 1998). If present in sea lice, components of the peritrophic matrix might be targeted as vaccine antigens. Proteins called peritrophins that are tightly associated with the peritrophic matrix have been demonstrated to have potential as vaccine antigens against the sheep blowfly (*Lucilia cuprina*) (cf. Wijffels *et al.*, 1999 and references therein).

4 Attachment and feeding mechanisms

4.1 *Attachment*

All of the developmental stages of sea lice use modified second antennae, maxillae and to a lesser extent the maxillipeds, for host attachment. These appendages are capable of piercing host tissues enabling a secure attachment. In the chalimus stages sea lice also rely upon the frontal filament to maintain their position on the host, especially between moults. In addition to the above-mentioned appendages and structures the mobile pre-adult and adult stages also use their post antennary processes, sternal furca and cephalothorax for attachment (Jønsdøttir *et al.*, 1992; Kabata, 1979; Kabata and Hewitt, 1971) (*Figure 2*). The cephalothorax or dorsal shield is formed by the fusion of the cephalon and thoraxic segments up to and including the segment that bears the third leg (Kabata, 1979; Kabata and Hewitt, 1971). In sea lice the third leg is modified to form a transverse barrier that closes off the posterior margin of the cephalothorax (Kabata and Hewitt, 1971). A thin strip of membrane fringes the margins of this barrier. Another thin flexible membrane called the marginal membrane fringes the margin of the cephalothorax (*Figure 2*). Both of these membranes enable a tight seal to be produced between the copepod's body and the skin of the host, allowing the cephalothorax to essentially function as a suction cup. Two openings on the posterior margin of the cephalothorax, the posterior sinuses, are equipped with membranes that serve as one-way valves allowing water to be expelled from beneath the cephalothorax. Co-ordinated movements of the first and second leg push jets of water through these valves, causing both downward and forward thrust. This thrust allows the copepod to move forward while at the same time pressing it against the surface of the host. Replacement water passes under the frontal plates that form the anterior margin of the cephalothorax (Kabata and Hewitt, 1971). In pre-adult and adult stages of *Caligus* spp. cup-like suckers, the lunnules, are present on the anterior margin of the cephalothorax. These structures are thought to play a secondary role in attachment although their importance as attachment organs has not been conclusively demonstrated (Kabata, 1979).

4.2 *Feeding*

All of the developmental stages of sea lice that are found on the host feed on host mucus, skin and blood. Both their attachment and feeding activities ultimately cause

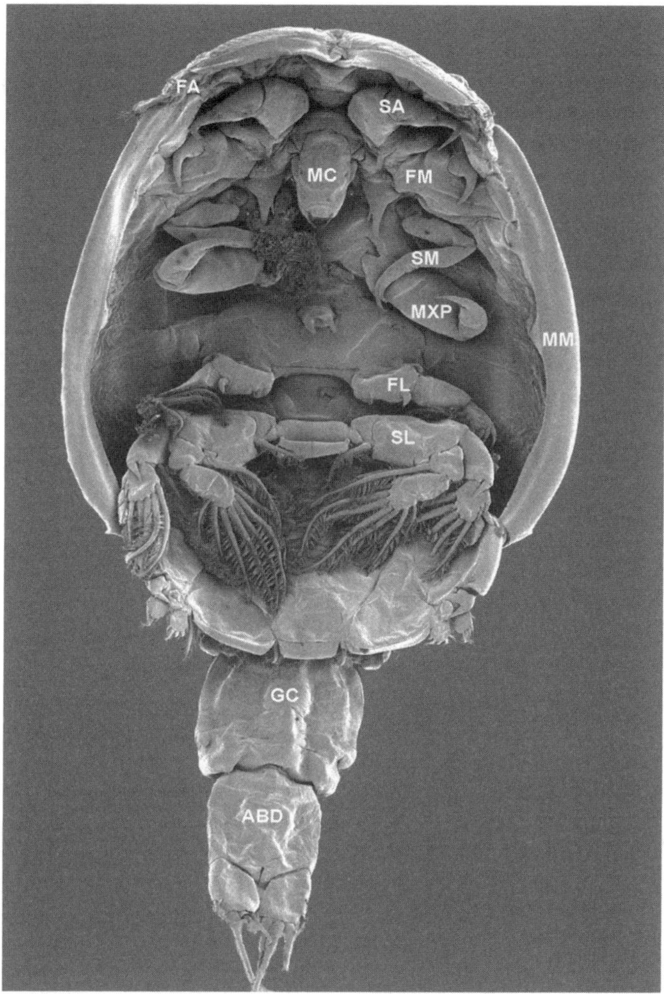

Figure 2 *Scanning electron micrograph of the ventral surface of a first pre-adult female* Lepeophtheirus salmonis. *Note the presence of the modified mouthparts, the marginal membrane and the modified third leg that are used to maintain its position on the host. ABD, abdomen; CR, caudal ramus; FA, first antennae; FL, first leg; FM, first maxilla; GC, genital complex; MM, marginal membrane; MXP, maxilliped; MC, oral cone; SA, second antennae; SL, second leg; SM, second maxilla.*

the development of lesions and in the case of heavy infections the development of disease. Kabata (1974) described in detail the structure of the mouth and mode of feeding of sea lice. Movements of the strigil scrape off host mucus and tissues and this material is then collected and moved by the mandibles into the oral cone. Contractions of the musculature of the labrum are thought to push this material towards the opening of the oesophagus. Peristaltic movements of the gut may also aid in the movement of this material into the digestive tract. Jónsdóttir *et al.* (1992) reported grooves on the skin surface of *L. salmonis*, in the vicinity of the oral cone that were consistent with the mode of feeding as described by Kabata (1974). Feeding activities may also be aided by the secretion of enzymes onto the surface of the host as has been reported for *L. salmonis* (Fast *et al.*, 2003; Firth *et al.*, 2000).

Although blood is the major component of the diet of some species of parasitic copepods its importance as a food item for sea lice is not well understood. Knowledge of the prevalence of blood feeding in the different developmental stages of sea lice is important to our understanding of the host–parasite relationship as different host factors will likely be present in the blood than in the mucus or skin. It is also very important with respect to the development of vaccines (Sections 7–10). To date the only species for which there are data on feeding is *L. salmonis*. Blood feeding by *L. salmonis* has been described by a number of authors including Brandal *et al.* (1976) who reported that 42% of adult females and 10% of adult males feeding on Atlantic salmon had blood in their guts. Haji Hamid *et al.* (1998) report that blood feeding occurs in both the pre-adult and adult stages, but only rarely in the chalimus larvae on Atlantic salmon. These authors report that blood-feeding females were larger than other females and that they had significantly longer egg strings, suggesting increased fecundity. However as Pike and Wadsworth (1999) point out it is unclear as to how the authors determined the past feeding history of these animals. There is little evidence to support the view that blood feeding is important with respect to the fecundity of *L. salmonis* or other sea lice species. Unlike the numerous arthropod ectoparasites that require a blood meal for reproduction there is no evidence to suggest that this is the case for any species of sea lice.

5 Pathological effects of sea lice attachment and feeding

The attachment and feeding activities of sea lice result in lesions that vary in their nature and severity depending upon: the species of sea lice, their abundance, the developmental stages present and the species of host. Sea lice produce physical and enzymatic damage at the site of their attachment and feeding to which their hosts may or may not mount an immune response. With respect to gross pathology lesions caused by copepodids and chalimus larvae are most often relatively minor with variable amounts of localized damage seen (Boxshall, 1977; Bron *et al.*, 1991; Johnson and Albright, 1992a, b; Jønsdøttir *et al.*, 1992; MacKinnon, 1993; Roubal, 1994). However, when chalimus larvae are abundant they can cause severe host damage including loss of fins and death (Dawson, 1998a, b; Parker and Margolis, 1964; Tully *et al.*, 1993). In most instances the pre-adult and adult stages of sea lice do not breach the epidermis of their hosts and cause only minor tissue damage (Johnson *et al.*, 1996; Ogawa, 1992; Roubal, 1994). However, in situations of severe disease such as seen in salmonids infected by high numbers of *L. salmonis*, extensive areas of skin erosion and haemorrhaging on the head and back, and a distinct area of erosion and sub-epidermal haemorrhage in the perianal region can be seen (Brandal and Egidius, 1979; Johnson *et al.*, 1996; Pike, 1989; Pike and Wadsworth, 1999). Similar skin and head lesions have also been reported in Atlantic halibut (*Hippoglossus hippoglossus*) infected with large numbers of *Caligus elongatus* and in the rabbit fish (*Siganus fuscescens*) infected with large numbers of *Caligus ovicipes* and *Lepeophtheirus atypicus* (cf. Bergh *et al.*, 2001; Lin *et al.*, 1996b). In addition to the feeding activities of sea lice the development of these large open lesions and extensive areas of gill or fin tissue damage is due in part to secondary infections and the resulting tissue necrosis.

A number of studies have examined sites of attachment and feeding using light and electron microscopy. Using light microscopy it has been demonstrated that tissue responses of naive chinook (*Oncorhynchus tshawytscha*), coho (*Oncorhynchus kisutch*) and Atlantic salmon to infection with *L. salmonis* are important with respect to establishment and maintenance of this species on its hosts (Johnson and Albright, 1992a, b). Although all of these species are natural hosts for *L. salmonis* they vary in their susceptibility to infection due to differences in the magnitude of their inflammatory and hyperplasic responses to the presence of the copepod. Atlantic salmon are the most susceptible species followed by chinook and then coho salmon (Johnson and Albright, 1992 a, b).

In naive Atlantic salmon the tissue response to *L. salmonis* is extremely limited irrespective of their developmental stage or the number of copepods that are present. Areas of tissue erosion, small amounts of haemorrhage, and mild inflammation have been reported at the attachment and feeding sites on the gills. In Atlantic salmon copepods are retained on the gills throughout the chalimus stages (Johnson and Albright, 1992a). With the exception of a mild inflammatory response in the dermis of some individuals, there was little evidence of fin tissue response to the presence of the copepodid or chalimus stages. In most cases the epidermis was breached and the underlying dermis and fin rays exposed to the external environment. In cases where inflammation was reported the inflammatory infiltrate consisted of abundant neutrophils and a few lymphocytes (Johnson and Albright, 1992a).

In contrast, attachment and feeding sites on gills of naive coho salmon were characterized by partial to complete erosion of the epithelium, minor haemorrhaging, and acute inflammation. In some cases mild epidermal hyperplasia was reported at the tips of the lamellae. In this species copepods are lost within days from the gill tissues (Fast *et al.*, 2002; Johnson and Albright, 1992a). Mild inflammation in the dermis of the fins was reported to occur in coho salmon as early as 1 day post-infection. Attachment and feeding sites of chalimus stages were characterized by well-developed epithelial hyperplasias that in some cases resulted in complete encapsulation of the copepods. In cases of partial or complete encapsulation the spaces surrounding the copepod were filled with tissue debris and a mixed inflammatory infiltrate. Neutrophils, macrophages and a few lymphocytes were present at sites of inflammation on both the gills and fins (Johnson and Albright, 1992a). Administration of hydrocortisol to naive coho salmon resulted in the suppression of these tissue responses and increased the survival rate of *L. salmonis* to a level similar to that seen on Atlantic salmon (Johnson and Albright, 1992b).

Jones *et al.* (1990) described the histopathology associated with the copepodid and early chalimus stages of *L. salmonis* on naturally infected Atlantic salmon obtained from commercial salmon farms on the coast of Scotland. As seen in laboratory studies these authors reported minor, if any, host response to the second antennae, maxillipeds or feeding activities of the copepodid stage. With respect to the chalimus stages there was no tissue response or only mild hyperplasia associated with the frontal filament. Interestingly these authors reported that upon the detachment of chalimus larvae from the frontal filament there was a host response to both the filament and the lesions caused by feeding activities. In the case of the frontal filament the tissue response consisted of epidermal hyperplasia, fibrosis and macrophage infiltration. In some sections fibrous nodules filled with inflammatory infiltrate were also evident. With respect to the feeding sites, epidermal hyperplasia with abundant areas of focal necrosis was evident along the margins of the lesions. Increased melanization of the stratum granulosum was evident beneath the hyperplasic tissues (Jones *et al.*, 1990).

Jónsdóttir *et al.* (1992) studied the histopathology associated with the pre-adult and adult stages of *L. salmonis* on naturally infected Atlantic salmon. The severity of lesions varied widely and this was thought to be due to the mobile nature of these developmental stages. In general, tissues were more heavily damaged under the cephalothorax than under other regions of the copepod's body. Cells beneath the cephalothorax showed a loss of cell surface structure as determined by SEM. Furthermore, the thickness of the epidermis varied relative to the thickness of the underlying layers and tissue swelling and in some cases splitting of the epidermal layer above the basal cells was reported. Basal cells were often reported to be poorly defined, rounded, and enlarged with large and granular nuclei. The basement membrane and pigment layer was thickened in some cases and in others it was ill-defined or broken (Jónsdóttir *et al.* 1992). These authors also reported that the inflammatory response to *L. salmonis* was more severe around the periphery of the attachment and feeding site than in the tissues directly beneath the cephalothorax.

Boxshall (1977) described the histopathology associated with attachment and feeding activities of *Lepeophtheirus pectoralis* on the fins of wild-caught flounder (*Platichthys flesus*). In this study it was unusual to find lesions caused by adult females that extended below the epidermis. As in *L. salmonis* host tissue responses including epithelial hyperplasia were more evident around the periphery of the attachment and feeding site then directly beneath it. Fibroblastic proliferation, fibre production and cellular infiltration within the dermis resulted in the formation of granulation tissues that resulted in the characteristic swelling of the fin tissues, observed beneath each copepod. It was reported that chalimus larvae due to their immobility often caused deep lesions that extended into the dermis, however the presence of an inflammatory or other responses to these stages was not mentioned.

Several studies have described the histopathology of attachment and feeding sites of *Caligus* species. MacKinnon (1993) examined the attachment and feeding sites of *C. elongatus* on naturally infected Atlantic salmon. Lesions associated with the chalimus stages extended through the epithelium to the basement membrane. In the vicinity around the lesion epithelial cells were detached from the basement membrane. Cells within lesions were necrotic and there was no evidence of inflammatory responses. In some of the lesions that were associated with fourth-stage chalimus larvae there was mild hyperplasia of the tissues surrounding the lesion. Similar lesions were reported for *Caligus epidemicus* on yellowfin bream (*Acantohopagrus australis*) (cf. Roubal, 1994). At feeding and attachment sites of both the copepodid and chalimus stages there was little evidence of tissue responses, although a few infiltrating cells were sometimes present. As reported for *L. salmonis*, frontal filaments from which copepods had detached provoked a strong tissue response that included epithelial hyperplasia, and infiltration of neutrophil-like cells, macrophages and lymphocytes. With respect to the pre-adult and adult stages only very minor tissue damage was reported and there was no evidence of a host response. *Caligus nanhaiensis* is a species that has been responsible for disease outbreaks in cultured banded grouper (*Epinephelus awoara*) (cf. Wu *et al.*, 1997). This species feeds on the gills of its host and is reported to cause severe damage of the gill filaments with little evidence of any host tissue responses.

With the exception of naive coho salmon, most species of fish have been reported to show little to no tissue response to the attachment and feeding activities of the copepodid and chalimus stages of sea lice. This lack of host response may be the result of suppression of the immune system caused by stress associated with the infection. However, the presence of well-developed host responses to frontal filaments and attachment and feeding sites after the copepods have become detached, as well as the host responses reported in tissues immediately outside of the sites of active attachment and feeding suggests that stress is not the only factor responsible for the limited tissue responses to sea lice. These observations have led to the belief that sea lice, like other parasites, have the ability to modulate their host's immune response at the sites of their attachment and feeding (Section 7). The lack of suppression of the tissue response in coho salmon may be related to the reduced level of secretions produced by *L. salmonis* in the presence of coho salmon mucus (Fast *et al.*, 2003).

Pre-adult and adult stages of sea lice have been reported to produce imprints of their cephalothoraxes on the surface of their hosts (Jónsdóttir *et al.*, 1992; Roubal, 1994). The secretion and retention of salivary or other components beneath the carapace, as well as the actions of the mouthparts and first and second thoracic legs may produce these imprints. The presence of host tissue responses around the periphery of the cephalothorax may indicate the physical limit of any immunomodulatory activity.

In addition to causing damage at their feeding and attachment site low numbers of pre-adult *L. salmonis* have been reported to cause subtle pathological changes at sites distant from their point attachment and feeding (Nolan *et al.*, 1999). These changes included increased apoptosis and necrosis of epithelial cells and decreased numbers of mucus cells in the skin; and in gills of infected fish swelling of the lamellae, detachment of the epithelium and apoptosis of chloride cells. These changes are thought to be the result of the stress response of the host, although it is also possible that enzymes secreted by sea lice onto the surface of the host may influence these changes (Fast *et al.*, 2003; Nolan *et al.*, 1999).

6 Physiological effects of sea lice infection

6.1 *The host stress response*

The magnitude of the physiological effects of sea lice infection on the host is related to sea lice abundance and the severity of lesions formed by their attachment and feeding activities. Several studies have explored how sea lice infection affects the stress response in fish. The stress response of teleosts can be divided into primary, secondary and tertiary responses (reviewed in Iwama *et al.*, 1999; Wendelaar Bonga, 1997; see also Chapter 6). The primary response is a neuroendocrine/endocrine response, in which increases in plasma levels of the stress hormones, such as cortisol occur, as well as cellular responses including heat shock protein (HSP) production. The secondary response comprises metabolic changes (plasma glucose and lactate levels), hydromineral disturbance (changes in chloride and sodium balance), haematological changes (haematocrit and haemoglobin), and changes in the immune system. The tertiary response is characterized by changes in the whole organism, such as reduced growth, swimming performance, reproductive success and disease resistance that can cause a reduction in survival (Iwama *et al.*, 1999). Of the hormones that make up the stress response, cortisol is the one that is most commonly measured and used as an assessment of stress in fish (see Chapter 6). Blood glucose levels and to a lesser extent serum sodium and chloride levels are also used. With respect to infection with *L. salmonis* a variety of authors have used challenge studies to examine the stress response using these parameters. These studies have used a variety of host species and host sizes but unfortunately they have not been conducted using standard methods. What is evident from these studies is that the level of stress caused by the infection is influenced by the intensity of infection, host size, host condition and possibly species of host. There are no data available on the host stress response for other species of sea lice. To enable some comparison between studies we have reported intensity of infection as the number of *L. salmonis* per gram of host body weight.

6.2 *Sea lice infection trials*

As mentioned previously a number of challenge studies have been undertaken to examine the physiological and immunological effects of sea lice on their hosts (*Table 1*; Tully and Nolan, 2002). Unfortunately, many of these studies have used high levels of infection that are more indicative of a disease state rather then the level that would occur when the host–parasite relationship is in balance. In these studies the stress response and especially the development of open lesions results in marked physiological and immunological changes and in some instances death. Ross *et al.* (2000) infected small Atlantic salmon (average weight 57 grams) with high (178 ± 67) and low (20 ± 13) numbers of *L. salmonis*. At the higher level of infection (average of 3.0 *L. salmonis*/g fish) 100% morbidity or mortality occurred within a 24-hour period of the moult to the first pre-adult stage (at 120–140 degree days (dd)). There was no gross evidence of open lesions in this study, however cortisol levels were significantly elevated in infected fish when compared to the controls in the first sample taken at 3 days post infection (dpi) (approximately 30 dd). High levels of cortisol that averaged between 100 and 170 nmol l^{-1} were maintained in the infected fish throughout the study. At the lower infection level (average 0.4 *L. salmonis*/g fish) no mortality occurred and cortisol levels did not increase significantly from the normal resting cortisol levels seen in the controls. Grimnes and Jakobsen (1996) also reported increased mortality upon the moult to the pre-adult stage. In their study, Atlantic salmon (average weight 40 g) infected with an average of 1.0 *L. salmonis*/g fish suffered 10% mortality at 220 dd and 90% mortality at 320 dd. Based upon their measurements of serum protein, serum albumin, plasma chloride and haematocrit they reported that chalimus stages even at this high level of infection had little effect on the physiology of their hosts. However, with the development to pre-adult stage significant differences in plasma chloride and serum protein and albumin and haematocrit were evident between infected and control fish. Unfortunately neither cortisol or glucose levels were measured in their study. At first sampling (7 dpi) Atlantic salmon infected with an average of 0.6 *L. salmonis* g^{-1} host showed elevated levels of cortisol when compared to controls (Finstad *et al.*, 2000). As seen previously other effects such as osmoregulatory problems (increased plasma chloride levels) and mortality occurred only after copepods moulted to the pre-adult stage.

Table 1 Lepeophtheirus salmonis *infection densities on salmonid hosts. The cortisol levels and mortality rates that are presented are those obtained at the moult from Chalimus IV to pre-adult stage I.*

Study	Lice per fish	Fish size (g)	Lice/g	Cortisol (nmol/L)	Mortality/ Morbidity (%)
Ross *et al.* (2000) [A]	178 ± 67	55.5	3.0	170	100
Ross *et al.* (2000) [A]	20 ± 13	56.9	0.4	<20	—
Grimnes and Jakobsen (1997) [A]	42	40	1.0	NM	10
Mustafa *et al.* (2000a) [A]	52	180	0.28	50–70	–
Bowers *et al.* (2000) [A]	106	680	0.16	150–200	–
Fast *et al.* (2002) [A]	18	115	0.16	20	–
Johnson and Albright (1992a) [A]	10	40–100	0.14	NM	
Dawson *et al.* (1997) [A]	40	413–499	0.09	NM	–
Nolan *et al.* (1999) [AB]	3, 6, 10	200–250	0.01–0.05	NM	–
Bjorn and Finstad (1996)[C]	38	93	0.41	>100	11
Nagasawa (2001)[E]	1.06–6.07	NM	NM	N/A	N/A
Tingley *et al.* (1997)[D]	5	980	<0.01	N/A	N/A
Mo and Heuch (1998) [D]	26.7	140	0.12	N/A	N/A
Mo and Heuch (1998) [D]	2.7	146	<0.01	N/A	N/A
Schram *et al.* (1998)[D]	3-8	440	<0.01	N/A	N/A

[A] Experimental infection of Atlantic salmon (*Salmo salar*);[B] Experimental infection conducted with pre-adults and adults rather than copepodids; [C] Experimental infection of sea trout (*Salmo trutta*); [D] Natural infection of sea trout; [E] Natural infection of Pacific salmonids (*Oncorhynchus kisutch, O. keta, O. gorbuscha, O. mykiss, O. tshawytscha*); NM – not measured; N/A – not applicable.

Bowers *et al.* (2000) infected Atlantic salmon with an average weight of 680 g with an average of 106 *L. salmonis* per fish (approximately 0.2 *L. salmonis*/g) copepods per fish and monitored plasma cortisol, glucose, electrolytes, thyroid hormones (T3 and T4) and haematocrit over a 290 dd period. At 210 dd both cortisol (mean 63.1 nmol l^{-1} controls: mean 178.9 nmol l^{-1} infected) and glucose (mean 3.54 nmol l^{-1} controls: mean 4.57 nmol l^{-1} infected) were elevated significantly due to the presence of *L. salmonis*. Although there was a trend of higher cortisol and glucose levels in individual fish with higher *L. salmonis* numbers this trend was not statistically significant. In later stages of the infection both plasma protein, chloride and sodium levels were elevated over that of the controls suggesting osmoregulatory failure, which the authors suggested might be caused by the breakdown of skin mucus. A marked reduction in host mucus consistency has been reported in Atlantic salmon that are heavily infected with *L. salmonis* (cf. Ross *et al.*, 2000).

Mustafa *et al.* (2000a) studied the stress response and macrophage function in Atlantic salmon infected with *L. salmonis*. The first trial that they reported on is the same trial presented in Bowers *et al.* (2000), although they reported only plasma cortisol and glucose data up to 210 dd. However, they did include data on respiratory burst and phagocytic activities of isolated head kidney macrophages collected over this period. In this trial they demonstrated a significant reduction in respiratory burst and phagocytic activity at 210 dd. In a second trial, salmon with average weight of 180 grams were infected with an average of 52 (0.5 *L. salmonis*/g fish) copepods per fish. This resulted in significantly elevated levels of cortisol and glucose in the infected fish when compared to controls at 70, 140 and 210 and 140 and 210 dd, respectively. At 210 dd when average cortisol levels exceeded 60 nmol l^{-1}, respiratory burst activity and phagocytic activity of isolated head kidney macrophages was significantly reduced in the infected fish. These authors suggest that the prolonged elevated levels of cortisol seen in these trials were responsible for the decrease in macrophage function. Fast *et al.* (2002) challenged Atlantic salmon, coho salmon and rainbow trout with a relatively low number of *L. salmonis* (0.16 *L. salmonis*/g fish). Over the 21-day study period (210 dd) there were no significant differences in plasma cortisol levels between infected and control fish for any of the host species with exception of 1 dpi (10 dd) in rainbow trout when cortisol levels were slightly elevated in the infected fish. These authors also reported a significant reduction in macrophage function as measured by respiratory burst and phagocytic activity for rainbow trout (at 210 dd) and Atlantic salmon (at 140 and 210 dd), which occurred in the absence of a cortisol response. At the latter time point the majority of copepods that were present were pre-adults.

Overall these studies have demonstrated that at relatively high levels of infection, similar to those responsible for disease, *L. salmonis* can cause a generalized stress response that may ultimately lead to decreased immune function, possibly increased susceptibility to *L. salmonis* or other disease-causing agents and death. Mustafa *et al.* (2000b) reported that rainbow trout were more susceptible to infection with the microsporidian *Loma salmonae* when infected with *L. salmonis*. The high levels of mortality reported within a short time after the moult to the pre-adult stage are more difficult to account for and may be related to substances produced by *L. salmonis* and secreted onto the surface of their hosts (see Section 9).

With the exception of a few instances when disease outbreaks due to *L. salmonis* have occurred in the wild most fish collected from the wild have much lower sea lice densities than the densities that have been used in laboratory studies. For example, sea trout (*Salmo trutta*) caught in Norwegian fjords and off the coast of England were infected at a level of 2–8 *L. salmonis* per fish (<0.01 *L. salmonis*/g of fish, *Table 1*) (Mo and Heuch, 1998; Tingley *et al.*, 1997). Similar levels of infection have been reported for Pacific salmon caught on the high seas and for Atlantic salmon and sea trout caught of the coast of Scotland and Norway (Berland, 1993; Jacobsen and Gaard, 1997; MacKenzie *et al.*, 1998; Nagasawa, 2001). Only a few studies have investigated the effects of low levels of sea lice infection on host physiology. Nolan *et al.* (1999) reported disruption of skin and gill epithelia over large areas of the body of Atlantic salmon infected with low numbers of *L. salmonis* (ranging from 3–10 pre-adult or adult *L. salmonis* per fish). Accompanying these changes was evidence of small disruptions of osmoregulation and increased turnover rates of chloride cells on the gills. Although plasma cortisol was not measured these changes are highly similar to those reported for stressors in general. Even in the absence of a cortisol or other physiological response, infection with low numbers of *L. salmonis* has been demonstrated to effect host recovery when exposed to additional stressors (Nolan *et al.*, 2000; Ruanne *et al.*, 2000). In the study by Ruane *et al.* (2000) it was demonstrated that although infection with low numbers of *L. salmonis* did not cause a significant increase in serum cortisol levels, the presence of copepods did result in higher levels of cortisol and glucose when compared to uninfected fish during a 24-hour recovery period post stress.

7 Immune responses to sea lice

There is a great deal of interest in the production of a vaccine against sea lice. However, there is very limited information with respect to the importance of acquired immunity in controlling the abundance of sea lice or aspects of their biology. One of the problems that have made studies of this difficult is the lack of a good infection model for sea lice. To date most studies have used single pulses of infection with high infection levels on a per individual basis (*Table 1*). Although this model might be argued to mimic the dynamics of infection seen in fish maintained under aquaculture conditions it is not likely to be a good model to use for studying host–parasite interactions that have evolved over time in response to infection pressures that are lower and spread out over time. Furthermore, single pulse infections of even relatively low numbers of sea lice result in increased levels of stress due to both the experimental conditions under which the fish are maintained as well as their response to infection (Nolan *et al.*, 1999).

There are only a few reports that discuss the possible roles that the immune system of fish plays in their relationship with parasitic copepods. With respect to sea lice there is some evidence that immunity may play a role in controlling aspects of reproduction in *L. salmonis*. It has been observed that the number of eggs carried by *L. salmonis* increases on mature coho salmon when compared to immature coho salmon raised at the same site (Johnson, 1993). Johnson (1993) suggested that increased egg numbers on mature hosts was related to a reduction in the function of their immune system that was brought about by sexual maturation. However it is unknown whether this change in egg number is related to aspects of the innate or acquired immune system or both.

There is only one published report of the serum antibody response of naturally infected salmonids to sea lice. Grayson *et al.* (1991) compared the immune responses of rainbow trout and Atlantic salmon that were naturally infected with *L. salmonis* to the responses obtained in rainbow trout and rabbits that were immunized with *L. salmonis* whole-body homogenates. In this study specific antibody responses against *L. salmonis* could not be identified in naturally infected rainbow trout. Naturally infected Atlantic salmon produced antibodies that recognize several antigens in unreduced samples, most notably one that had a molecular weight in excess of 200 kDa. Immunohistochemical techniques demonstrated the binding of serum from the naturally infected Atlantic salmon to the apices but not the crypts of the highly folded gut epithelium of *L. salmonis* (cf. Grayson *et al.*, 1991). Rainbow trout that were immunized with extracts of whole body homogenate were seen to produce antibodies that recognized a wider variety of antigens than seen for the naturally infected Atlantic salmon which is not surprising as many 'hidden antigens' would be present in the whole-body homogenates.

Although not very informative with respect to understanding sea lice host–parasite interactions several studies have demonstrated that Atlantic salmon produce antibodies against antigens derived from *L. salmonis*. Unfortunately, none of the antigens tested to date have been shown to be effective as a vaccine against sea lice. Reilly and Mulcahy (1993) investigated the humoral antibody response of Atlantic salmon to vaccination with extracts of *C. elongatus* and *L. salmonis*. This research was part of a larger project designed to identify antigens that might be suitable for use in vaccines. In this study the authors demonstrated that Atlantic salmon could make antibodies against several antigens present in the crude extracts from these two species of sea lice. They also demonstrated that there were many cross-reacting antigens between these species and their different life history stages. To date there is only one published vaccine trial that determined the efficacy of vaccination against *L. salmonis*. Grayson *et al.* (1995) immunized Atlantic salmon using a relatively crude preparation of soluble sea lice antigens that were partially purified using Con A affinity chromatography. Polyclonal antisera raised against these antigens in rats demonstrated that at least some of the antigens used were associated with the gut epithelium. There was no significant effect of immunization on *L. salmonis* abundance following laboratory challenge. However, it was noted that the average number of eggs produced by *L. salmonis* on immunized hosts was 26% less than produced by females attached to non-immunized hosts. There was no apparent difference in hatching success of the eggs from these two groups. Whether the fish would be naturally exposed to the antigens used in this study is unknown.

Several studies have investigated the importance of acquired immunity in the host–parasite relationships of a number of other species of parasitic copepods especially those that belong to the genus *Lernaea*. In this genus, adult females undergo a remarkable metamorphosis that results in the production of a root-like structure referred to as a holdfast. These species anchor themselves to the body of their hosts using the holdfast, which depending on the species can be imbedded deeply within a variety of different types of host tissues. Although such a structure should be expected to elicit a strong host response including antibody production the role of acquired immunity in the host–parasite relationships of species within this genus is not clear. Shields and Goode (1978) reported variable re-infection of goldfish, *Carassius auratus*, that were previously exposed to *Lernaea cyprinacea* suggesting the possibility that acquired immunity had developed. Shariff (1981) studied the distribution and abundance of adult female *Lernaea piscinae* growing on big head carp, *Aristichthys nobilis*. *Lernaea piscinae* was reported to be retained in the eyes for up to 3 weeks after all copepods were lost from other body areas. It was suggested that this could be an indication of an immune response as the copepods were retained only in the eyes, an immunologically privileged site. Woo and Shariff (1990) reported for *Lernaea cyprinacea* growing on kissing gourami, *Helostoma temmincki*, that previously infected fish lost copepods faster then naive hosts upon re-infection. They also noted that a higher proportion of egg sacs were lost from copepods growing on previously infected fish than naive fish. Furthermore, eggs from copepods growing on previously infected hosts either failed to develop or produced copepodids that had a low infectivity when compared to copepodids hatched from eggs of copepods growing on naive hosts. It was postulated that changes in the structure of the lesion or the production of antibodies that interfered with feeding or digestion might be responsible for their observations. In contrast to these other studies Kularatne *et al.* (1994) report that acquired immunity is not important in the interactions between *Lernaea minuta* and the Javanese carp, *Puntius gonionotus*. In their study they reported that the fish acquired *L. minuta* infections regardless of whether the fish had been previously infected, injected with plasma from recovered fish or injected with plasma from naive fish. They could also not statistically demonstrate a difference in infection intensities between the groups, although previously infected fish were reported to start losing their parasites a few days before that of the other groups. It was suggested that *L. minuta* does penetrate into the host tissues as deeply and that it does not cause as severe an inflammatory response as that reported for other species of *Lernaea*, thereby failing to illicit an acquired immune response. The failure of plasma from recovered fish to immobilize the first naupliar stage of this species was also considered as evidence for little or no specific antibody responses. Thoney and Burreson (1988) reported that there was no evidence of antibody production by the spot croaker (*Leiostomus xanthurus*) against the tissue dwelling copepod, *Lernaeenicus radiatus*. They attributed the lack of an antibody response to the presence of a thick capsule of host tissue that surrounds the head and the neck of this copepod that would limit antigenic stimulation of the host.

8 The effects of hosts on sea lice

There is only a small amount of published information on the effects of hosts on the biology of sea lice though one can imagine based on studies of other parasites that these are very important in the maintenance of the host–parasite relationship. In most instances available information is purely observational and has not been tested experimentally. Host effects on the biology of other types of parasitic copepods have been documented and include: changes in distribution on the host, interruption of egg sac production, loss of egg sacs, failure of eggs to develop and reduced infectivity of the copepodid stage (Paperna and Zwerner, 1982; Shariff, 1981; Woo and Sharrif, 1990). There is a body of evidence that indicates that *L. salmonis* grows at different rates on different species of hosts within its natural host range. Differences in growth rates on naive hosts correspond well with differences in susceptibility to infection under laboratory conditions with the highest growth rates reported on the most susceptible species, Atlantic salmon. In their study of the susceptibility of naive coho, chinook and Atlantic salmon to experimental infection with *L. salmonis* Johnson and Albright (1992a) reported that the development rate of the copepods differs on different regions of the body and between different species of naive hosts. They reported the presence of earlier developmental stages on the gills of Atlantic salmon when compared to the fins and other body surfaces. They suggested that this might be due to the presence of innate humoral factors affecting the developmental rate, noting that copepods feeding on the gills had blood in their guts whereas blood was less common in the guts of copepods from the fins and other body surfaces. They also reported that at 20 days post infection a marked difference in the age structure of *L. salmonis* between chinook and Atlantic salmon, which suggested that copepods developed at a slower rate on chinook salmon. This result was supported by the work of Johnson (1993) who compared the development and growth rates of *L. salmonis* on naive chinook and Atlantic salmon. In that study significant differences in *L. salmonis* developmental stage composition were evident in all samples with copepods developing at a slower rate on chinook salmon. Fast *et al.* (2002) conducted an experimental infection of naive rainbow trout, coho salmon and Atlantic salmon with *L. salmonis*. In their study *L. salmonis* was seen to mature at a slower rate on coho salmon, followed by rainbow trout and then Atlantic salmon.

In addition to growth, differences in the numbers of eggs carried by *L. salmonis* on different host species have been reported. Johnson (1993) reported that approximately twice as many eggs were carried on females growing on mature Atlantic salmon when compared to those collected from mature chinook salmon growing at the same site. Higher numbers of eggs were also reported on females collected from mature verses immature coho salmon reared at the same site. Immunization of Atlantic salmon with extracts of adult *L. salmonis* was reported not to affect the number of sea lice in a laboratory challenge. However, immunized fish had fewer ovigerous females that had lower numbers of eggs when compared to females from non-immunized fish (Grayson *et al.*, 1995).

The reason for differences in growth rates and the number of eggs produced on the different host species is not known and the relative importance of host nutritional and/or immunological factors in controlling these aspects of sea lice biology should be investigated. It would also be valuable to determine whether past exposure to sea lice would have an effect on sea lice growth rates or numbers of eggs. The effects of host stress and maturation state should also be investigated, as both of these are known to affect immune function. The presence of lower numbers of ovigerous females bearing lower numbers of eggs on immunized hosts is similar to results obtained for multiple tick species. Both *Boophilus microplus* (Tellam *et al.*, 2002; Willadsen, 1997; Willadsen *et al.*, 1995) and *Ornithodoros moubata* (Chinzei and Minoura, 1988) have exhibited reduced oviposition when feeding on immunized hosts.

9 Host immunomodulation by sea lice

In situations where the host–parasite interactions between sea lice and their host are in balance it is not in the best interest of the sea lice to cause large-scale immunomodulation of the host that might result in host mortality. The lack of a host response at the site of attachment and feeding, and more normal tissue responses at a distance from the parasite has been taken as evidence that sea lice, like other arthropod parasites, produce substances that have immunomodulatory effects on the host. The lack of an antibody response to sea lice may also be in part due to immunosuppression. However, many studies cannot be taken as definitive evidence for the presence of sea lice immunomodulatory compounds as the stress response that is seen in most challenge trials is in itself immunosuppressive. For example, elevated levels of serum cortisol are the most likely reason for the observed reduction of macrophage function in Mustafa *et al.* (2000a). Whereas, in the paper by Fast *et al.* (2002) a reduction of macrophage function could not be attributed to a cortisol response.

9.1 *Sea lice immunomodulatory substances*

There is a vast literature describing compounds produced by parasites that have been demonstrated to or are thought to be important in maintaining the host–parasite relationship. Arthropod parasites such as ticks, fleas and biting flies are particularly well studied due to their economic importance as disease-causing agents or vectors of other disease-causing agents of livestock and humans. Proteases, alkaline phosphatase, other enzymes, prostaglandins and low-molecular-weight proteins are major salivary constituents of parasites, including helminths and arthropods (Berasain *et al.*, 1997;

Gillespie *et al.*, 2000; Hotez and Cerami, 1983; Kerlin and Hughes, 1992; Rosenfeld and Vanderberg, 1998; Schoeler and Wikel, 2001). Trypsin-like proteases play important roles in the pathogenesis of a variety of parasitic diseases in which they assist in invasion of host tissues and evasion of host immune responses. They have been well studied for a variety of arthropod parasites including: the cattle tick (*Boophilus microplus*), the mosquito (*Aedes aegypti*), the sheep blowfly (*Lucilia cuprina*), the buffalo fly (*Haematobia irritans exigua*), the warble-fly (*Hypoderma lineatum*) and the cat flea (*Ctenocephalides felis*) (cf. Casu *et al.*, 1994, 1996; Gaines *et al.*, 1999; Kerlin and Hughes, 1992; Moiré, *et al.*, 1994).

Several studies have suggested that sea lice produce immunomodulatory compounds and that these compounds are at least in part responsible for the lack of host response seen at their sites of attachment and feeding (Bell *et al.*, 2000; Fast *et al.*, 2002, 2003; Johnson *et al.*, 2002; Ross *et al.*, 2000). It is expected that sea lice will produce similar compounds to those produced by other arthropod parasites and that these may be secreted in saliva or regurgitated onto the surface of the host to aid in feeding and avoidance of host immune response. With the exception of a few papers there is little published evidence to support the presence of such substances in sea lice and what work has been conducted is limited to studies on *L. salmonis*. The increased protease and alkaline phosphatase activity reported in mucus collected from infected hosts and in mucus samples of susceptible species that are incubated with live *L. salmonis* is good evidence for secretions being produced by sea lice. Ross *et al.* (2000) reported increased protease and alkaline phosphatase activity in the skin mucus of Atlantic salmon experimentally infected with *L. salmonis* compared to skin mucus from non-infected fish. This increase was primarily due to the appearance of low molecular weight (17–22 kDa) proteases with trypsin-like activity (Firth *et al.*, 2000). Fast *et al.* (2003) studied enzymes released by *L. salmonis* in response to mucus from different host species of salmonids and winter flounder, a non-salmonid that is not a host for this species. They report that a significantly higher number of *L. salmonis* released these 17–22-kDa proteases in the presence of rainbow trout and Atlantic salmon mucus than did in the presence of seawater, coho salmon or winter flounder mucus. Alkaline phosphatase activity was also significantly higher in Atlantic salmon mucus incubated with *L. salmonis* than in control mucus. Based upon this study it appears that these secretions are important with respect to maintaining the relationship between *L. salmonis* and its hosts. Naive coho salmon are capable of being infected with *L. salmonis* but quickly lose their parasites within 10–14 days due to a well-developed inflammatory response (Fast *et al.*, 2002; Johnson and Albright, 1992a, b). As mentioned previously other natural hosts such as Atlantic salmon show little if any host tissue response to *L. salmonis*. It is possible that a reduced or lack of *L. salmonis* secretory activity on coho salmon may reduce parasite feeding, allow host tissue responses, or both. Interestingly these authors noted that *L. salmonis* collected in the Pacific Ocean where coho salmon naturally occur had significantly higher mean protease activity and positive protease responses to coho salmon mucus than did *L. salmonis* from the Atlantic Ocean. This suggests that there may be differences between isolated populations of *L. salmonis* with respect to their responses to hosts.

Trypsin-like activities have previously been found in *L. salmonis* whole-body homogenate and gut epithelial cells using immunocytochemical techniques (Jenkins *et al.*, 1993; Roper *et al.*, 1995). Firth *et al.* (2000) characterized the LMW proteases released by *L. salmonis* in the presence of salmon mucus as trypsin. Since that study trypsin genes were identified in cDNA libraries from pre-adult *L. salmonis* and seven trypsin-like enzymes have been cloned and sequenced (Johnson *et al.*, 2000, 2002). These trypsins are very similar to other crustacean trypsins and insect hypodermins and trypsinogen expression can be identified in the three major cell types that line the midgut. Based on these studies and those conducted for other parasites it appears that trypsin decreases host phagocytic activity and immune responses following infection. As mentioned previously reduced phagocytic and respiratory burst responses were observed in the absence of elevated cortisol in *L. salmonis*-infected rainbow trout and Atlantic salmon at the same time that the multiple bands of LMW proteases occurred in the mucus (Fast *et al.*, 2002). Saliva from the cattle tick, *Rhipicephalus sanguineus*, and the sand fly, *Phlebotomus papatasi*, has been shown to impair host T cell proliferation and interferon (IFN)-γ-induced macrophage microbicidal activity and activation (Ferreira and Silva, 1998; Hall and Titus, 1995).

As mentioned previously Bell *et al.* (2000) reported the presence of glands in the vicinity of the oral cone of *C. elongatus* and *L. salmonis* that stain positive for peroxidase activity. They suggested that peroxidases produced and released by these glands might serve to protect the copepods from internal and/or external damage caused by reactive oxygen species produced by the host. Reactive oxygen species are produced by the host immune cells such as neutrophils in response to tissue damage and as a host defence response. In other blood-feeding arthropods salivary peroxidase activity is thought to prevent vasoconstriction, as well as have other roles such as the metabolism of prostaglandins and leukotrienes, thereby limiting the inflammatory response (Riberio and Valenzuela, 1999). Down-regulation of host inflammatory cytokines is often observed in other host–parasite relationships (Ferriera and Silva, 2001; Fuchsberger *et al.*, 1995; Gwakiska *et al.*, 2001; Hajnicka *et al.*, 2001; Kopecky *et al.*, 1999; Schoeler and Wikel, 2001; Schoeler *et al.*, 1999). Whether this is the case for sea lice is unknown.

Bell *et al.* (2000) also suggested that these peroxidases might also be involved in the production of prostaglandins. At present there is no evidence for the presence of prostaglandins or peroxidases in secretions of sea lice, although these compounds have been found in the saliva of other parasites (Gillespie *et al.*, 2000; Riberio and Valenzuela, 1999).

Based on what is known to date about sea lice and information from studies on other parasites it is highly likely that sea lice salivary secretions will contain a wide variety of substances, each of which may play a number of roles in host immunomodulation. Techniques utilized for the study of secretions of other parasites need to be applied to the study of sea lice (see Section 10).

10 Future directions

The ultimate goal of sea lice research has been to develop information that will be useful in the management, treatment and prevention of sea lice disease. To date a large amount of data has been collected; especially for *L. salmonis* and these data have been very useful in developing management plans and treatment strategies that have reduced but not eliminated the significant economic impact of sea lice on marine aquaculture. Within the last few years several research groups have started to investigate more closely the interactions between sea lice and their hosts with the ultimate goal being the development of a vaccine.

We have seen that there is good evidence that sea lice, like other parasites, maintain themselves on their hosts by using a variety of strategies including immunomodulation of the host responses. With respect to immunomodulation a major gap in our knowledge is in the identification, purification and characterization of the sea-lice-derived compounds that are responsible for it. There are several ways in which such compounds could be identified including the traditional method of isolation of secretions or the tissues responsible for their production; followed by extraction and fractionation of these secretions and identification of active compounds. This procedure requires the development of a sensitive and reproducible assay for the detection of activity, which does not exist for fish. This procedure also runs the risk that some or all of the biological activity may be lost during extraction and fractionation. This type of procedure has been used in some studies on *L. salmonis* (Fast *et al.*, 2003; Firth *et al.*, 2000; Ross *et al.*, 2000). However, these studies were limited to biochemical characterization of the secretions and did not attempt to determine biological activity for these fractions. Butler *et al.* (2000) attempted to characterize biological activity of sea lice compounds against host cells; however, it is unclear as to whether the activities reported were exclusively from the sea lice secretions. It has been difficult to identify biological activity of sea lice compounds using cell-based bioassays due to the highly variable cell responses between individual fish even when exposed to identical stimuli. Studies of mammalian host–parasite interactions do not have this problem due to the use of highly inbred strains of animals and/or well-characterized cell lines.

In a recent study by Valenzuella *et al.* (2002) gene and protein expression was studied in the salivary glands of the tick *Ixodes scapularis*. This species is a vector for Lyme disease. In this study the authors used Edman degradation of proteins from saliva and salivary gland extracts isolated by 1-D SDS-PAGE and compared amino acid sequences to translated protein sequences from gene expression studies. By comparing their data against the GenBank protein database and searching against the Conserved Domains Database these authors were able to describe the presence of a large number of novel proteins that may be important in tick feeding and disease transfer. Ultimately these authors feel that some of these proteins may serve as novel vaccine targets. The study by Valenzuella *et al.* (2002) is a good example of how a variety of techniques could be used to determine proteins present in sea lice secretions or frontal filaments. Once identified and cloned genes from sea lice could be expressed using one of several available expression systems to obtain the relatively large amounts of purified proteins required for further characterization and/or testing as antigens for use in vaccines.

Recently a number of sea lice research groups have started to use molecular techniques to determine which genes are up-regulated in sea lice in response to establishment and initiation of feeding. However, to date there are only two published reports on gene expression in *L. salmonis* (cf. Johnson *et al.*, 2000, 2002). Johnson *et al.* (2000) described the use of ESTs (expressed sequence tags) as a method to identify expressed genes in the whole body of the pre-adult stages of *L. salmonis*. A variety of genes for proteins that might be important in the host–parasite interaction were identified including trypsin that is secreted onto the surface of the host during infection (Firth *et al.*, 2000; Johnson *et al.*, 2002). Although such studies will give an accurate profile of all of the genes that are being expressed at the time of sampling a high proportion of the genes are likely to play little or no role in the host–parasite interaction. Although difficult due to the size of sea lice, a better strategy would have been to prepare tissue-specific cDNA libraries from tissues identified or thought to be responsible for production of substances important in the host–parasite interaction. Such tissues could include those responsible for the production of the frontal filament or the peroxidase-positive tissues that have been identified in association with the oral cone (Bell *et al.*, 2000). Another approach would be to use a subtractive library produced from sea lice collected prior to attachment and after some time of attachment and feeding on the host. Such studies for sea lice are presently underway in several laboratories.

A problem with molecular studies on poorly studied groups of organisms is the lack of homology of the obtained sequences with sequences available within published databases. Furthermore, for many significant matches the function of the gene is unknown. For example, in a small EST study of *L. salmonis* 47% of the sequences had no matches against sequences within public databases and of those that did 4% had significant matches with genes that are unidentified as to function (Johnson *et al.*, 2000). This lack of data in public databases can be overcome by combining molecular studies with studies of the organism's expressed proteins. In these instances the expressed proteins can be searched against the molecular data to obtain the genes that are responsible for their production.

In the study by Valenzuela *et al.* (2002) proteins were identified by Edman degradation, a process that requires relatively large amounts of the protein. To obtain these proteins from small animals is often a very laborious and time-consuming process. Another common approach used for large-scale protein identification is proteomics. Proteomics allows the generation of amino acid sequence data from very small amounts of protein using mass spectrometry. In the past proteomics has been used to identify proteins that have been separated by one- or two-dimensional SDS-PAGE. More recently techniques have been developed to separate and identify proteins directly from biological extracts. Proteomic studies on sea lice are likely to be limited by the availability of protein sequences within available databases. As mentioned above conducting molecular studies of the parasite and using those data as a source of sequence information for protein identification may overcome this limitation.

There are two important things to consider with respect to novel vaccine targets. These targets must be expressed in all of the developmental stages of interest and they need to be expressed at sites where they can be exposed to components of the host immune system. There are several ways in which proteins can be localized within different developmental stages or body regions of sea lice. In the past, studies such as Grayson *et al.* (1995) utilized immunohistochemical techniques to identify sites of protein production in *L. salmonis.* These techniques require the purification of proteins of interest and the production of antisera against them and are a very labour-intensive method for screening, especially when you have large numbers of potential antigens. If the nucleotide sequence of the protein of interest is known it is relatively easy to screen different developmental stages or regions of the body for expression of the gene using RT-PCR or real time PCR techniques. It is also possible to generate gene-specific probes and to use these probes to determine sites of gene expression in histological sections using *in-situ* hybridization. For example, an *in-situ* hybridization technique was used to locate sites of trypsinogen expression in sections of the gut of

L. salmonis (Johnson *et al.*, 2002). As more sequence data become available for sea lice it will become possible to produce sea lice cDNA microarrays that will allow the simultaneous study of expression of large numbers of genes.

Due to ongoing large-scale sequencing projects on fish we are presently experiencing a rapid increase in our knowledge of genes involved in both the innate and acquired immune responses. In some species such as the Atlantic salmon cDNA microarrays that contain large numbers of immunologically relevant genes are now becoming available. Using RT-PCR, real time PCR and/or microarrays it is now possible to examine the effects of sea lice secretions or recombinant sea lice proteins on host gene expression. These techniques and tools should enable a rapid expansion of our understanding of how sea lice interact with their hosts.

References

Bell, S., Bron, J.E. and Sommerville, C. (2000) The distribution of exocrine glands in *Lepeophtheirus salmonis* and *Caligus elongatus* (Copepoda: Caligidae). *Cont. Zool.* **69:** 9–20.

Berasain, P., Goni, F., McGonigle, S., Dowd, A., Dalton, J.P., Frangione, B. and Carmona. C. (1997) Proteinases secreted by *Fasciola hepatica* degrade extracellular matrix and basement membrane components. *J. Parasitol.* **83:** 1–5.

Bergh, O., Nilsen, F. and Samuelsen, O.B. (2001) Diseases, prophylaxis and treatment of the Atlantic halibut *Hippoglossus hippoglossus*: a review. *Dis. Aquat. Org.* **48:** 57–74.

Berland, B. (1993) Salmon lice on wild salmon (*Salmo salar*) in western Norway. In: Boxshall, G.A. and Defaye, D. (eds) *Pathogens of Wild and Farmed Fish: Sea Lice,* pp. 179–187. Ellis Horwood, Chichester, UK.

Bowers, J.M., Burka, J.F., Mustafa, A., Speare, D.J., Conboy, G.A., Sims, D.E. and Brimacombe, M. (2000) The effects of a single experimental challenge of sea lice, *Lepeophtheirus salmonis*, on the stress response of Atlantic salmon, *Salmo salar*. *J. Fish Dis.* **23:** 165–172.

Boxshall, G.A. (1977) The histopathology of infection by *Lepeophtheirus pectoralis* (Müller, 1776) (Copepoda: Caligidae). *J. Fish Biol.* **10:** 411–415.

Brandal, P.O., Egidius, E. and Romslo, I. (1976) Host blood: a major food component for the parasitic copepod *Lepeophtheirus salmonis* Krøyer, 1838 (Curstacean, Caligidae). *Nor. J. Zool.* **24:** 341–343.

Bron, J.E., Sommerville, C., Jones, M. and Rae, G.H. (1991) The settlement and attachment of early stages of the salmon louse, *Lepeophtheirus salmonis* (Copepoda: Caligidae) on the salmon host, *Salmo salar*. *J. Zool.* **224:** 201–212.

Bron, J.E., Sommerville, C. and Rae, G.H. (1993a) Aspects of the behaviour of copepodid larvae of the salmon louse *Lepeophtheirus salmonis* (Krøyer, 1837). In: Boxshall, G.A. and Defaye, D. (eds) *Pathogens of Wild and Farmed Fish: Sea Lice,* pp. 125–142. Ellis Horwood, Chichester, UK.

Bron, J.E., Sommerville, C. and Rae, G.H. (1993b) The functional morphology of the alimentary canal of larval stages of the parasitic copepod *Lepeophtheirus salmonis*. *J. Zool. London* **230:** 207–220.

Butler, R., Bowden, T.J., Bron, J.E., Bricknell, J.R. and Sommerville, C. (2000) The use of an in vitro model to investigate suppression of Atlantic salmon cellular immune responses during infection of *Lepeophtheirus salmonis*. *Acta Parasitol.* **45:** 271.

Casu, R.E., Pearson, R.D., Jarmey, J.M., Cadogan, L.C., Riding, G.A. and Tellam R.L. (1994) Excretory/secretory chymotrypsin from *Lucilia cuprina*: purification, enzymatic specificity and amino acid sequence deduced from mRNA. *Insect Mol. Biol.* **3:** 201–211.

Casu, R.E., Eisemann, C.H., Vuocolo, T. and Tellam, R.L. (1996) The major excretory/secretory protease from *Lucilia cuprina* larvae is also a gut digestive protease. *Int. J. Parasitol.* **26:** 623–628.

Chinzei, Y. and Minoura, H. (1988) Reduced oviposition in *Ornithodoros moubata* (Acari: Argasidae) fed on tick-sensitized and vitellin-immunized rabbits. *J. Med. Entomol.* **25:** 26–41.

Dawson, L.H.J. (1998a) Physiological, behavioural and pathological effects of sea lice, *Lepeophtheirus salmonis* on salmonids. Ph.D. thesis, University of Aberdeen, Aberdeen, UK.

Dawson, L.H.J. (1998b) The physiological effects of salmon lice (*Lepeophtheirus salmonis*) infections on returning post-smolt sea trout (*Salmo trutta* L.) in western Ireland. *ICES J. Mar. Sci.* **55:** 193–200.

Dawson, L.H.J., Pike, A.W., Houlihan, D.F. and McVicar, A.H. (1997) Comparison of the susceptibility of sea trout (*Salmo trutta* L.) and Atlantic salmon (*Salmo salar* L.) to sea lice (*Lepeophtheirus salmonis* (Krøyer, 1837)) infection. *ICES J. Mar. Sci.* **54:** 1129–1139.

Einszporn, T. (1965) Nutrition of *Ergasilus sieboldi* Nordmann. I Histological structure of the alimentary canal. *Acta. Parasitol. Polonica* **13:** 71–80.

Fast, M.D., Ross, N.W., Mustafa, A., Sims, D.E., Johnson, S.C., Conboy, G.A., Speare, D.J., Johnson, G. and Burka, J.F. (2002) Susceptibility of rainbow trout *Oncorhynchus mykiss*, Atlantic salmon *Salmo salar* and coho salmon *Oncorhynchus kisutch* to experimental infection with sea lice *Lepeophtheirus salmonis*. *Dis. Aquat. Org.* **52:** 57–68.

Fast, M.D., Burka, J.F., Johnson, S.C. and Ross, N.W. (2003) Enzymes released from *Lepeophtheirus salmonis* in the presence of salmonid mucus. *J. Parasitol.* **89:** 7–13.

Ferreira, B.R. and Silva, J.S. (1998) Saliva of *Rhipicephalus sanguineus* tick impairs T cell proliferation and IFN-β-induced macrophage microbicidal activity. *Vet. Immunol. Immunopathol.* **64:** 279–293.

Ferreira, B.R. and Silva, J.S. (2001) Successive tick infestations selectively promotes a Th_2 cytokine profile in mice. *Immunology* **96:** 434–439.

Finstad, B., Bjorn, P.A., Grimmes, A. and Hvidsten, N.A. (2000) Laboratory and field investigations of salmon lice *Lepeophtheirus salmonis* (Krøyer) infestation on Atlantic salmon (*Salmo salar* L.) post-smolts. *Aquacult. Res.* **31:** 795–803.

Firth, K.J., Johnson, S.C. and Ross, N.W. (2000) Characterisation of proteases in the skin mucus of Atlantic salmon (*Salmo salar*) infected with the salmon louse (*Lepeophtheirus salmonis*) and in *L. salmonis* whole-body homogenates. *J. Parasitol.* **86:** 1199–1205.

Fuchsberger, N., Kita, M., Hajnicka, V., Imanishi, J., Labuda, M. and Nuttall, P.A. (1995) Ixodid tick salivary gland extracts inhibit production of lipopolysaccharide-induced mRNA of several different human cytokines. *Exp. Appl. Acarol.* **19:** 671–676.

Gaines, P.J., Sampson, C.M., Rushlow, K.E. and Stiegler, G.L. (1999) Cloning of a family of serine protease genes from the cat flea *Ctenocephalides felis*. *Insect Mol. Biol.* **8:** 11–22.

Gillespie, R.D., Mbow, M.L. and Titus, R.G. (2000) The immunomodulatory factors of blood feeding arthropod saliva. *Parasit. Immunol.* **22:** 319–331.

Gonzalez-Alanis, P., Wright, G.M., Johnson, S.C. and Burka, J.F. (2001) Frontal filament morphogenesis in the salmon louse, *Lepeophtheirus salmonis*. *J. Parasitol.* **87:** 561–574.

Grayson, T.H., Jenkins, P.G., Wrathmell, A.B. and Harris, J.E. (1991) Serum responses to the salmon louse, *Lepeophtheirus salmonis* (Krøyer,1838), in naturally infected salmonids and immunised rainbow trout, *Oncorhynchus mykiss* (Walbaum), and rabbits. *Fish Shell. Immunol.* **1:** 141–155.

Grayson T.H., John, R.J., Wadsworth, S., Greaves, K., Cox, D., Roper, J., Wrathmell, A.B., Gilpin, M.L. and Harris, J.E. (1995) Immunisation of Atlantic salmon against the salmon louse: identification of antigens and effects on louse fecundity. *J. Fish Biol.* **47:** 85–94.

Grimnes, A. and Jakobsen, P.J. (1996) The physiological effects of salmon lice infection on post-smolt of Atlantic salmon. *J. Fish Biol.* **48:** 1179–1194.

Gwakiska, P., Yoshihara, K., To, T.L., Gotoh, H., Amano, F. and Momotani, E. (2001) Salivary gland extract of *Rhipicephalus appendiculatus* ticks inhibits the in vitro transcription and secretion of cytokines and production of nitric oxide by LPS-stimulated JA-4 cells. *Vet. Parasitol.* **99:** 53–61.

Hajnicka, V., Kocakova, P., Slavikova, M., Slovak, M., Gasperik, J., Fuchsberger, N. and Nuttall, P.A. (2001) Anti-interleukin-8 activity of tick salivary gland extract. *Parasite Immunol.* **23:** 483–489.

Haji Hamid, H.L., Bron, J.E., Shin, A.P. and Sommerville, C. (1998) The occurrence of blood feeding in *Lepeophtheirus salmonis*. Fourth International Crustacean Congress, Amsterdam, The Netherlands, July 20–24, 1998, p. 191.

Hall, L.R. and Titus, R.G. (1995) Sand fly vector saliva selectively modulates macrophage functions that inhibit killing of *Leishmania major* and nitric oxide production. *J. Immunol.* **155:** 3501–3506.

Hansen, U. and Peters, W. (1998) Structure and permeability of the peritrophic membranes of some small crustaceans. *Zool. Anz.* **236:** 103–108.

Heuch, P.A. (1995) Experimental evidence for aggregation of salmon louse copepodids (*Lepeophtheirus salmonis*) in step salinity gradients. *J. Mar. Biol. Assoc. UK* **75:** 927–939.

Heuch, P.A. and Karlsen, H.E. (1997) Detection of infrasonic water oscillations by copepodids of *Lepeophtherius salmonis* (Copepoda: Caligidae). *J. Plankton Res.* **19:** 735–747.

Heuch, P.A., Parsons, A. and Boxaspen, K. (1995) Diel vertical migration: A possible host-finding mechanism in salmon louse (*Lepeophtheirus salmonis*) copepodids. *Can. J. Fish. Aquat. Sci.* **52:** 681–689.

Ho, J-S. (2000) The major problem of cage aquaculture in Asia relating to sea lice. In: Liao, I.C. and Lin, C.K. (eds) *Proceedings of the First International Symposium on Cage Aquaculture in Asia,* pp. 13–19. Manila: Asian Fisheries Society, and Bangkok: World Aquaculture Society.

Hotez, P.J. and Cerami, A. (1983) Secretion of a proteolytic anticoagulant by *Anclystoma* hookworms. *J. Exp. Med.* **157:** 1594–1603.

Huys, R. and Boxshall G.A. (1991) *Copepod Evolution*. The Ray Society. Unwin Brothers Ltd., the Gresham Press, Old Woking, Surrey, UK.

Ingvarsdóttir, A., Birkett, M.S., Duce, I., Genna, R.L., Mordue, W., Pickett, J.A., Wadhams, L.J. and Mordue, A.J. (2002) Semiochemical strategies for sea louse control: host location cues. *Pest Manag. Sci.* **58:** 537–545.

Iwama, G.K., Vijayan, M.M., Forsyth, R.B. and Ackerman, P.A. (1999) Heat shock proteins and physiological stress in fish. *Am. Zool.* **39:** 901–909.

Jacobsen, J.A. and Gaard, E. (1997) Open ocean infection by salmon lice (*Lepeophtheirus salmonis*): comparison of wild and escaped farmed Atlantic salmon. *ICES J. Mar. Sci.* **42:** 843–849.

Jenkins, P.G., Grayson, T.H., Hone, J.V., Wrathmell, A.B. and Harris, J.E. (1993) The extraction and analysis of potential candidate vaccine antigens from the salmon louse *Lepeophtheirus salmonis* (Krøyer 1837). In: Boxshall, G.A. and Defaye, D. (eds) *Pathogens of Wild and Farmed Fish: Sea Lice,* pp. 311–322. Ellis Horwood, Chichester, UK.

Johnson, S.C. (1991) The biology of *Lepeophtheirus salmonis* (Krøyer, 1837) (Copepoda: Caligidae). PhD. thesis, Department of Biology, Simon Fraser University, Vancouver, BC, Canada.

Johnson, S.C. (1993) A comparison of development and growth rates of Lepophtheirus salmonis (Copepoda: Caligidae) on naïve Atlantic (*Salmo salar*) and Chinook (*Oncorhynchus tshawytscha*) salmon. In: Boxshall, G.A. and Defaye, D. (eds) *Pathogens of Wild and Farmed Fish: Sea Lice,* pp. 68–80. Ellis Horwood, Chichester, UK.

Johnson, S.C. and Albright, L.J. (1991) The developmental stages of *Lepeophtheirus salmonis* (Krøyer, 1837) (Copepoda: Caligidae). *Can. J. Zool.* **69:** 929–950.

Johnson, S.C. and Albright, L.J. (1992a) Comparative susceptibility and histopathology of the host response of naive Atlantic, chinook, and coho salmon to experimental infection with *Lepeophtheirus salmonis* (Copepoda: Caligidae). *Dis. Aquat. Org.* **14:** 179–193.

Johnson, S.C. and Albright, L.J. (1992b) The effects of cortisol implants on the susceptibility and histopathology of the responses of naive coho salmon *Oncorhynchus kisutch* to experimental infection with *Lepeophtheirus salmonis* (Copepoda: Caligidae). *Dis. Aquat. Org.* **14:** 195–205.

Johnson, S.C., Blaylock, R.B., Elphick, J. and Hyatt, K. (1996) Disease caused by the salmon louse *Lepeophtheirus salmonis* (Copepoda: Caligidae) in wild sockeye salmon (*Oncorhynchus nerka*) stocks of Alberni Inlet, British Columbia. *Can. J. Fish. Aquat. Sci.* **53:** 2888–2897.

Johnson, S.C., Ewart, K.V., Osborne, J.A., McIntosh, S.E., Stratton L. and Ross, N.W. (2000) Expressed sequence tags (ESTs) of the salmon louse, *Lepeophtheirus salmonis.* **Bull. Aqua. Soc. Can. 100-1:** 8–12.

Johnson, S.C., Ewart, K.V., Osborne, J.A., Delage, D., Ross N.W. and Murray, H.M. (2002) Molecular cloning of trypsin cDNAs and trypsin gene expression in the salmon louse *Lepeophtheirus salmonis* (Copepoda: Caligidae). *Parasitol. Res.* **88:** 789–796.

Johnson, S.C., Treasurer, J.W., Bravo, S., Nagasawa, K. and Kabata, Z. (in press). A review of the impacts of parasitic copepods on marine aquaculture. *Zool. Stud.*

Jones, M.W., Sommerville, C. and Bron, J. (1990) The histopathology associated with the juvenile stages of *Lepeophtheirus salmonis* on the Atlantic salmon, *Salmo salar* L. *J. Fish Dis.* 13: 303–310.

Jónsdóttir, H., Bron, J.E., Wootten, R. and Turnbull, J.F. (1992) The histopathology associated with the pre-adult and adult stages of *Lepeophtheirus salmonis* on the Atlantic salmon, *Salmo salar* L. *J. Fish Dis.* **15:** 521–527.

Kabata, Z. (1974) Mouth and mode of feeding of Caligidae (Copepoda), parasites of fishes, as determined by light and scanning electron microscopy. *J. Fish Res. Board Can.* **31:** 1583–1588.

Kabata, Z. (1979) *Parasitic Copepoda of British Fishes*. The Ray Society, London.

Kabata, Z. and Hewitt, G.C. (1971) Locomotory mechanisms in Caligidae (Crustacea: Copepoda). *J. Fish. Res. Board Can.* **28:** 1143–1151.

Kerlin, R.L. and Hughes, S. (1992) Enzymes in saliva from four parasitic arthropods. *Med. Vet. Entomol.* **6:** 121–126.

Kopecky, J., Kuthejlova, M. and Pechova, J. (1999) Salivary gland extract from *Ixodes ricinus* ticks inhibits production of interferon-γ by the upregulation of interleukin-10. *Parasite Immunol.* **21:** 351–356.

Kularatne, M., Subasinghe, R.P. and Shariff, M. (1994) Investigations on the lack of acquired immunity by the Javanese carp, *Puntius goniontus* (Bleeker), against the crustacean parasite, *Lernaea minuta* (Kuang). *Fish Shell. Immunol.* **4:** 107–114.

Lehane, M.J. (1997) Peritrophic matrix structure and function. *Ann. Rev. Entomol.* **42:** 525–550.

Lin, C-L., Ho, J-S. and Chen, S-N. (1996a) Developmental stages of *Caligus epidemicus* Hewitt, a copepod parasite of tilapia cultured in brackish water. *J. Natl Hist.* **30:** 661–684.

Lin, C-L., Ho, J-S. and Chen S-N. (1996b) Two species of Caligidae (Copepoda) parasitic on cultured rabbit fish (*Siganus fuscescens*) in Taiwan. *Fish Pathol.* **31:** 129–139.

Lin, C-L., Ho, J-S. and Chen, S-N. (1997) Development of *Caligus multispinosus* Shen, a caligid copepod parasitic on black sea bream (*Acanthopagrus schlegeli*) cultured in Taiwan. *J. Natl His.* **31:** 1483–1500.

MacKinnon, B.M. (1993) Host response of Atlantic salmon (*Salmo salar*) to infection by sea lice (*Caligus elongatus*). *Can. J. Fish. Aquat. Sci.* **50:** 789–792.

MacKenzie, K., Longshaw, M., Begg, G.S. and McVicar, A.H. (1998) Sea lice (Copepoda: Caligidae) on wild sea trout (*Salmo salar* L.) in Scotland. *ICES J. Mar. Sci.* **55:** 151–162.

Mo, T.A. and Heuch, P.A. (1998) Occurrence of *Lepeophtheirus salmonis* (Copepoda: Caligidae) on sea trout (*Salmo trutta*) in the inner Oslo Fjord, south-eastern Norway. ICES. *J. Mar. Sci.* **55:** 176–180.

Moiré, N., Bigot, Y., Periquet, G. and Boulard, C. (1994) Sequencing and gene expression of hypodermins A, B, C in larval stages of *Hypoderma lineatum*. *Mol. Biochem. Parasitol.* **66:** 233–240.

Monterosso, B. (1930) Struttura e funzione dell'intestino media di '*Peroderma cylindricum*' . *Atti della Reale Accademice (Nazionale) dei Lincei, Rendiconti* **(6)11:** 433–437.

Mustafa, A., MacWilliams, C., Fernandez, N., Matchett, K., Conboy, G.A. and Burka, J.F. (2000a) Effects of sea lice (*Lepeophtheirus salmonis* Kröyer, 1837) infestation on macrophage functions in Atlantic salmon (*Salmo salar* L.). *Fish Shell. Immunol.* **10:** 47–59.

Mustafa, A., Speare, D.J., Daley, J., Conboy, G.A. and Burka, J.F. (2000b) Enhanced susceptibility of seawater cultured rainbow trout, *Oncorhynchus mykiss* (Walbaum), to the microsporidian *Loma salmonae* during a primary infection with the sea louse, *Lepeophtheirus salmonis*. *J. Fish Dis.* **23:** 337–341.

Mustafa A., Rankaduwa, W. and Campbell, P. (2001a) Estimating the cost of sea lice to salmon aquaculture in eastern Canada. *Can. J. Vet.* **42:** 54–56.

Mustafa A., Conboy, G.A. and Burka, J.F. (2001b) Life-span and reproductive capacity of sea lice, *Lepeophtheirus salmonis*, under laboratory conditions. *Bull. Aquacult. Soc. Can.* **2000 (4):** 113–114.

Nagasawa, K. (2001) Annual changes in the populations size of the salmon louse *Lepeophtheirus salmonis* (Copepoda: Caligidae) on high-seas Pacific salmon (*Oncorhynchus* spp.), and relationship to host abundance. *Hydrobiology* **453/454:** 411–416.

Nolan, D.T., Reilly, P. and Wendelarr Bonga, S.E. (1999) Infection with low numbers of the sea louse *Lepeoptheirus salmonis* (Krøyer) induces stress-related effects in post-smolt Atlantic salmon (*Salmo salar* L.). *Can. J. Fish. Aquat. Sci.* **56:** 947–959.

Nolan, D.T., Ruane, N.M., van der Heijden, Y., Quabius, E.S., Costelloe, J. and Wendelaar Bonga, S.E. (2000) Juvenile *Lepeophtheirus salmonis* (Krøyer) affect the skin and gills of rainbow trout *Oncorhynchus mykiss* (Walbaum) and the host response to a handling procedure. *Aquacult. Res.* **31:** 823–833.

Nylund, A., Økland, S. and Bjørknes, B. (1992) Anatomy and ultrastructure of the alimentary canal in *Lepeophtheirus salmonis* (Copepoda: Siphonostomatioda). *J. Crust. Biol.* **12:** 423–437.

Ogawa, K. (1992) *Caligus longipedis* infection of cultured striped jack, *Pseudocaranx dentex* (Teleostei: Carangidae) in Japan. *Fish Pathol.* **27:** 197–205.

Parker, R.R., and Margolis, L. (1964) A new species of parasitic copepod, *Caligus clemensi* sp. nov. (Caligoida: Caligidae), from pelagic fishes in the coastal waters of British Columbia. *J. Fish. Res. Bd. Can.* **21:** 873–889.

Paperna, I. and Zwerner, D.E. (1982) Host–parasite relationship of *Ergasilus labracis* Krøyer (Cyclopidea, Ergasilidae) and the striped bass, *Morone saxatilis* (Walbaum) from the lower Chesapeake Bay. *Ann. Parasitol. (Paris)* **57:** 393–405.

Piasecki, W. (1996) The developmental stages of *Caligus elongatus* von Nordmann, 1832 (Copepoda, Caligidae). *Can. J. Zool.* **74:** 1459–1478.

Piasecki, W. and MacKinnon, B.M. (1993) Changes in structure of the frontal filament in sequential developmental stages of *Caligus elongatus* von Nordmann, 1832 (Crustacea, Copepoda, Siphonostomatoida). *Can. J. Zool.* **71:** 889–895.

Piasecki W. and MacKinnon, B.M. (1995) Life cycle of a sea louse, *Caligus elongatus* von Nordmann, 1832 (Copepoda, Siphonostomatoida, Caligidae). *Can. J. Zool.* **73:** 74–82.

Pike, A.W. (1989) Sea lice – major pathogens of farmed Atlantic salmon. *Parasitol. Today* **5:** 291–297.

Pike, A.W., MacKenzie, K. and Roland, A. (1993) Ultrastructure of the frontal filament in chalimus larvae of *Caligus elongatus* and *Lepeophtheirus salmonis* from Atlantic salmon, *Salmo salar*. In: Boxshall, G.A. and Defaye, D. (eds) *Pathogens of Wild and Farmed Fish: Sea Lice,* pp. 99–113. Ellis Horwood, Chichester, UK.

Pike, A.W. and Wadsworth, S.L. (1999) Sea lice on salmonids: Their biology and control. *Adv. Parasitol.* **44:** 233–337.

Rae, G.H. (2002) Sea louse control in Scotland, past and present. *Pest Manag. Sci.* **58:** 515–520.

Raynard, R.S., Bricknell, I.R., Billingsley, P.F., Nisbet, A.J., Vigneau, A. and Sommerville, C. (2002) Development of vaccines against sea lice. *Pest Manag. Sci.* **58:** 569–575.

Reilly, P. and Mulcahy, M.F. (1993) Humoral antibody response in Atlantic salmon (*Salmo salar* L.) immunised with extracts derived from the ectoparasitic caligid copepods, *Caligus elongatus* (Nordmann, 1832) and *Lepeophtheirus salmonis* (Kroyer, 1838). *Fish Shell. Immunol.* **3:** 59–70.

Ribeiro, J.M.C. and Valenzuela, J.G. (1999) Purification and cloning of the salivary peroxidase/catechol oxidase of the mosquito *Anopheles albimanus*. *J. Exp. Biol.* **202:** 809–816.

Roper, J., Grayson, T.H., Jenkins, P.G., Hone, J.V., Wrathmell, A.B., Russell, P.M. and Harris, J.E. (1995) The immunocytochemical localisation of potential candidate vaccine antigens from the salmon louse *Lepeophtherius salmonis* (Krøyer 1837). *Aquaculture* **132:** 221–232.

Rosenfeld, A. and Vanderberg, J.P. (1998) Identification of electrophoretically separated proteases from midgut and hemolymph of adult *Anopheles stephansi* mosquitoes. *J. Parasitol.* **84:** 361–365.

Ross, N.W., Firth, K.J., Wang, A., Burka, J.F. and Johnson, S.C. (2000) Changes in hydrolytic enzyme activities of naïve Atlantic salmon (*Salmo salar*) skin mucus due to infection with the salmon louse (*Lepeophtheirus salmonis*) and cortisol implantation. *Dis. Aquat. Org.* **41:** 43–51.

Roubal, F.R. (1994) Histopathology caused by *Caligus epidemicus* Hewitt (Copepoda: Caligidae) on captive *Acanthopagrus australis* (Guenther) (Pisces: Sparidae). *J. Fish Dis.* **17:** 631–640.

Ruane, N.M., Nolan, D.T., Rotllant, J., Costelloe, J. and Wendelaar Bonga, S.E. (2000) Experimental exposure of the rainbow trout *Oncorhynchus mykiss* (Walbaum) to the infective stages of the sea louse *Lepeophtheirus salmonis* (Krøyer) influences the physiological response to an acute stresser. *Fish Shellfish Immunol.* **10:** 451–463.

Schram, T.A. (1993) Supplemental descriptions of the developmental stages of *Lepeophtheirus salmonis* (Krøyer, 1837) (Copepoda: Caligidae). In: Boxshall, G.A. and Defaye, D. (eds) *Pathogens of Wild and Farmed Fish: Sea Lice,* pp. 30–50. Ellis Horwood, Chichester, UK.

Schram, T.A., Knutsen, J.A., Heuch, P.A. and Mo, T.A. (1998) Seasonal occurrence of *Lepeophtheirus salmonis* and *Caligus elongatus* (Copepoda: Caligidae) on sea trout (*Salmo trutta*) off Norway. *ICES J. Mar. Sci.* **55:** 163–175.

Schoeler, G.B. and Wikel, S.K. (2001) Modulation of host immunity by haematophagous arthropods. *A. Trop. Med. Parasitol.* **95:** 755–771.

Schoeler, G.B., Manweiler, S.A. and Wikel, S.K. (1999) *Ixodes scapularis*: effects of repeated infestations with pathogen-free nymphs on macrophage and T lymphocyte cytokine responses of BALB/c and C3H/HeN mice. *Exp. Parasitol.* **92:** 239–248.

Shao, L., Devenport, M. and Marcelo, J-L. (2001) The peritrophic matrix of hematophagous insects. *Arch. Insect Biochem. Physiol.* **47:** 119–125.

Shariff, M. (1981) The histopathology of the eye of big head carp, *Aristichthys noblis* (sic.) (Richardson), infested with *Lernaea piscinae* Harding, 1950. *J. Fish Dis.* **4:** 161–168.

Shields, R.J. and Goode, R.P. (1978) Host rejection of *Lernaea cyprinacea* L. (Copepoda). *Crustaceana* **35:** 301–307.

Sinnott, R. (1999) Cost of sea lice to Scottish salmon farmers. *Trouw Outlook* **11:** 8–10.

Spates, G. (1979) Fecundity of the stable fly: effect of soybean trypsin inhibitor and phospholipase A inhibitor on the fecundity. *Ann. Entomol. Soc. Am.* **72:** 845–849.

Spates, G. and Harris, R.L. (1984) Reduction of fecundity, egg hatch and survival in adult horn flies fed protease inhibitors. *Southwest. Entomol.* **9:** 399–403.

Terra, W.R. (2001) The origin and functions of the insect peritrophic membrane and peritrophic gel. *Arch. Insect Biochem. Physiol.* **47:** 47–61.

Tellam, R.L., Kemp, D., Riding, G., Briscoe, S., Smith, D., Sharp, P., Irving, D. and Willadsen, P. (2002) Reduced oviposition of *Boophilus microplus* feeding on sheep vaccinated with vitellin. *Vet. Parasitol.* **103:** 141–156.

Thoney, D. and Burreson, E.M. (1988) Lack of a specific humoral antibody response in *Leiostomus xanthurus* (Pisces; Sciaenidae) to parasitic copepods and monogeans. *J. Parasitol.* **74:** 1919–1924.

Tingley, G.A., Ives, M.J. and Russell, I.C. (1997) The occurrence of lice on sea trout (*Salmo trutta* L.) captured in the sea off the East Anglian coast of England. *ICES J. Mar. Sci.* **54:** 1120–1128.

Tully, O. and Nolan, D.T. (2002) A review of population biology and host-parasite interactions of the sea louse *Lepeophtheirus salmonis* (Copepoda: Caligidae). *Parasitology* **124:** S165–S182.

Tully, O., Poole, W.R., Whelan, K.F. and Merigoux, S. (1993) Parameters and possible causes of epizootics of *Lepeophtheirus salmonis* (Krøyer) infesting sea trout (*Salmo trutta* L.) off the west coast of Ireland. In: Boxshall, G.A. and Defaye, D. (eds) *Pathogens of Wild and Farmed Fish: Sea Lice,* pp. 202–218. Ellis Horwood, Chichester, UK.

Valenzuela, J.C., Francischetti, I.M.B., Pham, V.M., Garfield, M.K., Mather, T.N. and Ribeiro, M.C. (2002) Exploring the sialome of the tick *Ixodes scapularis*. *J. Exp. Biol.* **205:** 2843–2864.

Wendelaar Bonga, S.E. (1997) The stress response in fish. *Physiol. Rev.* **77:** 591–625.

Wijffels, G., Hughes, S., Gough, J., Allen, J., Don, A., Marshall, K., Kay, B. and Kemp, D. (1999) Perotrophins of adult dipteran ectoparasites and their evaluation as vaccine antigens. *Int. J. Parasitol.* **29:** 1363–1377.

Willadsen, P. (1997) Novel vaccines for ectoparasites. *Vet. Parasitol.* **71:** 209–222.

Willadsen, P., Cobon, G.S., Hungerford, J. and Smith, D. (1995) Role of vaccination in current and future strategies for control. In: Rodríguez C.S. and Fragoso S.H. (eds) *Memorias III Seninario Internacional de Parasitologia Animal,* Acapulco, Mexico, 1995: 88-100.

Woo, P.T.K. and Shariff, M. (1990) *Lernea cyprinacea* L. (Copeoda: Caligidae) in *Helostoma temmincki* Cuvier & Valenciennes: the dynamics of resistance in recovered and naïve fish. *J. Fish Dis.* **13:** 485–493.

Wu, Z., Jinpei, P. and Qiwei, Q. (1997) The lice disease in cultured banded grouper *Epinephelus awoara*: Pathology. *Acta Hydrobiol. Sinica* **21:** 207–212. (In Chinese with English abstract)

Yoshikoshi, K. and Kô, Y. (1988) Structure and function of the peritrophic membranes of copepods. *Bull. Japan. Soc. Sci. Fish.* **54:** 1007–1082.

8

Interactive associations between fish hosts and monogeneans

Kurt Buchmann, Thomas Lindenstrøm and Jose Bresciani

1 Introduction

1.1 *Monogeneans*

Helminth parasites of fishes comprise several thousands of described species belonging to well-characterized taxonomic entities such as nematodes, acanthocephalans, trematodes (digeneans and aspidogastreans), cestodes and monogeneans. Thus, it has been estimated that a total of more than 30 000 helminth species parasitizing fish have been described (Williams and Jones, 1994). Trematodes, cestodes and monogeneans have traditionally been grouped together as platyhelminths based on their external morphology, the conspicuous slightly dorsoventrally flattened body. Among these flatworms the monogeneans depart due to their outstanding morphology and ecology. These worms are mostly ectoparasites with preferences for fins, body skin, gills, gill chamber, buccal cavity, cornea or nostrils.

Diversity

Although a number of monogeneans parasitize amphibians, reptiles and mammals the group is principally known to be ectoparasites of fish. Monogeneans are considered to be among the most host-specific parasites existing and most fish species are parasitized by one or more specific species. This knowledge can be used to estimate the total number of monogenean species. Thus, the number of known fish species exceeds 24 000 and a qualified guess is that each of those fishes are parasitized by at least one monogenean species. It can therefore be estimated that the total number of existing species is far higher than 24 000. However, literature is at present only acquainted with less than 4000 descriptions of monogenean species (Whittington, 1998). Gyrodactylids are a special entity within monogeneans and have a very high species richness with 402 species recorded from 19 fish orders and are useful for studying parasite–host interactions (Bakke *et al.*, 2002). The viviparity of most representatives

Host–Parasite Interactions, edited by G. F. Wiegertjes & G. Flik.

of this group makes the worms excellent for specificity investigations. Thus, micropopulations based on one founder parasite can be produced on susceptible hosts. Further, it is possible to perform controlled infections of individual fish simply by transferring isolated worms to a fish.

External anatomy

These ectoparasitic platyhelminths possess a posterior attachment organ which is designated the opisthaptor. It is equipped with varying types and numbers of accessory hooks or anchors (hamuli), hooklets and suckers or clamps, the combination of which has been used for further subdivision (Chisholm *et al.*, 1998) of these helminths into two main groups, the monopisthocotyleans (*Figures 1, 2* and *3*) and the polyopisthocotyleans (*Figures 4* and *5*). Another classification can be applied and in this case the two groups are to a certain degree and with a few exceptions identical with Polyonchoinea, Oligonchoinea and Polystomatoinea respectively. The haptor is considered responsible for major pathological impact on host structures especially in monopisthocotyleans.

Gut and feeding

Representatives of the Monopisthocotylea group primarily feed by browsing host epithelia (skin, fins and gills) whereas the Polyopisthoctylea group comprise blood-feeders. The mouths of the worms are located anteriorly or antero-ventrally. The ingested host material passes through the more or less muscular pharynx to the intestinal caeca. No anus is found and non-digested material is regurgitated. Thus,

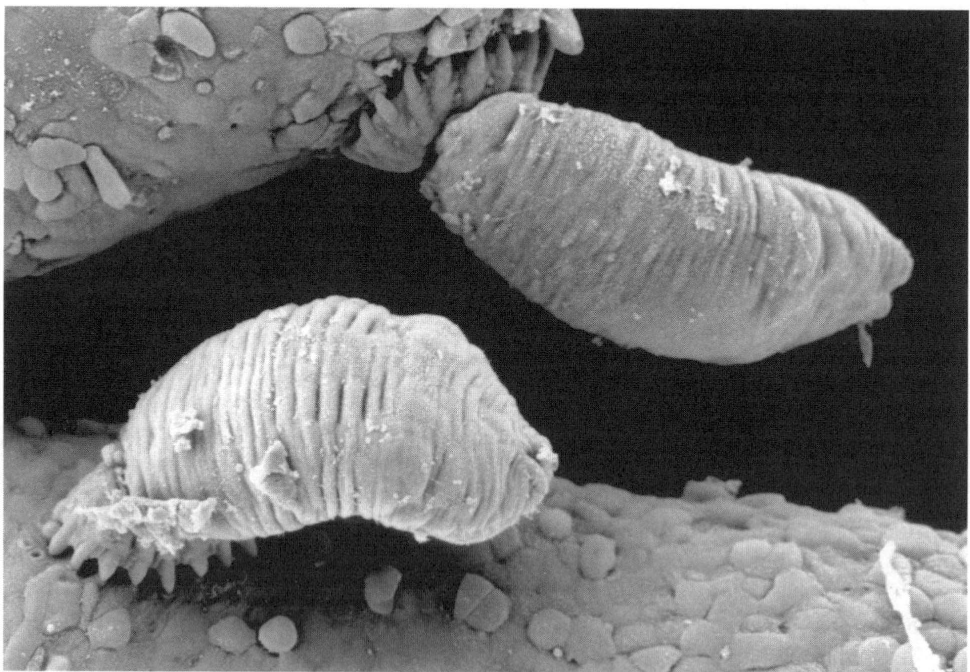

Figure 1 Gyrodactylus derjavini, *a monopisthocotylean viviparous monogenean parasitizing fins (and rarely gills) of rainbow trout. The opisthaptor is equipped with 16 marginal hooklets inserted into the epidermis.*

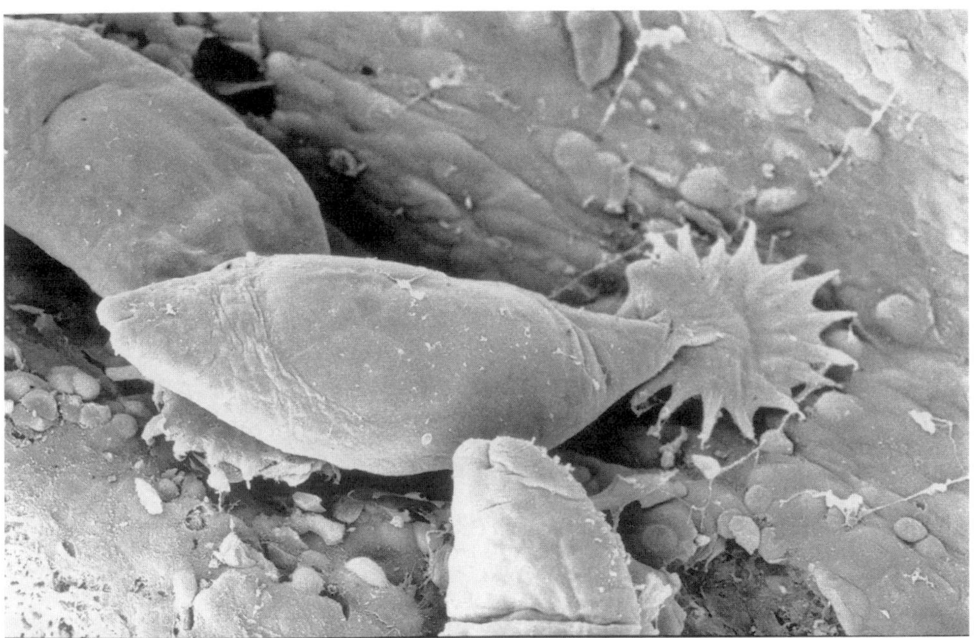

Figure 2 Gyrodactylus salaris, *a monopisthocotylean viviparous monogenean parasitizing fins (and rarely gills) of Atlantic salmon.*

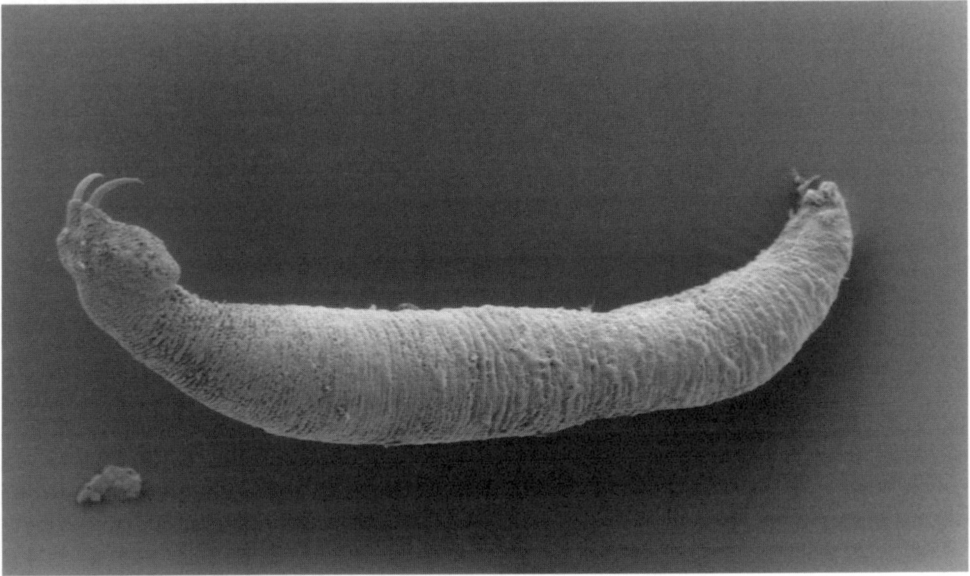

Figure 3 Pseudodactylogyrus bini, *a monopisthocotylean oviparous monogenean parasitizing gills of the European eel. The opisthaptor has two large hamuli and 14 hooklets which penetrate the gill tissue.*

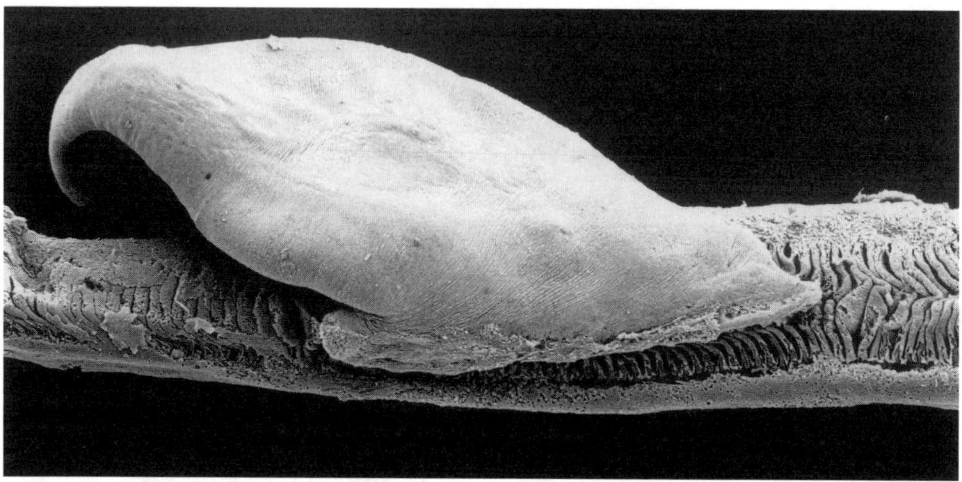

Figure 4 Axine belones, *a blood-feeding polyopisthocotylean from the gills of garfish. The opisthaptor is equipped with numerous clamps securing attachment to the host gill filaments.*

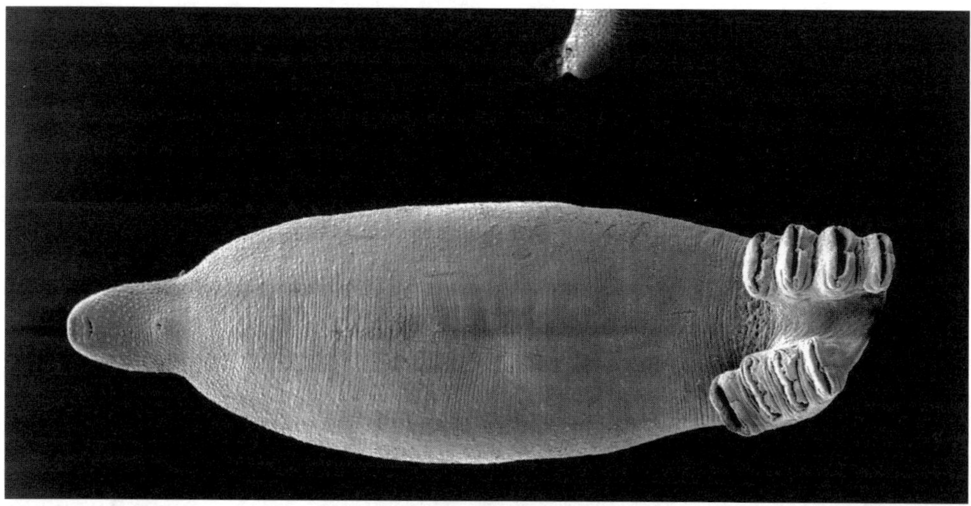

Figure 5 Discocotyle sagittata, *a blood-feeding polyopisthocotylean from the gills of rainbow trout. The opisthaptor is attaching to the host gills with eight clamps.*

digestive enzymes associated with this material will probably get in contact with host epithelia. In addition, digestion is intracellular in gut cells of the monogenean gastrodermis but enzymes may be secreted to initiate external digestion of host surfaces.

Nervous system

Monogeneans have a relatively well-developed orthogonal nervous system. This means that from the two cerebral ganglia with a circum-oral nerve ring longitudinal paired nerve trunks are leading anteriorly and posteriorly. These are interconnected by transverse commissures and connected with the somatic musculature and sense organs (mechano- and chemoreceptors) which are dipersed over the tegument (Lyons, 1969; Watson and Rohde, 1994). Thus, monogeneans have the capability to react to external stimuli with fast neuromusculatory actions. It has been indicated that a range of neurotransmitters is responsible for the co-ordinated neuromuscular system (Halton *et al.*, 1993).

Glands

Several gland types have been recognized in monogeneans. Both in connection with the anterior and cephalic part (Kearn and Evans-Gowing, 1998; Kritsky, 1978), the digestive system (Buchmann, 1997a) and the opisthaptor (Whittington and Cribb, 2001). Secretions from these are liable to reach and affect the host epidermis and thereby stimulate reactions in the host.

Excretion and osmoregulation

These functions are based on protonephridia (flame cells) which are interconnected with longitudinal collecting ducts. These empty in an antero-lateral to antero-dorsal excretory pore. The arrangement of protonephridia has mostly been used for taxonomical purposes (Malmberg, 1998). However, the physiological and immunological importance of the excretory products should be considered. To what extent the excreted material affect the host has been poorly studied but should have priority in future studies.

Parasite surface

The outer surface of the monogenean is called the tegument. It is a syncytium, richly equipped with mitochondria, extending from nucleated epidermal cells located below a basement lamina and the longitudinal, circular and diagonal muscle layer. These muscles create the basis for extensive dynamic body changes (stretching, bending and contraction), which allow the worm to move in a leech-like manner.

Life cycle and reproduction

The life cycle of monogeneans is direct without involvement of intermediate hosts. They are hermaphroditic but conduct copulation by the use of cirrus (penis) and in some cases vagina. These structures are characteristically developed which allows their use for taxonomical purposes. Pheromones as an active ingredient in mating, which is known in other systems e.g. among digeneans (Armstrong, 1965; Fried, 1986), is possible but less well studied.

Most monogenean species are oviparous, producing eggs which release infective larvae (oncomiracidia) upon hatching. Following a relatively short free-living phase the larvae settle on the host surface and develop into adult stages depending on the prevailing abiotic (temperature, oxygen, pH, salinity, water currents, light) and biotic (host structure, physiology and responsiveness) conditions. If the host is appropriate and the environmental conditions satisfactory infections can build up to devastating levels within a relatively short time. Other monogeneans are viviparous, giving birth to live and almost fully developed offspring. Most of the members of the family Gyrodactylidae are viviparous. In one adult worm it is possible to discern three generations at a time. In the uterus of the parent worm a fully developed daughter is present and in this a developing embryon can be observed. Following birth the newborn (which is not equipped with a penis) is able to give birth to one daughter without copulation. Following the first birth the penis starts developing and the male copulatory organ is fully matured after the second birth. At this stage the worm can take part in copulation. Such a reproductive strategy allows a rapid population increase on the host surface (Cable and Harris, 2002). Monogeneans are equipped with well-developed testes and sperm structure has been studied in some detail. Temperature is the main abiotic factor affecting monogenean populations. Thus, reproduction of monogeneans, both oviparous and viviparous forms, is highly dependent on temperature. For both monopisthocotylean and polyopisthocotylean oviparous species it has been observed that egg production rate, embryonation and hatching of eggs and post-larval development are all events which will increase with temperature within certain limits (Buchmann, 1997a; Gannicott and Tinsley, 1997). Likewise, in viviparous gyrodactylids the reproductive rate is positively correlated with temperature until an upper limit (Andersen and Buchmann, 1998; Janssen and Bakke, 1991).

Taxonomy, evolution and diagnostics based on molecular studies

Sequencing of ribosomal RNA genes for both lsr (large subunit of ribosome) and ssr (small subunit of ribosome) has been the primary interest up until now. The internal transcribed spacers and intergenic spacers associated with these genes have been isolated and studied intensively (Cunningham, 1997). This has not only led to development of diagnostic tools such as PCR-RFLP but also clearly indicated that monopisthocotyleans and polyopisthocotyleans differ at this level (Olson and Littlewood, 2002). Further, genes encoding cytochrome oxidase subunit I have been sequenced. Studies on genes encoding factors of direct influence on parasite–host interactions, such as virulence genes, have not been conducted yet but should have priority in the future. In this context establishment of cDNA libraries from closely related pathogenic and non-pathogenic parasite species (e.g. *G. salaris* and *G. thymalli*) would be relevant.

Abiotic factors versus biotic factors and the monogeneans

The macro-environment occupied by the fish plays a major role in parasite–host interactions. Both the physiological, immunological and behavioural responses of the fish are influenced by parameters such as temperature, oxygen, light, water currents, salinity and pH. Likewise, the ectoparasite itself is directly exposed to and highly influenced by these factors (*Figure 6*). In addition, further abiotic elements such as metals (Al, Zn, Cu, Fe) have a strong impact on not only fish but also monogeneans (Larsen and Buchmann, 2003; Soleng *et al.*, 1999).

Variations within the monogenean group

As mentioned above not only feeding strategies but also internal and external anatomy, sperm structure and rDNA sequences differ significantly between monopisthocotyleans and polyopisthocotyleans. It is therefore evident that monogeneans comprise a vast amount of highly differing forms which makes it difficult to create a picture that covers all interactive associations between these parasites and their hosts. The following section will outline the main aspects of the question with special reference to specific parasite–host systems, which have been studied in some detail.

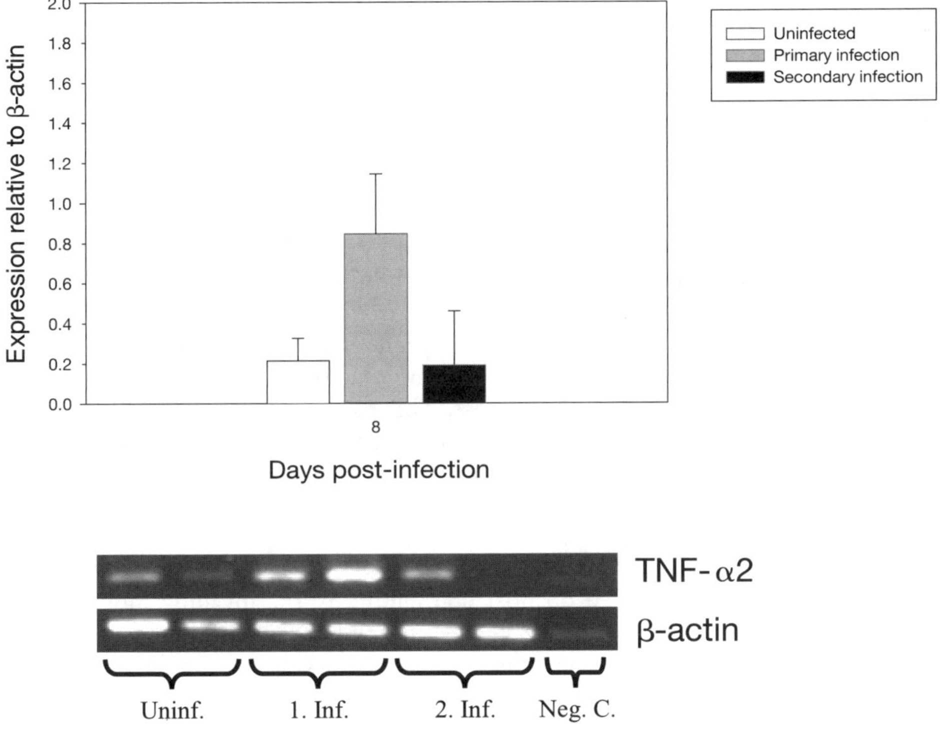

Figure 6 *Methodology to study activation of specific genes following a monogenean infection. Skin of trout was collected and RNA isolated. RT-PCR procedures on genes expressing TNF-alpha in rainbow trout skin during both primary and secondary* Gyrodactylus derjavini *infections were subsequently performed. TNF-alpha is shown as relative expression to the constitutively expressed house-keeping beta-actin gene.*

2 Monogenean interactions with the host

Interactions between host and parasite do not only include the mechanical, physical and chemical signals emitted by the parasite and the various physical and chemical factors emitted by the host. Also, the intricate reaction patterns elicited in the host following infection are part of this picture.

The microhabitats on fish occupied by monogeneans comprise the entire surface including fins, body, gill apparatus, buccal cavity, nostrils, cornea and opercula. The characteristics of these anatomical units will evidently limit the possibilities for certain parasites due to their specialized anatomy.

Microhabitats

One characteristic issue by monogeneans is their pronounced predilection for particular microhabitats on the fish. Thus, even within the same genus different species of these platyhelminths often select well-defined parts of the fish (Buchmann, 1997a; Simkhov *et al.*, 2000). It has been suggested that this is due to differences in water currents over gills whereby worms less suited to stronger current will seek sheltered compartments of the gill apparatus. Also, differential mortality has been suggested as an explanation. This means that some parts of the gills simply would create dangerous microhabitats for particular monogeneans leaving these areas empty of the parasite in question. Direct intra- and interspecific competition between the worms is also a possible explanation of the observed differential distribution of monogenean species. Inherent strategies to prevent hybridization (Rohde, 1977) or inherent strategies to prevent predation (Euzet and Combes, 1998) have also been suggested as explanations. Further, host reactions may play a part in the micro-ecology on the host. Thus, tissue reactions of cells in eel gills induced by *Pseudodactylogyrus bini* is associated with a marked decline of *P. anguillae* from the gills (Buchmann, 1988). A similar phenomenon was observed with *Dactylogyrus vastator* and *D. extensus* by Paperna (1964). Likewise, *Gyrodactylus derjavini* has a preference for pectoral fins of trout in the initial phase of infection. However, during the response phase the parasites aggregate on the corneal surface of the host. This part is considered as an immunoprivileged site due to lack of vascularization, mucous cells and mast cells (Buchmann and Bresciani, 1998; Sigh and Buchmann, 2000).

2.1 *Effects of monogeneans on the host*

The mechanical injuries induced on host epithelium by the invading monogeneans vary considerably between species. Heavy devastating burdens of some gyrodactylids on the fins and skin of teleosts (even in natural conditions) have been documented. In contrast light infections even by blood-feeders of delicate gill filaments do not always lead to a pathogenic state of the host. Action of hamuli, marginal hooklets and clamps on the opisthaptor may induce wounds on the fish surface. The mouth and feeding activities may play an even larger role among mechanical stimuli. However, these may not be the only signals from the parasite. Excretion, regurgitation of gut content and gland release are events which must affect the host epithelium to some extent.

Pathological changes of host epidermis have been studied in some detail for a number of monogeneans (*Table 1*).

2.2 *Fish host molecules detected by monogeneans*

Fishes excrete or release a range of different compounds. Although nitrogen-containing molecules such as ammonia and to a lesser extent urea are important players, release of host materials is not only via the kidney–urinary system. Ammonia is mainly excreted through the gills and additional molecules may be emitted via the skin. The entire surface of the fish is covered with a high-density layer of mucous cells discharging mucus with associated compounds. Mucus is composed of mucopolysaccharides and steroids, bile salts, aliphatic acids, nucleotides, amino acids and carbon dioxide are some of the various additional chemical substances released from fishes. N-compounds released comprise uric acid, purines, methylamine, taurine, imidazoles, creatine and creatinine. It is noteworthy that these molecules are part of a chemical communication system between fishes based on their well-developed chemoreceptors (Hara, 1993). It would be surprising if ectoparasites, such as monogeneans, were not able to detect at least some of these substances. Therefore, it is suggested that these factors act as chemo-attractants or at least influence parasite behaviour whereby

Table 1 *Examples of recorded pathological reactions in fish hosts due to monogenean infections*

Parasite	Host	Organ affected	Author
Dactylogyrus vastator	Carp	Gills	Wunder (1929) Paperna (1963)
Dactylogyrus anchoratus	Carp	Gills	Wunder (1937) Prost (1963)
Dactylogyrus extensus	Carp	Gills	Prost (1963) Buchmann *et al.* (1993)
Dacytylogyrus minutus	Carp	Gills	Buchmann *et al.* (1993)
Dactylogyrus macracanthus	Tench	Gills	Wilde (1935, 1937)
Dactylogyrus lamellatus	Grass carp	Gills	Molnar (1972)
Dactylogyrus corporalis	Fallfish	Gills	Putz and Hoffmann (1964)
Pseudodactylogyrus anguillae and P. bini	European eel	Gills	Buchmann (1988)
Pseudodactylogyrus anguillae and P. bini	Japanese eel	Gills	Chan and Wu (1984)
Cleidodiscus robustus	Blue-gill	Gills	Thune and Rogers (1981)
Linguadactyla molvae	Ling	Gills	Bychowsky (1957)
Callorhyncicola multitesticulatus	Elephant fish	Gills	Llewellyn and Simmons (1984)
Dermophthirius carcharhini	Galapagos shark	Skin	Rand *et al.* (1986)
Neobenedenia melleni	Several species	Skin	Jahn and Kuhn (1932)
Benedenia monticelli	Mugilid fish	Skin	Paperna *et al.* (1984)
Allomurraytrema	Sparids	Gills	Roubal (1995)
Gyrodactylus salaris	Atlantic salmon	Skin and fins	Malmberg (1993)
Gyrodactylus derjavini	Rainbow trout	Skin and fins	Buchmann and Bresciani (1997)
Gyrodactylus salmonis	Rainbow trout	Skin and fins	Cone and Odense (1984)
Gyrodactylus anguillae	European and Japanese eel	Gills	Ogawa and Egusa (1980)
Heterobothrium okamotoi	Tiger puffer	Branchial cavity	Ogawa (2002)
Neoheterobothrium hirame	Japanese flounder	Buccal cavity	Ogawa (2002)

they would play a role in parasite–host interactions. Host substances are also known to elicit hatching of monogenean eggs. Eggs from *Entobdella soleae* and *Acanthocotyle lobianchi* needed host extracts for hatching (Kearn and MacDonald, 1976). Urea was suggested as the causative factor. Likewise, Xia *et al.* (1996) found hatching-inducing factors in mandarin fish mucus and Gannicott and Tinsley (1997) studying *D. sagittata* pointed to a hatching factor in rainbow trout mucus. Host hormones are known to play a major role in the life cycle of certain monogeneans. Circumstantial evidence for host hormone regulation of oviposition in polystomatids parasitizing amphibia has been presented (Tinsley, 1990) and testosterone levels in rainbow trout affect the susceptibility to gyrodactylid infection (Buchmann, 1997b). The extent to which these host factors are detected by monogeneans will be treated below.

2.3 *Host-specificity*

Different species of monogeneans are relatively specific showing preference for narrowly related host fishes. The reason for this is only poorly understood. Anatomical and chemical elements in the parasites may play a role by fitting into corresponding factors in the host. Likewise host structures and substances may bind to receptors in parasites eliciting appropriate behaviour and physiological activity. Detailed descriptions of this are lacking. However, several investigations have been conducted which contribute to the notion that both chemo-attraction and immunological host elements are active in this question.

2.4 *Chemo-attraction*

A few studies have shown that monogeneans are able to detect host-derived molecules. Early observations by Frankland (1955) showed that *Diclidophora denticulata* oncomiracidia preferred its natural host *Pollachius virens* to *Gadus morhua*. The classical work by Kearn (1967) with focus on *Entobdella soleae* parasitizing *Solea solea* clearly indicated that oncomiracidia of this monogenean are attracted by host substances in scales or mucus from the specific host. Fish skin extracts was also found to influence behaviour of oncomiracidia and induce their attachment response to a host when studying *Neobenedenia girellae* (Yoshinaga *et al.*, 2000) and *Benedenia seriolae* (Yoshinaga *et al.*, 2002). However, these authors could not confirm the specificity in these systems where extracts from a range of hosts did not differ in their induction capabilities. The mechanism of induction was suggested to rely on lectins in the worms recognizing carbohydrates on the fish host. It can be suggested that the mechanism in some cases is based on at least two components: (1) induction of attachment and (2) survival and postlarval development. If the second part is affected negatively by the un-appropriate host, then host specificity will be evident within a short time irrespective of initial attachment. A somewhat corresponding system operates in the digenean cercariae of *Diplostomum*. Infection of the fish host is based on an initial settling by the cercaria due to certain host molecules. This is followed by penetration induced by other substances (Haas *et al.*, 2002). Due to the finding of lectins in the penetration glands of the parasite (Mikes and Horak, 2000) it could be suggested that these events operate with lectin–carbohydrate interactions. This was also indicated by the finding of corresponding lectins and carbohydrates in trout skin and tegument of *G. derjavini* (Buchmann, 2001). The exact compositions of all the gland types found in monogeneans have not been established yet, but these materials are evident candidates for a role in host specificity (Whittington *et al.*, 2000).

2.5 *Immune factors and host specificity*

Evidence has been presented which suggests involvement of immunological elements in determination of host specificity. Corticosteroids are generally assumed to inhibit a range of immunological elements in mammals. That a similar situation exists for salmonids has indeed been indicated by the experimental inhibition of multiplication and metabolism of rainbow trout macrophages (Pagniello *et al.*, 2002). These authors found that cortisol and dexamethasone are compounds with a marked influence on these macrophage functions in rainbow trout and Zou *et al.* (2000) detected impairment of IL-1 production in cortisol-treated rainbow trout macrophages. These observations do indeed suggest a connection between host specificity and immunity. Studies using treatment of salmonid hosts with immuno-suppressants, such as hydrocortisone and dexamethasone, showed that unsuitable parasites could colonize the treated host with greater success compared to untreated controls. Thus, two races of Atlantic salmon, *Salmo salar*, which are unsuitable hosts for *Gyrodactylus derjavini*, allowed increased population growth following dexamethasone administration (Olafsdottir *et al.*, 2003). Further, these parasites often aggregate on the cornea of unsuitable hosts. The cornea is regarded as an immuno-privileged site due to lack of mucous cells and mast cells in this cell layer (Buchmann and Bresciani, 1998; Sigh and Buchmann, 2000). Interestingly, in immuno-suppressed hosts the unsuitable parasites did not aggregate on the corneal surface to the same extent as seen in responding hosts (Olafsdottir *et al.*, 2003). Other studies have shown that brown trout *Salmo trutta*, which is an unsuitable host for *Gyrodactylus salaris* (Bakke *et al.*, 1999) become significantly more susceptible to this species following hydrocortisone acetate administration (Harris *et al.*, 2000). The finding that trout complement exists in various forms with different binding abilities (Sunyer *et al.*, 1997, 1998; Zarkadis *et al.*, 2001) suggests that these molecules may be part of the immunologically relevant recognition molecules. Although these matters are relatively well described it should be noted that genes for a range of additional acute-phase molecules are expressed upon infection of a fish (Bayne *et al.*, 2001). It cannot be excluded that such substances will play a role in the interactions between monogeneans and fishes.

3 Host responses

Several investigations have demonstrated that the fish host is able to mount a protective response against monogenean infestation. Early studies by Jahn and Kuhn (1932) and Nigrelli and Breder (1934) indicated that a number of fish hosts are able to reduce infection intensities with *Neobenedenia* (*Epibdella*) *melleni*. Common carp (*Cyprinus carpio*) was later found to combat infections with *Dactylogyrus vastator* (Paperna, 1963) and to *Dactylogyrus minutus* and *D. extensus* (Buchmann *et al.*, 1993). Thus, following heavy primary infections the number of gill monogeneans was found to fall significantly in the later stages of infection. The European eel *Anguilla anguilla* showed a significantly lower number of *Pseudodactylogyrus* spp. on the gills of previously infected eels, which was interpreted as acquired immunity to infection (Slotved and Buchmann, 1993).

Gyrodactylids induce a clear protection in many host species. Early work by Lester and Adams (1974) showed that sticklebacks are able to eliminate or control *Gyrodactylus alexanderi* populations on a short-term basis. Scott and Robinson (1984) provided evidence that guppies (*Poecilia reticulata*) partially resist re-infections with *Gyrodactylus bullatarudis*. This was further studied and confirmed by Richards and Chubb (1996) who also described a cross resistance to *G. turnbulli*. English sole mounts a relative response against *Gyrodactylus stellatus* (Moore *et al.*, 1994) and extensive studies (see Bakke *et al.*, 1990, 2002) have shown clear reactions in a range of salmonids against *G. salaris*. Thus, even after a steady build up of parasites on the host surface some factors in the host skin affect the population adversely, resulting in a rapid decline in the infection intensity. In a similar way the closely related *G. thymalli* has been shown to be controlled by a host response of Atlantic salmon (Sterud *et al.*, 2002). This was also demonstrated in various salmonids infected with *G. derjavini* (Bakke *et al.*, 2002; Buchmann and Uldal, 1997; Lindenstrøm and Buchmann, 2000).

These reactions were further shown to be impaired by dexamethasone treatment (Lindenstrøm and Buchmann, 1998). The Japanese flounder has also a capability of counteracting heavy build up of *Neobenedenia girellae* (Bondad-Reantaso *et al.*, 1995) and similar events were also indicated in related host–parasite system, the common sole, *Solea solea,* harbouring *Entobdella soleae* (Kearn, 2002). Infection of rainbow trout with *Discocotyle sagittata* elicits a host response to subsequent infections which, however, is weak (Rubio-Godoy and Tinsley, 2002). Protection against challenge infections of fish hosts following artificial immunization (vaccination) was found by Vladimirov (1971), Kim *et al.* (2000) and Nakane (2000) (cited by Ogawa, 2002) and will be described below.

3.1 *Immune evasion mechanisms*

Despite the well-developed protection mechanisms present in fish a vast majority of pathogens are able to infect the host with varying success. Evidently the parasites must possess some elements of immune evasion. In fact, the estimate of a total of more than 24 000 monogenean species worldwide is an indicator of the successful evolution of the group and the powerful evasion of the anti-parasitic host armament. In fact, histopathological studies of fish tissue infected with monogeneans do often show a limited damage to parasite surfaces. Although surrounded by a dense layer of reactive cells the monogenean tegument appear intact. It has even been suggested that the parasites exploit the tissue reaction by using it for improved attachment on the host. It is especially noteworthy that the blood-feeding polyopisthocotyleans are able to resist direct intake of blood containing complement factors, immunoglobulin, lectins, leucocytes and other reactive elements and substances. This will indicate well-developed evasion mechanisms. The subject of immune evasion has especially been treated for helminth groups such as digeneans, cestodes and nematodes. The various evasion mechanisms in these groups range from protection against complement, immunoglobulins, diversion of cytokine expressions to behavioural responses.

Monogeneans evading the inflammatory reactions in the host skin have been reported. Thus, the behavioural response of the ectoparasite to localized skin reactions is mainly escape reactions and search for non-reacting skin areas. At the molecular level it can be suggested that factors limiting effects of the anti-parasitic armament of the host must occur. Thus, despite a high antibody level in the serum of the trout host the polyopisthocotylean *Discocotyle sagittata* survives the direct feeding on host blood (Rubio-Godoy *et al.*, 2002, 2003). Various mechanisms in the parasite make it fit for coping with complement, antibodies, reactive oxygen and nitrogen intermediates, lectins, leucocytes, mast cells and macrophages. The nature of these substances is still poorly studied. Therefore priority should be given to studies on production of protective layers in monogeneans corresponding to the lamina containing complement-inhibiting molecules in *Echinococcus granulosus* cysts in mammals (Ferreira *et al.*, 2000) or in a range of other micro-organisms such as fungi and bacteria (Cooper, 1991). Further, knowing that cytokines are expressed following monogenean infection (Lindenstrøm *et al.*, 2002) it would be appropriate to analyse if parasite factors inhibit inflammatory cytokines, such as TNF, as was seen in murine macrophages primed with *Spirometra* products (Miura *et al.*, 2000).

3.2 *Mechanical barriers*

The fish skin and the gill epithelia create mechanical barriers to external invading organisms. The fish skin is evidently much more resistant compared to the delicate gill tissue. It requires certain abilities of a monogenean to penetrate the epithelial cell layer, especially of the skin, and besides this a dense layer of mucous cells produce mucus, which consists of sticky muco-polysaccharides. The physical/mechanical effect of mucus alone on monogeneans is uncertain. Hostile factors in or isolated with mucus are often ascribed to local cellular production although leakage or active secretion of serum components could be involved.

3.3 *Innate immune reactions*

It has been suggested that some of the factors responsible for host protection against monogeneans are non-specific. Responses against *Gyrodactylus* in guppies protect to the same degree against other species in the same genus (Richards and Chubb, 1996). The response activated in trout against *G. derjavini* did confer a relative protection of the host towards the parasitic ciliate *Ichthyophthirius multifiliis* (Buchmann *et al.*, 1999). Trout infected by live anisakid nematodes eliminated *G. derjavini* clearly better than controls (Larsen *et al.*, 2002). However, rainbow trout vaccinated and protected against the bacterium *Yersinia ruckeri* did not cope with *G. derjavini* better than controls when exposed 3, 4 and 5 months post vaccination (Buchmann *et al.*, 2003). This suggests that the factors in the innate immune system, such as lectins, pattern recognition receptors, complement and phagocytes (often considered non-specific players) do show some specificity.

3.4 *Specific immune reactions*

Specific reactions in fish are connected to the production of specific antibodies, B-lymphocytes and T-lymphocytes. The former cell population is well described due to the presence of immunoglobulin on the plasma membrane which can be detected immuno-cytochemically. In contrast the T-lymphocyte populations have been more problematic to characterize but recently the gene for the T-cell receptor in rainbow trout was described and monoclonal antibodies reacting with the T-cell receptor on rainbow trout lymphocytes were subsequently raised (Timmusk *et al.*, 2002). This will create access to future direct studies on the role of these lymphocytes in host responses against monogeneans parasitizing rainbow trout.

3.5 *Humoral reactions*

Humoral immunity comprises both elements of the specific and non-specific system. Thus, antibodies with relatively high specificity, complement factors and lectins with a limited repertoire of binding abilities.

Antibodies
Production of specific antibodies in teleosts directed against monogeneans occurs in some cases following natural infection. It is primarily gill-parasitizing monogeneans which elicit detectable serum antibodies. Probably the delicate gill tissue with a minimal separation between blood and parasite allows priming of systemic immunity

in fish against monogeneans. Thus, carp infected with *Dactylogyrus vastator* and *D. extensus* produce detectable serum antibodies (Vladimirov, 1971). Likewise, eels infected with the congeneric gill parasites *Pseudodactylogyrus bini* and *P. anguillae* contain enough serum antibody to produce a clear signal in western blot (Buchmann, 1993; Mazzanti *et al.*, 1999, Monni and Cognetti-Variale, 2002). In addition, Tiger puffer (naturally infected) produces detectable specific antibodies to *Heterobothrium okamotoi* (Wang *et al.*, 1997) and rainbow trout, both artificially and naturally infected, produce specific antibodies against *Discocotyle sagittata* (Rubio-Godoy, 2003). In contrast, *G. derjavini* infecting fins and body of rainbow trout do not elicit specific serum antibody production against surface epitopes on the parasite when analysed immuno-cytochemically (Buchmann, 1998). Likewise, no specific antibodies against *G. derjavini* antigens were detected in serum from this host species by using western blot (Jens Sigh, unpublished). A similar negative result was done using ELISA by Thoney and Burreson (1988) working with the fish *Leiostomus xanthurus* infected with the monogenean *Heteraxinoides xanthophilis*. A lack of antibody titre rise was also found by Bondad-Reantaso *et al.* (1995) in Japanese flounder reacting against *Neobenedenia girellae*.

Complement

Complement reactions comprise the classical, alternative and lectin activation pathways, also in fish. In fact, complement factors seem to be highly effective against monogeneans, at least *in vitro*. Both *G. derjavini* (Buchmann, 1998) and *G. salaris* (Harris *et al.*, 1998) are highly susceptible to complement exposure *in vitro*. In addition, during infection with *G. salaris* a higher rise in complement concentration in serum and mucus occurs in resistant salmon compared to susceptible salmon (Bakke *et al.*, 2000). It has been demonstrated by immuno-cytochemistry that complement factor C3 as well as factor B from rainbow trout serum bind to *G. derjavini* tegumental epitopes in association with *in vitro* killing (Buchmann, 1998; Lindenstrøm and Buchmann, 2001). Several studies strongly suggest that the complement-mediated killing in this system is independent of parasite-specific antibodies, whereas the alternative or lectin activation pathway is responsible (Buchmann and Lindenstrøm, 2002).

Lectins

Lectins in fish are likely to play a significant role in monogenean–host relations. Lectins with specificities for mannose, galactose and other carbohydrates occur in fish. In addition, the tegument of monogeneans display various carbohydrates on the surface membrane. *Gyrodactylus derjavini* shows a specific arrangement of mannose around the cephalic gland openings and a general distribution of galactose residues on the tegument (Buchmann, 2001). *Gyrodactylus salaris* seems to show a different distribution (Steen Jørndrup, unpublished). More forms of the same lectin in one fish species exist and due to their implication in complement activation it can be speculated that these interactions do not only work immunologically but also contribute to host specificity (Buchmann, 2001; Yoshinaga *et al.*, 2000, 2002).

3.6 Cellular reactions

Numerous investigations have indicated that epithelial proliferation and tissue reactions following infection of fish gills by various monogeneans are of common occurrence. Thus, cyprinids infected by dactylogyrids show extensive gill reactions (Buchmann *et al.*, 1993; Molnar, 1972; Paperna, 1963; Prost, 1963; Wilde, 1935; Wunder, 1929). Likewise, diplectanids infecting *Dicentrarchus labrax* gills elicit heavy reactions which will in some cases embed part of the parasite (Gonzales-Lanza *et al.*, 1991; Oliver, 1977). *Pseudodactylogyrus bini* causes extensive gill tissue proliferation resulting in clubbing and partial over-growth of the monogenean (Buchmann, 1988, 1997a). *Anoplodiscus tai* infection of red sea bream *Pagrus major* is associated with tissue reactions (Ogawa, 1994) and this is also the case for diclidophorids on tiger puffer and Japanese flounder (Ogawa, 2002).

No studies have previously elucidated the role of T-lymphocytes in the fish host reaction to monogenean infection. Following the development of immunological tools such as monoclonal antibodies to T-cell receptors this line should have priority in future research. Further, developments of molecular techniques such as RT-PCR will make it feasible to study the expression of T-cell-related cytokines in infected tissue. What is well documented is the involvement of leucocytes in interactions (*Table 1*). These are clearly seen in histopathological studies of fish tissue injured or altered by monogeneans. *In vitro* studies have also shown that macrophages from rainbow trout will colonize the tegument of *G. derjavini* and subsequently kill the parasite (Buchmann and Bresciani, 1999). It is noteworthy that the rainbow trout host response is dexamethasone labile (Lindenstrøm and Buchmann, 1998) and that dexamethasone is known to affect proliferation and metabolism of host macrophages (Pagniello *et al.*, 2002). Further, this corticosteroid inhibits IL-1 gene expression in rainbow trout leucocytes (Zou *et al.*, 2000). This association will suggest that leucocytes are involved in the response against *G. derjavini*. The production in the macrophages of complement factors and reactive oxygen metabolites was suggested as an active element of the killing process (Buchmann and Bresciani, 1999). One cell type that also is likely to play a role in the cellular reactions is the mucous cell (goblet cell) which is prevalent on most external surfaces of fish (Buchmann and Bresciani, 1998; Sterud *et al.*, 1998; Wells and Cone 1990). Due to the fact that this cell is present around monogeneans on fish in densities of several hundreds per square mm and that mucus released from this cell type has a profound effect on parasites *in vitro* indicates these cells as an important cell population in interactive associations between fish and monogeneans. Thionin-positive cells (putative mast cells) were detected in the skin of trout by Sigh and Buchmann (2000). Degranulation following infection of the fish with *G. derjavini* suggested the action of these cells in host reactions to gyrodactylids. Recent studies have identified the so-called Toll-like receptors which occur in both invertebrate and vertebrate taxons. These receptors endow the host cells with a capability to initiate a range of important physiological and immunological events (Fallon *et al.*, 2001). The role of these molecules in monogenean–host association should be studied.

3.7 *Cytokine-mediated regulation of anti-parasitic immunity in fish*

The well-established and recognized immune responses in fish against monogeneans urged the formulation of a model for these interactions. Thus, it was suggested that the epidermis of teleosts comprises a micro-environment below the surface epithelium where leucocytes, mucous cells, mast cells, putative lymphocytes and epithelial cells communicate through various cytokines, which results in the production and release of a range of hostile molecules eliminating the ectoparasites (Buchmann, 1999). Recent experiments using RT-PCR have confirmed that expression of the gene encoding the pro-inflammatory cytokine interleukin 1 (IL-1 beta) is initiated in rainbow trout skin following infection with *G. derjavini*. In addition it was found that the production is at least co-regulated by a decoy receptor for IL-1 (Lindenstrøm *et al.*, 2003). Parasite-induced expression of genes for several other immunologically relevant molecules has also been found. Thus, expression of genes for another pro-inflammatory cytokine, TNF-alpha, has been detected following *G. derjavini* infection (*Figure 7*). Several other immunologically relevant genes are likely to be expressed as well. It is therefore suggested that a range of reactions are activated following attachment of gyrodactylid monogeneans to the host (*Figure 8*). The 16 marginal hooklets of each worm penetrate the epithelial cells and the parasite starts feeding by placing the mouth opening on the epithelium. Gland secretions from the cephalic glands, the intestine-associated glands and opisthaptoral glands may contribute to epithelial stimulation. It is known that epithelial cells are able to produce IL-1 in mammals. Preformed IL-1 can be emitted following injury (mechanical or chemical) and thereby drive its own expression. It could be suggested that the same is the case in teleosts whereby the first host reaction would be expression of the pro-inflammatory cytokine. As IL-1 is a mucous secretagogue it could in turn initiate secretion and discharge of mucous cells and activate resident leucocytes at the site of inflammation. A rather fast degranulation of thionine-positive cells (putative mast cells) has been observed following infection and the release of potent substances from these cells may further stimulate this process. The various leucocytes may initiate expression of other cytokines, leukotrienes and prostaglandins and thereby attract additional leucocytes to the site of inflammation. The events will lead to emission of a number of hostile molecules ranging from reactive oxygen and nitrogen metabolites, lectins, defensins, complement factors, C-reactive protein and more mucus. Whether or not immunoglobulins or T-cells take part in these interactions is at present unknown but may depend on processing and presentation of parasite-derived antigens. The host defence substances may affect the tegument of the monogenean to some extent. However, the gut of the parasite is especially exposed due to the ingestion of the reactive host surface. Future work should focus on the action of immune factors and their action in the monogenean gut.

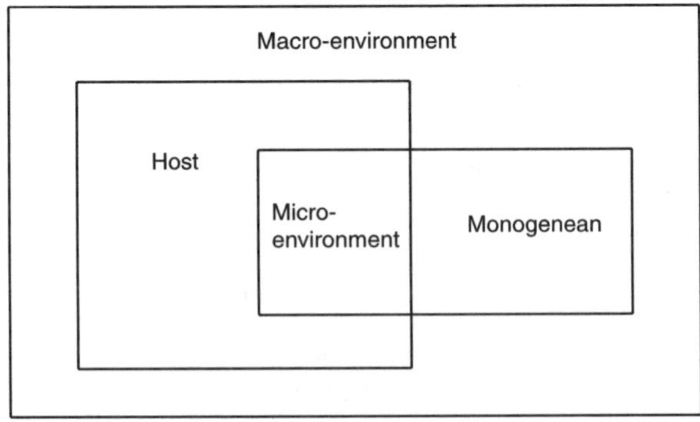

Figure 7 *Diagrammatic illustration of the interactive associations between macro-environment, host, monogenean and the microenvironment.*

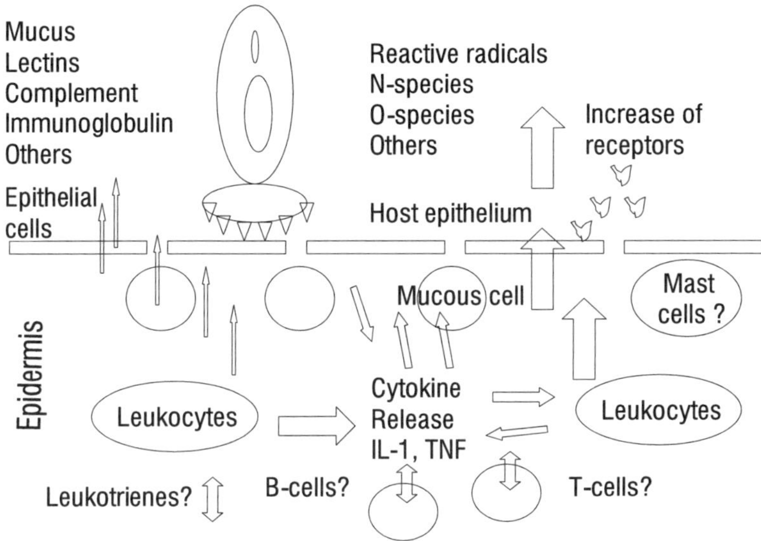

Figure 8 *Interactions between a gyrodactylid monogenean and the fish host. Molecular and cellular events elicited by an infection with a monogenean on fish skin.*

4 Control measures against monogeneans based on interactive associations

4.1 *Vaccination*

Vaccination studies in fish have been conducted with various pathogens ranging from virus, bacteria and protozoans. Especially, the antibacterial vaccines have obtained great success and are currently used in the aquaculture industry. A few limited studies have been performed on vaccination of fish against monogenean infestation. Vladimirov (1971) injected carp with an emulsion containing *Dactylogyrus* parasites and found a relative protection against infection.

Kim *et al.* (2000) vaccinated rockfish *Sebastes schlegeli* with a homogenate of *Microcotyle sebastis* which resulted in a significant protection. Intra-peritoneal injection conferred higher protection compared to immersion vaccination.

Injection of tiger puffer with homogenate of *Heterobothrium okamotoi* does result in a relative protection seen by a lower number of adult parasites on the branchial wall (Nakane, 2000, cited by Ogawa, 2002). It has been attempted to confer immunity to fish by passive immunization (Lindenstrøm and Buchmann, 2000). These authors collected serum from rainbow trout responding to *G. derjavini* infection and transferred it (with or without complement) by intraperitoneal injection to a naive host. However, no immunity was transferred by any of these processes compared to fish receiving non-immune serum. Rubio-Godoy (2003) immunized rainbow trout with different fractions of the blood-feeding *Discocotyle sagittata* and obtained some protection which was associated with specific antibody protection.

5 Conclusion

Monogeneans are one of the largest helminth groups existing and the total number may exceed 24 000. Most of the worms are ectoparastic on fish and have shown a high degree of host specificity. This is difficult to explain adequately but various factors may be associated with this fascinating part of monogenean biology. Immune factors may be of importance although it was previously considered that monogeneans were isolated from the host immune system. However, recent research has elucidated clear host responses against monogeneans in many fish orders. It has even been demonstrated that vaccination of these hosts is possible. What remains to be described in detail is the factors involved in the protective response. Isolated experiments have indicated host leucocytes as a main character in this play. Humoral factors such as complement factors, lectins, and to some extent immunoglobulins, may affect the parasites. The cytokine network in fishes seems to be connected to the regulation of these reactions. Besides the immunological issues a number of basic recognition systems may be involved. Host substances are recognized by the monogeneans and evidence for chemotactic mechanisms in these parasite–host relations have been demonstrated.

Acknowledgements

This work was conducted as a part of the EU-project QLRT-2000-01631: The genetic basis of *Gyrodactylus* resistance in Atlantic salmon (*Salmo salar)*. It is connected to the research network SCOFDA under the Danish Agricultural and Veterinary Research Council and the Danish Ministry of Food, Agriculture and Fisheries.
The *Gyrodactylus salaris* population was donated to us by Dr. Tor Atle Mo, Norway, and the specimens of *Discocotyle, sagittata* were kindly donated by Dr. Miguel Rubio-Godoy and Professor Dr. Richard C. Tinsley, UK. Mr. Eric Buchmann, Denmark, provided Axine belones infected garfish.

References

Andersen, P.S. and Buchmann, K. (1998) Temperature dependent population growth of *Gyrodactylus derjavini* on rainbow trout, *Oncorhynchus mykiss*. *J. Helminthol.* **72:** 9–14.

Armstrong, J.C. (1965) Mating behaviour and development of schistosomes in the mouse. *J. Parasitol.* **51:** 605–616.

Bakke, T.A., Jansen, P.A. and Hansen, L.P. (1990) Differences in host resistance of Atlantic salmon, *Salmo salar* L., stocks to the monogenean *Gyrodactylus salaris* Malmberg, 1957. *J. Fish Biol.* **37:** 577–587.

Bakke, T.A., Soleng, A. and Harris, P.D. (1999) The susceptibility of Atlantic salmon (*Salmo salar* L.) x brown trout (*Salmo trutta* L.) hybrids to *Gyrodactylus salaris* Malmberg and *Gyrodactylus derjavini* Mikailov. *Parasitology* **119:** 467–481.

Bakke, T.A., Soleng, A., Lunde, H. and Harris, P.D. (2000) Resistance mechanisms in *Salmo salar* stocks infected with *Gyrodactylus salaris*. Abstract. Proc. EMOP VIII, Poznan, Sept. 2000. *Acta Parasitol.* **45:** 272.

Bakke, T.A., Harris, P.D. and Cable, J. (2002) Host specificity dynamics: observations on gyrodactylid monogeneans. *Int. J. Parasitol.* **32:** 281–308.

Bayne, C.J., Gerwick, L., Fujiki, K., Nakao, M. and Yano, T. (2001) Immune relevant (including acute phase) genes identified in the liver of rainbow trout *Oncorhynchus mykiss* by means of suppression subtractive hybridisation. *Dev. Comp. Immunol.* **25:** 205–217.

Bondad-Reantaso, M.G., Ogawa, K., Yoshinaga, T. and Wakabayashi, H. (1995) Acquired protection against *Neobenedenia girellae* in Japanese flounder. *Fish Pathol.* **30:** 233–238.

Buchmann, K. (1988) Interactions between the gill parasitic monogeneans *Pseudodactylogyrus anguillae* and *P. bini* and the fish host *Anguilla anguilla*. *Bull. Eur. Ass. Fish Pathol.* **8:** 98–100.

Buchmann, K. (1993) A note on the humoral immune response of infected *Anguilla anguilla* against the gill monogenean *Pseudodactylogyrus bini*. *Fish Shellfish Immunol.* **3:** 397–399.

Buchmann, K. (1997a) Infection biology of gill parasitic monogeneans with special reference to the congeners *Pseudodactylogyrus bini* and *P. anguillae* (Monogenea: Platyhelminthes) from European eel. Dissertation. Royal Veterinary and Agricultural University, Frederiksberg, Denmark.

Buchmann, K. (1997b) Population increase of *Gyrodactylus derjavini* on rainbow trout induced by testosterone treatment of the host. *Dis. Aquat. Org.* **30:** 145–150.

Buchmann, K. (1998) Binding and lethal effect of complement from *Oncorhynchus mykiss* on *Gyrodactylus derjavini* (Platyhelminthes, Monogenea). *Dis. Aquat. Org.* **32:** 195–200.

Buchmann, K. (1999) Immune mechanisms in fish skin against monogeneans – a model. *Folia Parasitol.* **46:** 1–9.

Buchmann, K. (2001) Lectins in fish skin: Do they play a role in monogenean–fish host interactions? *J. Helminthol.* **75:** 227–232.

Buchmann, K. and Bresciani, J. (1997) Parasitic infections in pond-reared rainbow trout *Oncorhynchus mykiss* in Denmark. *Dis. Aquat. Org.* **28:** 125–138.

Buchmann, K. and Bresciani, J. (1998) Microenvironment of *Gyrodactylus derjavini* on rainbow trout *Oncorhynchus mykiss*: association between mucous cell density in skin and site selection. *Parasitol. Res.* **84:** 17–24.

Buchmann, K. and Bresciani, J. (1999) Rainbow trout leukocyte activity: influence on the ectoparasitic monogenean *Gyrodactylus derjavini*. *Dis. Aquat. Org.* **35:** 13–22.

Buchmann, K. and Lindenstrøm, T. (2002) Interactions between monogenean parasites and their fish hosts. *Int. J. Parasitol.* **32:** 309–319.

Buchmann, K. and Uldal, A. (1997) *Gyrodactylus derjavini* infections in four salmonids: comparative host susceptibility and site selection of parasites. *Dis. Aquat. Org.* **28:** 201–209.

Buchmann, K., Slotved, H.C. and Dana, D. (1993) Epidemiology of gill parasite infections in *Cyprinus carpio* in Indonesia and possible control methods. *Aquaculture* **118:** 9–21.

Buchmann, K., Lindenstrøm, T. and Sigh, J. (1999) Partial cross-protection against *Ichthyophthirius multifiliis* infection in *Gyrodactylus derjavini* immunized rainbow trout. *J. Helminthol.* **73:** 189–195.

Buchmann, K., Nielsen, C.V. and Nielsen, M.E. (2003) Immune responses against *Yersinia ruckeri* have no effect on colonization of rainbow trout *Oncorhynchus mykiss* (Walbaum) by *Gyrodactylus derjavini* Mikailov, 1975. *J. Fish Dis.* **26:** 183–186.

Bychowsky, B.E. (1957) *Monogenetic trematodes, their systematics and phylogeny*. Leningrad: Izdatelstvo Akademiya Nauk SSSR. (English translation American Institute of Biological Sciences, 1961.)

Cable, J. and Harris, P.D. (2002) Gyrodactylid developmental biology: historical review, current status and future trends. *Int. J. Parasitol.* **32:** 255–280.

Chan, B. and Wu, B. (1984) Studies on the pathogenicity, biology and treatment of Pseudodactylogyrus in fish farms. *Acta Zool. Sin.* **30:** 173–180.

Chisholm, L.A., Whittington, I.D. and Schokaert, E. (1998) Morphology and development of the haptors among the monocotylidae (Monogenea). *Hydrobiologia* **383:** 251–261.

Cone, D.K. and Odense, P.H. (1984) Pathology of five species of *Gyrodactylus* Nordmann, 1832 (Monogenea). *Can. J. Zool.* **62:** 1084–1088.

Cooper, N.R. (1991) Complement evasion strategies of microorganisms. *Immunol. Today* **12:** 327–331.

Cunningham, C.O. (1997) Species variation within the internal transcribed spacer (ITS) region of Gyrodactylus (Monogenea; Gyrodactylidae). *J. Parasitol.* **83:** 215–219.

Euzet, L. and Combes, C. (1998) The selection of habitats among the monogenea. *Int. J. Parasitol.* **28:** 1645–1652.

Fallon, P.G., Allen, R.L. and Rich, T. (2001) Primitive Toll signalling: bugs, flies, worms and man. *Trends Immunol.* **22:** 63–66.

Ferreira, A.M., Irigoin, F., Breijo, M., Sim, R.B. and Diaz, A. (2000) How *Echinococcus granulosus* deals with complement. *Parasitol. Today* **16:** 168–172.

Frankland, H.M.T. (1955) The life history and bionomics of *Diclidophora denticulata* (Trematoda: Monogenea). *Parasitology* **45:** 313–351.

Fried, B. (1986) Chemical communication in hermaphroditic digenetic trematodes. *J. Chem. Ecol.* **12:** 1659–1677.

Gannicott, A.M. and Tinsley, R.C. (1997) Egg hatching in the monogenean gill parasite *Discocotyle sagittata* from the rainbow trout (*Oncorhynchus mykiss*). *Parasitology* **114:** 569–579.

Gonzales-Lanza, C., Alvarez-Pellitero, P. and Sitja-Bobbadilla, A. (1991) Diplectanidae (Monogenea) infestations of sea bass, *Dicentrarchus labrax* (L.), from the Spanish mediterranean area. *Parasitol. Res.* **77:** 307–314.

Haas, W., Stiegler, P., Keating, A., Kullmann, R., Rabenau, H., Schönamsgruber, E. and Haberl, B. (2002) *Diplostomum spathaceum* cercariae respond to a unique profile of cues during recognition of their fish host. *Int. J. Parasitol.* **32:** 1145–1154.

Halton, D.W., Maule, A.G. and Shaw, C. (1993) Neuronal mediators in monogenean parasites. *Bull. Fr. Peche Piscic.* **328:** 82–104.

Hara, T.J. (1993) Chemoreception. In: Evans, D.H. (ed.) *The Physiology of Fishes*, pp. 191–218. CRC Press, Boca Raton, FL, USA.

Harris, P.D., Soleng, A. and Bakke, T.A. (1998) Killing of *Gyrodactylus salaris* (Platyhelminthes, Monogenea) mediated by host complement. *Parasitology* **117:** 137–143.

Harris, P.D., Soleng, A. and Bakke, T. A. (2000) Increased susceptibility of salmonids to the monogenean *Gyrodactylus salaris* following administration of hydrocortisone acetate. *Parasitology* **120:** 57–64.

Jahn, T.L. and Kuhn, L.R. (1932) The life history of *Epibdella melleni* Maccallum 1927, a monogenetic trematode parasitic on marine fishes. *Biol. Bull.* **62:** 89–111.

Janssen, P.A. and Bakke, T.A. (1991) Temperature-dependent reproduction and survival of *Gyrodactylus salaris* Malmberg, 1957 (Platyhelminthes: Monogenea) on Atlantic salmon (*Salmo salar*). *Parasitology* **102:** 105–112.

Kearn, G.C. (1967) Experiments on host-finding and host-specificity in the monogenean skin parasite *Entobdella soleae*. *Parasitology* **57:** 585–605.

Kearn, G.C. (2002) *Entobdella soleae* – pointers to the future. *Int. J. Parasitol.* 32: 367–372.

Kearn, G.C. and Evans-Gowing, R. (1998) Attachment and detachment of the anterior adhesive pads of the monogenean (platyhelminth) parasite *Entobdella soleae* from the skin of the common sole (*Solea solea*). *Int. J. Parasitol.* **28:** 1583–1593.

Kearn, G.C. and Macdonald, S. (1976) The chemical nature of host hatching factors in the monogenean skin parasites *Entobdella soleae* and *Acanthocotyle lobianchi*. *Int. J. Parasitol.* **6:** 457–466.

Kim, K.H., Hwang, Y.J., Cho, J.B. and Park, S.I. (2000) Immunization of cultured rockfish *Sebastes schlegeli* against *Microcotyle sebastis* (Monogenea). *Dis. Aquat. Org.* **40:** 29–32.

Kritsky, D. (1978) The cephalic glands and associated structures in *Gyrodactylus eucaliae* Ikezaki and Hoffmann, 1957 (Monogenea: Gyrodactylidae). *Proc. Helminthol. Soc. Wash.* **45:** 37–49.

Larsen, A.H., Bresciani, J. and Buchmann, K. (2002) Interactions between ecto- and endoparasites in trout *Salmo trutta*. *Vet. Parasitol.* **103:** 167–173.

Larsen, T.B. and Buchmann, K. (2003) Effects of aqueous aluminium chloride and zinc chloride on survival of the gill parasitizing monogenean Pseudodactylogyrus anguillae from European eel Anguilla anguilla. *Bull. Eur. Ass. Fish Pathol.* **23:** 123–127.

Lester, R.J.G. and Adams, J.R. (1974) A simple model of a *Gyrodactylus* population. *Int. J. Parasitol.* **4:** 497–506.

Lindenstrøm, T. and Buchmann, K. (1998) Dexamethasone treatment increases susceptibility of rainbow trout, *Oncorhynchus mykiss* (Walbaum), to infections with *Gyrodactylus derjavini* Mikailov. *J. Fish Dis.* **21:** 29–38.

Lindenstrøm, T. and Buchmann, K. (2000) Acquired resistance in rainbow trout against *Gyrodactylus derjavini*. *J. Helminthol.* **74:** 155–160.

Lindenstrøm, T. and Buchmann, K. (2001) Importance of alternative activating pathways in complement mediated killing of *Gyrodactylus derjavini*. *Bull. Scand. Soc. Parasitol.* **11:** 21–22.

Lindenstrøm, T., Buchmann, K. and Secombes, C. J. (2002) *Gyrodactylus derjavini* elicit expression of IL-1 beta genes in rainbow trout skin. *Fish Shellfish Immunol.* **15:** 107–115.

Llewellyn, J. and Simmons, J. E. (1984) The attachment of the monogenean parasite *Callorhyncicola multitesticulatus* to the gills of its holocephalan host *Callorhynchus millii*. *Int. J. Parasitol.* **14:** 191–196.

Lyons, K.M. (1969) Sense organs of monogenean skin parasites ending in a typical cilium. *Parasitology* **59:** 625–636.

Malmberg, G. (1993) Gyrodactylidae and gyrodactylosis of salmonidae. *Bull. Fr. Peche Piscic.* 328: 5–46.

Malmberg, G. (1998) On the evolution within the family Gyrodactylidae (Monogenea). *Int. J. Parasitol.* **28:** 1625–1635.

Mazzanti, C., Monni, G. and Varriale, A.M.C. (1999) Observations on antigenic activity of *Pseudodactylogyrus anguillae* (Monogenea) on the European eel (*Anguilla anguilla*). *Bull. Eur. Ass. Fish Pathol.* **19:** 57–59.

Mikes, L. and Horak, P. (2000) Purification and characterization of a beta-1,3-glucan binding protein from penetration glands of *Diplostomum pseudospathaceum* cercariae. *Dev. Comp. Immunol.* **24:** S92.

Miura, K., Fukumoto, S., Dirgahayu, P. and Hirai, K. (2000) Excretory/secretory products from plerocercoids of *Spirometra erinaceieuropaei* suppress gene expressions and production of tumour necrosis factor – alpha in murine macrophages stimulated with lipopolysaccharide or lipoteichoic acid. *Int. J. Parasitol.* **31:** 39–47.

Molnar, K. (1972) Studies on gill parasitosis of grass-carp (*Ctenopharyngodon idella*) caused by *Dactylogyrus lamellatus* Achmerow, 1952. IV.: Histopathological changes. *Acta Vet. Acad. Scient. Hung.* **22:** 9–24.

Monni, G. and Cognetti-Varriale, A.M. (2002) Antigenicity of *Pseudodactylogyrus anguillae* and *P. bini* (Monogenea) in the European eel (*Anguilla anguilla* L.) under different oxygenation conditions. *Fish Shellfish Immunol.* **13:** 125–131.

Moore, M.M., Kaattari, S.L. and Olson, R.E. (1994) Biologically active factors against the monogenetic trematode *Gyrodactylus stellatus* in the serum and mucus of infected juvenile English soles. *J. Aquat. Anim. Health* **6:** 93–100.

Nigrelli, R.F. and Breder, C.M. (1934) The susceptibility and immunity of certain fishes to *Epibdella melleni*, a monogenetic trematode. *J. Parasitol.* **20:** 259–269.

Ogawa, K. (1994) *Anoplodiscus tai* sp.nov. (Monogenea: Anoplodiscidae) from cultured red sea bream *Pagrus major*. *Fish Pathol.* **29:** 5–10.

Ogawa, K. (2002) Impacts of diclidophorid monogeneans on fisheries in Japan. *Int. J. Parasitol.* **32:** 373–380.

Ogawa, K. and Egusa, S. (1980) *Gyrodactylus* infestations of cultured eels (*Anguilla japonica* and *A. anguilla*) in Japan. *Fish Pathol.* **15:** 95–99.

Olafsdottir, S.H., Lassen, H.P.O. and Buchmann, K. (2003) Labile resistance of Atlantic salmon, *Salmo salar* L., to infections with *Gyrodactylus derjavini* Mikailov, 1975: implications for host specificity. *J. Fish Dis.* **26:** 51–54.

Oliver, G. (1977). Effet pathogene de la fixation de *Diplectanum aequans* (Wagener, 1857) Diesing, 1858 (Monogenea, Monopisthocotylea, Diplectanidae) sur led branchies de *Dicentrarchus labrax* (Linnaeus, 1758), Pisces, Serranidae). *Zeitschr. Parasitenk.* **53:** 7–11.

Olson, P.D. and Littlewood, D.T.J. (2002) Phylogenetics of the Monogenea – evidence from a medley of molecules. *Int. J. Parasitol.* **32:** 233–244.

Pagniello, K.B., Bols, N.C. and Lee, L.E. (2002) Effect of corticosteroids on viability and proliferation of the rainbow trout monocyte/macrophage cell line, RTS11. *Fish Shellfish Immunol.* **13:** 199–214.

Paperna, I. (1963) Dynamics of *Dactylogyrus vastator* Nybelin (Monogenea) populations on the gills of carp fry in fish ponds. *Bamidgeh* **15:** 31–50.

Paperna, I. (1964) Competitive exclusion of *Dactylogyrus extensus* by *Dactylogyrus vastator* (Trematoda, Monogenea) on the gills of reared carp. *J. Parasitol.* **50:** 94–98.

Paperna, I., Diamant, A. and Overstreet, R.M. (1984) Monogenean infestations and mortality in wild and cultured Red Sea fishes. *Helgol. Meeresunters.* **37:** 445–462.

Prost, M. (1963) Investigations on the development and pathogenicity of *Dactylogyrus anchoratus* (Duj., 1845) and *D. extensus* Mueller et van Cleave, 1932 for breeding carps. *Acta Parasitol. Pol.* **11:** 17–47.

Putz, R.E. and Hoffmann, G.L. (1964) Studies on *Dactylogyrus corporalis* n.sp. (Trematoda: Monogenea) from the fallfish *Semotilus corporalis. Proc. Helminthol. Soc. Washington* **31:** 139–143.

Rand, T.G., Wiles, M. and Odense, P. (1986) Attachment of *Dermophthirius carcharhini* (Monogenea: Microbothriidae) to the Galapagos shark *Carcharhinus galapagensis. Trans. A. Microscop. Soc.* **105:** 158–169.

Richards, G.R. and Chubb, J.C. (1996) Host responses to initial and challenge infections, following treatment of *Gyrodactylus bullatarudis* and *G. turnbulli* (Monogenea) on the guppy (*Poecilia reticulata*). *Parasitol. Res.* **82:** 242–247.

Rohde K. (1977) A non-competitive mechanism responsible for restricting niches. *Zool. Anz. Jena.* **199:** 164–172.

Roubal, F.R. (1995) Microhabitats, attachment of eggs and histopathology by the monogenean *Allomurraytrema robustum* on *Acanthopagrus australis* (Pisces: Sparidae). *Int. J. Parasitol.* **25:** 293–298.

Rubio-Godoy, M., Sigh, J., Buchmann, K., and Tinsley, R.C. (2003) Immunization of rainbow trout Oncorhynchus mykiss against Discocotyle sagittata (Monogenea). *Dis. Aquat. Org.* **55:** 23–30.

Rubio-Godoy, M. and Tinsley, R. (2002) Trickle and single infection with *Discocotyle sagittata* (Monogenea: Polyopisthocotylea): effect of exposure mode on parasite abundance and development. *Folia Parasitologica* **49:** 269–278.

Scott, M.E. and Robinson, M.A. (1984) Challenge infections of *Gyrodactylus bullatarudis* (Monogenea) on guppies (*Poecilia reticulata*) following treatment. *J. Fish Biol.* **24:** 581–586.

Sigh, J. and Buchmann, K. (2000) Associations between epidermal thionin-positive cells and skin parasitic infections in brown trout *Salmo trutta*. *Dis. Aquat. Org.* **41:** 135–139.

Simková, A., Desdevises, Y., Gelnar, M. and Morand, S. (2000) Co-existence of nine gill ectoparasites (*Dactylogyrus*: Monogenea) parasitizing the roach (*Rutilus rutilus* L.): history and present ecology. *Int. J. Parasitol.* **30:** 1077–1088.

Slotved, H.-C. and Buchmann, K. (1993) Acquired resistance of the eel *Anguilla anguilla* L. to challenge infections with gill monogeneans. *J. Fish Dis.* **16:** 585–591.

Soleng, A., Poleo, A.B.S., Alstad, N.E.W. and Bakke, T.A. (1999) Aqueous aluminium eliminates *Gyrodactylus salaris* (Platyhelminthes, Monogenea) infections in Atlantic salmon. *Parasitology* **119:** 19–25.

Sterud, E., Harris, P.H. and Bakke, T. A. (1998) The influence of *Gyrodactylus salaris* Malmberg, 1957 (Monogenea) on the epidermis of Atlantic salmon, *Salmo salar* L., and brook trout, *Salvelinus fontinalis* (Mitchill), experimental studies. *J. Fish Dis.* **21:** 257–263.

Sterud, E., Mo, T.A., Collins, C.M. and Cunningham, C.O. (2002) The use of host specificity, pathogenicity and molecular markers to differentiate between *Gyrodactylus salaris* Malmberg, 1957 and *G. thymalli* Zitnan, 1960 (Monogenea, Gyrodactylidae). *Parasitology* **124:** 203–214.

Sunyer, J.O., Tort, L. and Lambris, J.D. (1997) Diversity of the third form of complement C3 in fish: Functional characterization of five forms of C3 in the diploid fish *Sparus aurata*. *Biochem. J.* **326:** 877–881.

Sunyer, J.O., Zarkadis, I., Sarrias, M.R., Hansen, J.D. and Lambris, J.D. (1998) Cloning, structure and function of two trout Bf molecules. *J. Immunol.* **161:** 4106–4114.

Thoney, D.A. and Burreson, E.M. (1988) Lack of specific humoral antibody response in *Leiostomus xanthurus* (Pisces: Serranidae) to parasitic copepods and monogeneans. *J. Parasitol.* **74:** 191–194.

Thune, R.L. and Rogers, W.A. (1981) Gill lesions in bluegill, *Lepomis macrochirus* Rafinesque, infested with *Cleidodiscus robustus* Mueller, 1934 (Monogenea: Dactylogyridae). *J. Fish Dis.* **4:** 277–280.

Timmusk, S., Jansson, E. and Pilström, L. (2002) The generation of monoclonal antibodies by genetic immunisation: antibodies against trout TCR alpha and IgL isotypes. *Fish Shellfish Immunol.* **12:** 1–20.

Tinsley, R.C. (1990) Host behaviour and opportunism in parasite life cycles. In: Barnard, C.J. and Behnke, J.M. (eds) *Parasitism and Host Behaviour*, pp. 158–192. Taylor and Francis, London.

Vladimirov, V.L. (1971) The immunity of fishes in the case of dactylogyrosis. *Parasitologiya* **5:** 51–58 (in Russian). English translation: *Parasitology Riverdale* **1:** 58–68.

Wang, R., Kim, J.-H., Sameshima, M. and Ogawa, K. (1997) Detection of antibodies against the monogenean *Heterobothrium okamotoi* in Tiger puffer by ELISA. *Fish Pathol.* **32:** 179–180.

Watson, N.A. and Rohde, K. (1994) Two new sensory receptors in *Gyrodactylus* sp. (Platyhelminthes, Monogenea, Monopisthocotylea). *Parasitol. Res.* 80: 442–445.

Wells, P.R. and Cone, D.K. (1990) Experimental studies on the effect of *Gyrodactylus colemanensis* and *G. salmonis* (Monogenea) on the density of mucous cells in the epidermis of fry of *Oncorhynchus mykiss*. *J. Fish Biol.* **37:** 599–603.

Whittington, I.D. (1998) Diversity down under: monogeneans in the antipodes Australia with a prediction of monogenean biodiversity worldwide. *Int. J. Parasitol.* **28:** 1481–1493.

Whittington, I.D. and Cribb, B.W. (2001) Adhesive secretions in the Platyhelminthes. *Adv. Parasitol.* **48:** 101–224.

Whittington, I.D., Cribb, B.W., Hamwood, T.E. and Halliday, J.A. (2000) Host-specificity of monogenean (platyhelminth) parasites: a role for anterior adhesive areas? *Int. J. Parasitol.* **30:** 305–320.

Wilde, J. (1935) Der Schleiendactylogyrus (*Dactylogyrus macracanthus*) und die Schädigung der Schleienkieme diesen Parasiten. *Fischerei-Zeit.* **38:** 661–663.

Wilde, J. (1937) *Dactylogyrus macracanthus* Wegener als Krankheitserreger auf den Kiemen der Schleie (*Tinca tinca*). *Zeitschr. Parasitenk.* **9:** 201–236.

Williams, H. and Jones, A. (1994) *Parasitic Worms of Fish.* Taylor and Francis, London.

Wunder, W. (1929) Die *Dactylogyrus*-krankheit der Karpfenbrut, ihre Ursache und ihre Bekämpfung. Zeitchr. *Fischerei* **27:** 511–545.

Xia, X., Nie P. and Yao, W. (1996) Effects of light, temperature and host mucus on the egg hatching of *Ancyrocephalus mogurndae* (Monogenea). *Acta Hydrobiol. Sin.* **20:** 195–196.

Yoshinaga, T., Nagakura, T., Ogawa, K. and Wakabayashi, H. (2000) Attachment-inducing capacities of fish tissue extract on oncomiracidia of *Neobenedenia girellae* (Monogenea, Capsalidae). *J. Parasitol.* **86:** 214–219.

Yoshinaga, T., Nagakura, T., Ogawa, K., Fukuda, Y. and Wakabayashi, H. (2002) Attachment inducing capacities of fish skin epithelial extracts on oncomiracidia of *Benedenia seriolae* (Monogenea: Capsalidae). *Int. J. Parasitol.* **32:** 381–384.

Zarkadis, I.K., Sarrias, M.R., Sfyroera, G., Sunyer, J.O. and Lambris, J. D. (2001) Cloning and structure of the rainbow trout C3 molecules: a plausible explanation for their functional diversity. *Dev. Comp. Immunol.* **25:** 11–24.

Zou, J., Holland, J., Pleguezuelos, O., Cunningham, C. and Secombes, C.J. (2000) Factors influencing the expression of interleukin-1 beta in cultured rainbow trout (*Oncorhynchus mykiss*) leucocytes. *Dev. Comp.Immunol.* **24:** 575–582.

9

Comparative aspects of the tick–host relationship: immunobiology, genomics and proteomics

Francisco J. Alarcon-Chaidez and Stephen K. Wikel

1 Introduction

Infectious agents transmitted by blood-feeding arthropods are responsible for millions of cases of disease annually (Gubler, 2000). In addition, the emergence and resurgence of new and previously recognized vector-borne diseases continues to rise due to a number of factors that range from changes in public health policies to changes in the genetic makeup of infectious agents (Gratz, 1999).

The number of vector-borne transmitted diseases that affect humans has increased significantly since the identification of *Borrelia burgdorferi* as the causative agent of Lyme disease (Parola and Raoult, 2001). At present, control of ticks and tick-borne diseases depends almost exclusively on the use of acaricides. However, the widespread occurrence of tick resistance to these chemicals and the growing concern of environmental contamination have prompted research into development of alternative methods of controlling tick populations (Mitchell, 1996). Vaccination against tick infestation has shown great potential and progress has been made in the search for efficient tick vaccine candidates (Walker, 1998). Research efforts are now focused on the immunological events occurring at the tick–host interface through a combination of genomic and proteomic approaches, which will help develop new strategies for the control of tick and tick-borne infectious diseases. This chapter addresses some of the relevant aspects of tick biology that are important for transmission of infectious agents to humans. We also summarize recent findings in the immunological analysis of host response to tick feeding, and describe recent advances in the application of proteomic and genomic strategies, which continue to be increasingly attractive alternatives to expedite identification of potential vector-blocking vaccine candidates.

Host–Parasite Interactions, edited by G. F. Wiegertjes & G. Flik.

2 Life cycles and ecology

This section briefly describes selected aspects of the ecology and physiology of tick vectors of disease and readers are referred to the primary literature in these areas which have been extensively reviewed (Anderson, 2002; Parola and Raoult, 2001; Sonenshine, 1991).

Ticks are obligate haematophagous ectoparasites that belong to the Class Arachnida, a group of arthropods including mites and spiders that are able to infest a wide variety of vertebrate hosts in most habitats of the world (Klompen *et al.*, 1996; Sonenshine, 1991). Ticks are generally classified into three major families: the Ixodidae or 'hard ticks', the Argasidae or 'soft ticks', and the Nuttalliellidae comprised of a single species. Argasid ticks, characterized by the absence of a dorsal cuticular shield (scutum) (*Figure 1*), are generally represented by the genus *Ornithodoros* which includes species of medical importance such as *Ornithodoros moubata* found in the African continent, and *Ornithodoros hermsii*, common in the western United States. Argasids are normally found in the immediate vicinity of their hosts (endophilic) and are known for their ability to resist desiccation and starvation. Developmental stages of soft ticks may include several nymphal stages, and characteristically both nymphs and adults feed frequently for short periods of time (Anderson, 2002; Parola and Raoult, 2001).

Ixodid ticks have a scutum covering the entire dorsal surface in males but only the more anterior dorsal surface in females (*Figure 2*), which allows for greater expansion of the cuticle during feeding (*Figure 3*). Representative genera of the Ixodidae are *Ixodes, Haemaphysalis, Aponomma, Amblyomma,* and *Hyalomma*, which probably emerged as obligate ectoparasites of reptilian hosts during the late Mesozoic era. On the other hand, important members of the subfamily *Rhipicephalinae,* which includes *Boophilus, Rhipicephalus* and *Dermacentor*, among others, appear to have evolved more

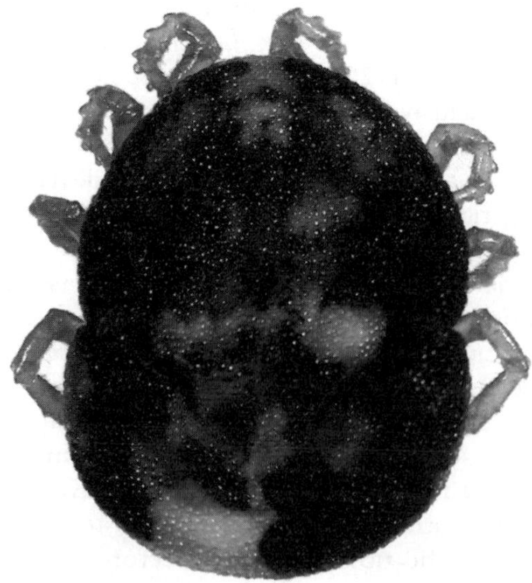

Figure 1 Ornithodoros moubata *adult soft tick. (Photo courtesy of K. Bouchard.)*

recently as avian and mammalian host species became more abundant (Black *et al.*, 1997; Klompen *et al.*, 1996; Sonenshine, 1991). Since ixodid ticks are sensitive to hostile environments, they occur in regions with more temperate climates, and tend to have exophilic habits. Ixodids can feed on one (*Boophilus*), two (*Rhipicephalus*) or three hosts (*Ixodes* and *Haemaphysalis*) during their life cycle. Their life cycle feeding stages include larvae, a single nymphal stage, and adult (*Figure 4*) and during feeding, they typically remain attached to the host for hours to days at a time.

Figure 2 *Female (left) and male (right)* Dermacentor andersoni *ticks also known as wood ticks. (Photo courtesy of K. Bouchard.)*

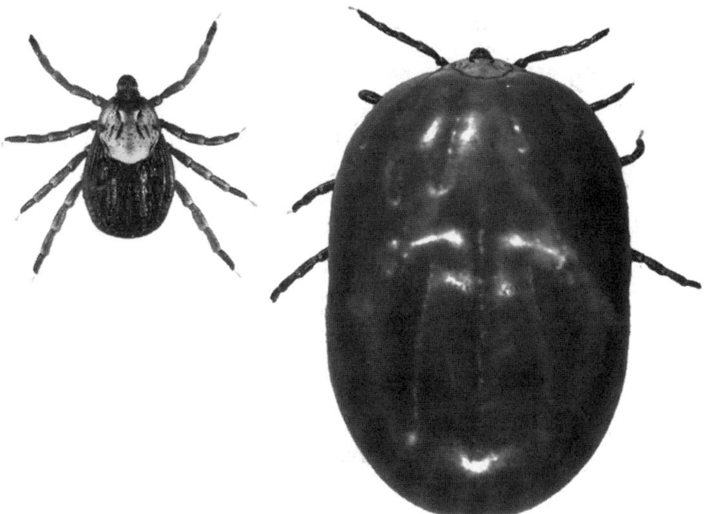

Figure 3 Dermacentor andersoni *female ticks unfed and engorged. (Photo courtesy of K. Bouchard.)*

Figure 4 *Three stages of the deer tick* Ixodes scapularis. *From left to right: larvae, nymph, and adult male and female. (Photo courtesy of K. Bouchard.)*

Morphological aspects and life cycle differences among ixodid ticks allow for further classification into two major groups, the metastriate and the prostriate. Among the 13 genera of this family, the prostriate *Ixodes* and the metastriate *Amblyomma, Dermacentor* and *Rhipicephalus* are considered to be the most important tick ectoparasites of livestock and are also vectors of many zoonotic diseases, indicating their medical importance (Parola and Raoult, 2001; Spach *et al.*, 1993; Walker, 1998).

3 Tick-borne pathogens

The public health importance of ticks is due to their active role in the transmission of a significantly greater number of disease-causing agents to humans and domestic animals than any other type of blood-feeding arthropod. Aetiologic agents known to be transmitted by ticks include a variety of bacteria, viruses and protozoa (Gayle and Ringdahl, 2001; Parola and Raoult, 2001) (*Table 1*).

Typically, tick larval stages acquire pathogens while feeding on infected rodents. The micro-organisms are then transmitted from one life stage to the next (trans-stadially) or from generation to generation (transovarially), thus acting as reservoirs for that pathogen (Walker, 1998). After larvae moult into nymphs, they transmit pathogens when they feed on other rodents or humans. The first report of a tick-borne infectious disease was the description of the transmission of the aetiologic agent of Texas cattle fever, *Babesia bigemina*, by the metastriate tick *Boophilus annulatus* (Smith and Kilbourne, 1893). *Babesia* organisms are protozoan parasites that replicate in host erythrocytes causing a malaria-like illness. In eastern and midwestern United States, human babesiosis is caused predominantly by *Babesia microti*, a piroplasm transmitted by *Ixodes scapularis*, also the vector tick for Lyme borreliosis and human granulocytic ehrlichiosis (HGE). In Europe, the major cause of human babesiosis is the bovine

Table 1 *Major tick-borne diseases, their vectors, and causative agents*

Genus in which primary vector species occurs	Diseases	Infectious agent	Reference
Ixodes	Lyme borreliosis	*Borrelia burgdorferi* *B. afzelii* *B. garinii*	(Steere, 2001)
	Human granulocytic ehrlichiosis (HGE)	*Anaplasma phagocytophilium*	(Dumler and Bakken, 1998)
	Tularaemia	*Francisella tularensis*	(Parola and Raoult, 2001)
	Rickettsiosis	*Rickettsia spp.*	(Parola and Raoult, 2001)
	Babesiosis	*Babesia microti*	(Krause, 2002)
		Babesia divergens	(Krause, 2002)
	Tick-borne encephalitis (TBE)	*Flavivirus* (Flaviviridae)	(Suss, 2003)
Dermacentor	Rocky Mountain spotted fever (RMSF)	*Rickettsia rickettsii*	(Walker, 1998)
	Tularaemia	*F. tularensis*	(Choi, 2002)
	Colorado tick fever	*Coltivirus* (Reoviridae)	(Klasco, 2002)
Amblyomma	Human monocytic ehrlichiosis (HME)	*Ehrlichia chaffeensis*	(Childs and Paddock, 2003)
	Tularaemia	*F. tularensis*	(Choi, 2002)
	Rickettsiosis (uncertain)	*Rickettsia* spp.	(Childs and Paddock, 2003)
Hyalomma	Crimean–Congo hemorrhagic fever	*Nairovirus* (Bunyaviridae)	(Hoogstraal, 1979)
Ornithodoros	Relapsing fever	*Borrelia hermsii* *Borrelia turicate* *Borrelia duttonii*	(Dworkin *et al.*, 2002b)
Rhipicephalus	Thogoto virus	*Thogotovirus* (Orthomyxoviridae)	(Jones *et al.*, 1992a)
	Rickettsiosis	*Rickettsia conorii*	(Walker, 1998)

pathogen *Babesia divergens*, and the tick *Ixodes ricinus* has been identified as the vector for this human infection (Kjemtrup and Conrad, 2000).

Lyme disease, caused by species of the spirochete *B. burgdorferi* sensu lato complex, is currently the leading tick-borne disease in the United States, with more than 17 000 cases reported in 2000 (Center for Disease Control, 2002). The prostriate ticks *I. scapularis*, and *Ixodes pacificus* in the US, *I. ricinus* in Europe and *I. persulcatus* in Asia (Yanagihara and Masuzawa, 1997) are the primary vectors of *B. burgdorferi* to humans. Relapsing fever is another tick-borne disease caused by *Borrelia* species that are transmitted by members of the argasid tick genus *Ornithodoros*, namely *O. moubata* in Africa and *O. hermsii* in western United States and Canada (Dworkin *et al.*, 2002a; Schwan and Piesman, 2002).

Ixodid ticks also transmit *Ehrlichia* species, which are Gram-negative intracellular organisms responsible for cases of HGE and also human monocytic ehrlichiosis (HME). *Ehrlichia chaffeensis* and *Ehrlichia ewingii* have been identified as the agents of HME, while *Anaplasma phagocytophilum* is responsible for HGE (Dumler and Bakken, 1998; Telford *et al.*, 1996). Most cases of HME have been shown to occur in areas where the metastriate *Amblyomma americanum* (Lone Star Tick) predominates although this illness has also been detected within the geographic distribution of the American dog tick, *Dermacentor variabilis*. These two species of metastriate ticks are also known to transmit spotted fever, a disease caused by obligate intracellular bacteria belonging to the genus *Rickettsia*, and tularaemia, an illness caused by the Gram-negative coccobacillus *Francisella tularensis*. In addition to tularaemia, the Rocky Mountain wood tick *Dermacentor andersoni* transmits *Rickettsia rickettsii* (spotted fever), and the Colorado tick fever virus.

Given the wide range of infectious agents that a particular species of tick may transmit, there is growing concern regarding the incidence of co-infection with more than one pathogen (Belongia, 2002). Although relatively unappreciated, simultaneous human infections with more than one tick-borne pathogen do occur as it has been described for the concurrent transmission of *B. burgdorferi* and *Babesia microti* as well as Lyme borreliosis and HGE by *I. scapularis* (Belongia, 2002; Krause *et al.*, 1996; Levin and Fish, 2000) and *I. ricinus* (Belongia, 2002; Skotarczak *et al.*, 2003). In clinical studies on disease manifestations in patients with dual tick-borne infections, co-infection with *Babesia* parasites was shown to increase the severity of the symptoms and prolong the persistence of the illness when compared to patients that had been infected with *B. burgdorferi* alone (Belongia, 2002; Krause *et al.*, 1996). Although the risks associated with concurrent infection by tick-borne infectious agents appear to be low, reports of double and even triple infections of single ticks with the pathogens described above continue to increase steadily and more research is needed to determine the potential for multiple transmission to humans (Skotarczak *et al.*, 2003; Stafford *et al.*, 1999; Varde *et al.*, 1998).

4 Host-acquired immunity to tick infestation

The host immune response against ectoparasites and many other infectious agents is regulated by $CD4^+$ T helper (Th) lymphocytes via cytokine secretion and activation of antigen-presenting cells (Jankovic *et al.*, 2001). As an innate immune response is induced, the profile of cytokine expression following the first encounter with a specific antigen is crucial in determining the adaptive immune response and the outcome of an infection. A Th1-dominated cytokine production pattern is generally induced by intracellular parasites, while a dominant Th2 cytokine profile is triggered by extracellular infections (Yazdanbakhsh *et al.*, 2001). Upon antigen presentation, naive precursor T helper cells (Thp) undergo differentiation from Th0 uncommitted cells to at least one of these two subsets, which will be dependent on the mediators present at the time of induction of an immune response (*Table 2*).

Th1 cells are associated with delayed-type hypersensitivity, increased phagocytic activity, and inflammatory responses (Szabo *et al.*, 2003). Cytokines secreted by this subset of T-cells include interferon γ (IFN-γ), interleukin-2 (IL-2), tumour necrosis factor β (TNF-β), and lymphotoxin-α (LT-α). Conversely, Th2 cells produce IL-4,

Table 2 *Characteristics of relevant cytokines secreted by the different Subsets of CD4⁺ T-helper (Th) cells.*

T-cell subset	Effector cytokines	Function	References
Naive Thp	IL-2	Precursor of T helper cell	(Liew, 2002)
Th0	IL-4, INF-γ	Uncommitted T helper cell	(Liew, 2002)
Th1	IL-2, TNF, INF-γ, LT-α	Cell-mediated immunity, control of intracellular pathogens	(Szabo *et al.*, 2003)
Th2	IL-4, IL-5, IL-9, IL-10, IL-13	Extra-cellular immunity, host defence against helminths, IgG class switching	(Finkelman and Urban, 2001)

IL-5, IL-9, IL-10 and IL-13, which stimulate growth and differentiation of mast cells and eosinophils, and support IgG1 subclass and IgE production (Finkelman and Urban, 2001). Other forms of $CD4^+$ T helper lymphocytes, which are beyond the scope of this chapter, are Th3 and Tr1 (regulatory) cells which appear to be involved in mucosal immunity and suppression of immune responses respectively (Groux *et al.*, 1997; Weiner, 2001).

Host-acquired immunity to tick infestation significantly impairs the ability of the ectoparasites to feed, resulting in reduced engorgement weight, decreased ova production, and egg and tick mortality (Wikel and Bergman, 1997). Depending on the tick–host association, resistance to infestation can result in extensive infiltration of basophils and eosinophils into the tick attachment sites inducing cutaneous basophil hypersensitivity, a form of delayed-type hypersensitivity mediated by Th1 lymphocytes (Brossard and Wikel, 1997). Tick-specific homocytotrophic antibodies bind and degranulate infiltrating basophils and mast cells resulting in the release of mediators of inflammation and chemotactic factors for eosinophils (Wikel, 1996a). Infiltration of basophils at tick attachment sites in animals resistant to tick infestation results in significantly increased levels of histamine, which correlate with reduced tick salivation and engorgement weight (Paine *et al.*, 1983; Wikel, 1982a).

Among the major antigen-presenting cells involved in triggering adaptive immunity, dendritic cells are critical in inducing an immune response (Liu *et al.*, 2001; Reise Sousa, 2001). Dendritic cells are widely distributed in both lymphoid and non-lymphoid tissues and they are capable of detecting foreign antigens during the very early stages of infection. Tick antigens' first contact with the host immune system occurs at the cutaneous interface. It is here that dendritic epidermal Langerhans cells play an important role in resistance to tick feeding by a process that involves tick antigen trapping, transportation of cell-bound antigens to draining lymph nodes, and delivery to T cells (Allen *et al.*, 1979a, b; Nithiuthai and Allen, 1985). Using guinea pigs as a model of acquired resistance to infestation with *D. andersoni*, Nithiuthai and Allen, (1984a) observed a decrease of up to a 65% in the density of ATPase-positive Langerhans cells at tick attachment sites on the third day of a primary infestation. Sensitized animals subjected to secondary infestations developed a significant increase in the numbers of Langerhans cells at the bite sites, resulting in epidermal hyperplasia and a granulocyte infiltration characteristic of an immune response directed towards tick feeding. Additional evidence came from a study involving depletion of Langerhans cells from the skin of resistant guinea pigs by treatment with UVC irradiation (Nithiuthai and Allen, 1984b). In this study, guinea pigs whose skin had been irradiated prior to infestation with *D. andersoni* larvae, showed a substantial decrease in the acquisition of resistance to tick feeding, which was determined from the number of larvae engorging over periods of 5 days (Nithiuthai and Allen, 1984b). Furthermore, by studying the role of dendritic cells in host immune modulation by tick-transmitted *B. burgdorferi*, Mbow and colleagues were first to describe that murine dendritic and Langerhans cells previously exposed to *B. burgdorferi* were able to induce protective immunity against tick-transmitted spirochetes after adoptive transfer into normal mice (Mbow *et al.*, 1997). Taken together, these results provided valuable insight into the role of dendritic cells in acquired resistance to tick infestation and demonstrated that Langerhans cells at the tick–host interface are central to mounting a successful immune response to tick infestation. However, answers regarding the mechanisms of protective immunity are still elusive, requiring a more extensive analysis of the events that take place at the host–tick interface during the early stages of tick infestation.

5 Bioactive compounds in tick saliva

During feeding, the cheliceral digits of hard ticks lacerate host tissues, enabling the hypostome to be inserted and anchored to the host skin and, in some species, a cement-like substance is secreted by the tick into the feeding cavity (Sonenshine, 1991). During the acquisition of a blood meal, ticks alternate feeding of blood with secretion of saliva into the wound. Tick saliva contains an array of pharmacological compounds that include immunomodulators, inhibitors of pain/itch response, anticoagulants, inhibitors of platelet aggregation, and vasodilatory molecules, all of which contribute to the evasion of the host immune and haemostatic defences (Ribeiro, 1995; Ribeiro and Francischetti, 2003; Wikel, 1999; Wikel and Alarcon-Chaidez, 2001).

5.1 *Anti-haemostatic factors*

Haematophagous arthropods have evolved to overcome host haemostasis by secreting a myriad of compounds in their saliva that act to ensure a continuous flow of blood to the bite sites during acquisition of a blood meal (Ribeiro, 1995; Wikel, 1996a). Blood coagulation is initiated when tissue factor (TF) is exposed to circulating factors VII or VIIa as a result of vascular damage. Proteolytic activation of factor X into Xa quickly follows, triggering a cascade of events that lead to the formation of thrombin, activation of platelet aggregation, and fibrin-clot formation at the site of injury (Broze *et al.*, 1990; Davie *et al.*, 1991). Stages of this process that could be the target of anti-haemostatic factors in arthropod saliva may include platelet adhesion/aggregation, activation of the intrinsic pathway of coagulation, and thrombin formation (Champagne and Valenzuela, 1996).

Molecules with anti-haemostatic properties were first described for the ixodid tick *I. scapularis* (=*I. dammini*) (Ribeiro *et al.*, 1985) and the soft Argasid tick *O. moubata* (Keller *et al.*, 1993; Waxman *et al.*, 1990), the vector of tick-borne relapsing fever (Parola and Raoult, 2001). A 60 amino acid peptide from *O. moubata* designated tick anti-coagulant peptide (TAP), was described as a serine protease inhibitor which showed limited homology to the family of Kunitz-type protease inhibitors and was able to specifically inhibit factor Xa (Waxman *et al.*, 1990). Tick-encoded serine protease inhibitors are considered to be important factors not only in blood coagulation, but also in complement activation, and inflammatory and immune responses (Mulenga *et al.*, 2001; Polanowski and Wilusz, 1996; Rubin, 1996).

Moubatin, a 17-kDa protein purified from *O. moubata* soluble extracts, was shown to inhibit aggregation of human platelets when stimulated with collagen, adenosine diphosphate (ADP) or thrombin (Keller *et al.*, 1993). The mechanism of inhibition by Moubatin resembles that of RPAI-1 (*Rhodnius prolixus* aggregation inhibitor 1), a 19-kDa lipocalin recently purified from the salivary glands of *R. prolixus* (Francischetti *et al.*, 2000). Once platelets are activated, they secrete factors such as thromboxane A2 and ADP that promote further platelet aggregation (Champagne and Valenzuela, 1996). Therefore, many haematophagous arthropods, including ticks, counteract aggregation by secreting enzymes known as apryases which hydrolyse adenosine triphosphate (ATP) and ADP into adenosine monophosphate (AMP) to inhibit blood coagulation (Champagne and Valenzuela, 1996). ADP is a potent inducer of platelet aggregation while ATP is pro-inflammatory.

Other inhibitors of platelet aggregation that have been characterized include Variabilin, an antagonist of fibrinogen receptor purified from salivary glands of *D. variabilis* (Wang *et al.*, 1996); anti-thrombin activities found in the saliva of the cattle tick *Boophilus microplus* (Horn *et al.*, 2000) and also in the salivary glands of *A. americanum* (Zhu *et al.*, 1997). Recently, a tissue factor pathway inhibitor (Francischetti *et al.*, 2002), along with a metalloproteinase-like protein (Francischetti *et al.*, 2003), and a group of related proteins with anti-coagulant properties (Narasimhan *et al.*, 2002) were purified from the saliva of *I. scapularis* and characterized.

5.2 *Anti-complement factors*

A critical component of host-acquired resistance subject to modulation by tick salivary antigens is the alternate pathway of complement activation, which is activated in the absence of specific antibody (Nielsen *et al.*, 2000). The complement system is a major effector mechanism of antibody-mediated as well as innate immunity. A central element in all complement pathways is activation induced by proteolysis of C3 which results in deposition of C3b on antigen surfaces or antibody bound to the antigen followed by phagocytosis of opsonized particles (Nielsen and Leslie, 2002). In addition, proteolytic complement fragments C3a, C4a and C5a increase vascular permeability and mediate mast cell degranulation contributing to inflammation at the sites of complement activation, whereas the membrane attack complex (MAC) C5b-9 causes osmotic lysis of cells (Kohl, 2001). An anti-complement activity has been detected in the saliva of *I. scapularis* the vector of Lyme borreliosis (Ribeiro, 1987; Valenzuela *et al.*, 2000). *Ixodes scapularis* saliva was able to inhibit deposition of complement components C3b and C5b on activating surfaces, and the generation of anaphylatoxin C3a, which reduces neutrophil chemotaxis and induces degranulation of mast cells localized at tick attachment sites with the release of vasoactive mediators. Further inhibition of anaphylatoxin production has also been attributed to a carboxypeptidase-like activity found in tick saliva (Ribeiro and Mather, 1998; Ribeiro *et al.*, 1985). This enzymatic activity was attributed to a kininase, a metalloenzyme that degrades bradykinin released in response to vascular injury and a known mediator of pain (Ribeiro and Mather, 1998). A similar activity was also detected in chromatographic fractions from the salivary glands of the cattle tick *Boophilus microplus*. However, involvement of this enzyme in facilitating tick feeding was not established (Bastiani *et al.*, 2002).

Anti-complement activity was recently detected in saliva from *I. ricinus*, which was able to prevent haemolysis of sheep red blood cells (SRBC) by the human alternative pathway of complement (Mejri *et al.*, 2002). Inhibition was shown to proceed in a dose-dependent manner with up to 70% lysis inhibition observed at the highest dose of saliva protein. Additional evidence supporting tick anti-complement was provided by the work of Rathinavelu *et al.* (2003) who observed inhibition of C3 deposition on *B. burgdorferi* spirochetes by incubation with the contents of *I. scapularis* nymphal tick guts.

The first line of defence against extracellular parasites resides in the alternative pathway of complement activation and recent findings from a number of studies suggest that ticks appear to have evolved mechanisms to impair this complement activity. Therefore, components of these mechanisms represent potential candidates for the design of transmission-blocking vaccines.

6 Tick modulation of host immune function

The prolonged periods of attachment required for full engorgement of ixodid ticks give the host ample opportunity to activate immune responses that can impair blood feeding and result in the development of acquired host immunity to tick infestation. Therefore, in order to overcome rejection, ixodid ticks have evolved ways to modulate the effector pathways involved in host innate and acquired immunity to infestation. As a result, suppression/modulation of host defences leads to reduced host reactions against the tick and renders the host less capable of responding to tick-borne infectious agents (Gayle and Ringdahl, 2001; Parola and Raoult, 2001; Schoeler and Wikel, 2001; Wikel and Alarcon-Chaidez, 2001).

Since T-lymphocytes play important roles in adaptive immune responses against foreign antigens, they are targets of tick-induced immune modulation. A number of studies have shown that tick infestation does not alter the lymphocyte proliferative response to B-cell mitogens, whereas responsiveness to T-cell mitogens is significantly suppressed (Wikel, 1996a). Furthermore, extensive studies dealing with tick-induced modification of cytokine expression profiles in murine cell systems support the hypothesis of tick saliva antigens playing an active role in suppression of T-cell proliferation (Wikel and Alarcon-Chaidez, 2001).

Immunomodulation of T-cell responsiveness by ixodid tick feeding was first observed following infestation of BALB/c mice with female *D. andersoni* ticks (Ramachandra and Wikel, 1992; Wikel, 1982b). In those studies, murine T-lymphocyte proliferation in response to the T-cell mitogen Concanavalin A (Con A) was markedly reduced by tick salivary gland extracts (SGE). At the same time, the levels of the macrophage-derived cytokines IL-1β and TNF-α along with the T-lymphocyte cytokines IL-2 and INF-γ were significantly reduced (Ramachandra and Wikel, 1992). INF-γ is a potent Th1 effector cytokine that is involved in activation of macrophages, and in combination with IL-2 and TNF-β, promotes differentiation of $CD8^+$ T-cells into active cytotoxic cells and recruitment and activation of inflammatory leucocytes respectively (Rao and Avni, 2000).

Increased expression of IL-4, a marker of differentiation of Th cells into the Th2 subset, has also been reported for murine models of tick infestation (Schoeler and Wikel, 2001). Tick saliva and SGE up-regulation of IL-4 were demonstrated for the brown tick *Rhipicephalus sanguineus*, a vector of ehrlichiosis and babesiois (Ferreira and Silva, 1999). In that study, the cytokines IL-10, and transforming growth factor-β (TGF-β) were also up-regulated, whereas the opposite was observed for the Th1 cytokines IL-2, IFN-γ and IL-12 (Ferreira and Silva, 1999). IL-10 and TGF-β are important anti-inflammatory cytokines which are known to exert suppressing effects on T-cell proliferation and APC activation in parasitic infections (Yazdanbakhsh *et al.*, 2001). Furthermore, IL-10 can block key components of the effector phase, such as mast cell degranulation, restricting tissue reactivity and impairing the ability of the host to mount a protective immune response to tick feeding and pathogen transmission. Although up-regulation of IL-4 and IL-10 expression has been reported for a number of ticks including *I. scapularis* (Schoeler *et al.*, 1999) *I. pacificus* (Schoeler *et al.*, 2000) and *I. ricinus* (Kopecky *et al.*, 1999; Mbow *et al.*, 1994a; Mejri *et al.*, 2001), contrasting results from a recent study utilizing *I. ricinus* saliva and SGE showed strong down-regulation of IL-10 production by murine splenocytes with concurrent reduction in ConA and LPS-induced splenocyte proliferation (Hannier *et al.*, 2003). In that study, inhibition of IL-10 production was shown to be stronger in LPS-stimulated splenocytes, suggesting a direct and T-cell-independent effect of saliva components on B cells. Tick saliva or SGE factors responsible for B-cell unresponsiveness were not identified but results were consistent with other studies showing an impaired ability of tick-infested hosts to mount appropriate antibody responses to invading antigens (Christe *et al.*, 2000; Wikel, 1985).

Host–tick associations in which acquired resistance develops are characterized by the activation of infiltrating $CD4^+$ T-cells and the generation of delayed hypersensitivity responses at the site of tick attachment. Examples of these events have been observed during tick infestation of a number of laboratory animals including guinea pigs, rabbits and mice (Brossard and Wikel, 1997; Wikel, 1996a). Interestingly, *I. ricinus* (Christe *et al.*, 1999; Mbow *et al.*, 1994b), *I. scapularis* and *I. pacificus* (Schoeler *et al.*, 2000) appear to impair the ability of the host to elicit acquired resistance to feeding. Indeed, in experiments with BALB/c mice infested with *I. ricinus*, $CD45R^+$ B-lymphocytes which may be involved in the induction of protective humoral response at the skin level were absent from tick attachment sites (Mbow *et al.*, 1994b). On the other hand, acquired resistance to *D. andersoni* involves a significant antibody component (Whelen and Wikel, 1993).

Natural killer (NK) cells are another population of lymphocytes that participate in innate immunity and early defence against viruses and other intracellular microbes (Biron *et al.*, 1999). These lymphocytes have also been the target of tick-induced immunosuppression (Kopecky and Kuthejlova, 1998; Kubes *et al.*, 1994, 2002). SGE from female *Dermacentor reticulatus* (Kubes *et al.*, 1994), and *Amblyomma variegatum* and *Haemaphysalis inermis* to a lesser extent (Kubes *et al.*, 2002), inhibited the activity of human NK cells when compared with that observed for SGE preparations from unfed ticks. Although NK cells are not known to have direct effector functions in tick-acquired immunity, they could potentially contribute to a Th1 immune response by production of IFN-γ, a Th1 effector cytokine that stimulates cell-mediated immunity, such as delayed-type hypersensitivity, macrophage activation and inflammatory responses. Also, their role in responding to tick-borne infectious agents can be potentially important.

The immunological basis for differences in the levels of expression of acquired resistance between different host and tick species is not thoroughly defined (Lawrie and Nuttall, 2001; Wikel, 1996a; Willadsen and Jongejan, 1999). Much of the accumulated evidence about the cellular and immunological basis for acquired resistance is derived from *in vitro* or *in vivo* studies with a laboratory animal species and extrapolation to other tick–host relationships needs to be addressed with caution. The complexity of the tick-induced immunological responses stresses the importance of developing a more detailed understanding of the mechanisms underlying resistance before strategies for enhanced protection against tick feeding can be implemented.

7 Ticks as vectors of disease

Molecules present in tick saliva that possess immunomodulatory and other pharmacological activities are known to be essential not only for tick survival but also for transmission of infectious agents (Ribeiro, 1995; Wikel, 1996b). Tick saliva enhances host infection with Thogoto virus (Jones *et al.*, 1992a), *Theileria parva* (Shaw *et al.*, 1993), tick-borne encephalitis virus (Labuda *et al.*, 1993), and vesicular stomatitis virus (Hajnicka *et al.*, 2000). Studies to evaluate the potential of a number of different tick species as vectors of Thogoto virus showed that *Rhipicephalus appendiculatus* was more competent in its ability to enhance transmission of the virus than a number of other ixodid or argasid ticks (Jones *et al.*, 1992a, b).

Borrelia burgdorferi sensu lato, the causative agent of Lyme borreliosis, is transmitted by the ixodid tick *I. scapularis* and is the most frequently reported tick-borne disease in the United States (Center for Disease Control, 2002). Modulation of the host immune defences by *I. scapularis* appears to significantly influence the transmission of the spirochete *B. burgdorferi* (Wikel *et al.*, 1997; Zeidner *et al.*, 1996, 1997). Infestation of Lyme disease-susceptible C3H/HeN mice with infected *I. scapularis* has been shown to increase the levels of the Th2 cytokine IL-4, whereas Th1 cytokines such as IL-2 and IFN-γ are down-regulated (Schoeler *et al.*, 1999). Conversely, BALB/c mice resistant to this disease do not seem to experience significant changes in their cytokine expression pattern.

Evidence supporting the importance of tick modulation of host cytokines in the transmission of infectious agents came from the work of Zeidner *et al.* (1996), in which passive transfer of the Th1 cytokines TNF-α, IL-2, and IFN-γ into disease-susceptible C3H/HeJ mice prevented transmission of the spirochete from infected *I. scapularis*. Infection rates ranged from 5% for TNF-α-treated mice to 30–45% for mice that received IL-2 or IFN-γ. In contrast, an 83% infection rate was observed in unreconstituted controls that were infested under similar conditions. In a related study, sensitization of BALB/c mice by repeated infestations with pathogen-free *I. scapularis* nymphs prior to infestation with *B. burgdorferi*-infected ticks, resulted in development of resistance to tick transmission of the spirochete with only 17% of the challenged mice being positive for *B. burgdorferi*, while control mice showed an infection rate of 100% (Wikel *et al.*, 1997). Initial infestation with pathogen-free ticks has also been shown to induce host responses capable of blocking subsequent tick transmission of *Francisella tularensis* (Bell *et al.*, 1979) and *B. burgdorferi* (Wikel *et al.*, 1997).

8 Anti-inflammatory and immunomodulatory molecules

Earlier studies on the characterization of tick-acquired resistance demonstrated that inflammatory responses could be reduced by the concurrent administration of histamine type I and type II receptor antagonists which bound histamine and reduced tick rejection (Brossard, 1982; Paine *et al.*, 1983; Wikel, 1982a). Histamine, a mediator of inflammation as well as itch responses, is secreted by mast cells and infiltrating basophils in response to tissue damage. Therefore, ticks have developed strategies to suppress inflammation by secreting histamine-binding proteins (HBPs) into the attachment site. HBPs function by sequestering secreted histamine molecules, preventing them from eliciting an inflammatory response that would result in tick rejection. Histamine-binding proteins have been reported in the saliva of *R. apendiculatus* and *D. reticulatus* (Paesen *et al.*, 1999, 2000; Sangamnatdej *et al.*, 2002).

Prostaglandins (PG) are metabolites derived from arachidonic acid which are normally found in high concentrations in tick saliva (Sauer *et al.*, 2000). Some of the activities attributed to tick saliva prostaglandins are inhibition of platelet aggregation through inhibition of ADP secretion, vasodilation, and impairment of T-cell function by suppression of the production of IL-2 and IFN-γ. However, their role in tick feeding and pathogen transmission has not been established (Bowman *et al.*, 1996; Champagne and Valenzuela, 1996; Sauer *et al.*, 2000).

Attempts to characterize tick salivary components that may potentially influence tick–host interactions and pathogen transmission, have led to the identification of a number of molecules, mainly proteins, that are specifically induced by tick salivary glands during feeding. Immunomodulation upon exposure to tick SGE and/or saliva has been observed for a number of ixodid ticks including *D. andersoni* (Wikel, 1982b), *B. microplus* (Inokuma *et al.*, 1993) *I. ricinus* (Dusbabek *et al.*, 1995) and *R. sanguineus* (Ferreira and Silva, 1998). Salivary gland extracts from feeding *D. andersoni* were able to suppress mitogen-induced *in vitro* proliferation of murine T-lymphocytes (Ramachandra and Wikel, 1992), and this suppressive activity was attributed in part to a 36-kDa protein (p36) present in the saliva of female *D. andersoni* (Alarcon-Chaidez *et al.*, 2003; Bergman *et al.*, 2000). In a recent report, the recombinant form of an *I. ricinus* salivary gland protein, Iris, suppressed *in vitro* proliferation of Con A stimulated naive spleen cells by up to 81% (Leboulle *et al.*, 2002a). Iris was also able to down-regulate production of IFN-γ by T-lymphocytes and antigen-presenting cells, while leaving the levels of IL-5 and IL-10 unchanged or reduced (Leboulle *et al.*, 2002a).

A homologue of the human pro-inflammatory macrophage inhibitory factor (MIF) was isolated from a cDNA library from *A. americanum* ticks (Jaworski *et al.*, 2001). Using anti-MIF serum, this protein was localized to the midgut and salivary gland tissues of the tick and it is the first cytokine-like factor identified from ticks. However, the functional role of tick MIF, concerning its potential role as an immunosuppressant, remains to be determined.

An IL-2-binding activity present in *I. scapularis* saliva has been shown to inhibit mouse spleen cell *in vitro* proliferation and CD4+ T-cell activation in response to anti-CD3 monoclonal antibodies (Anguita *et al.*, 2002; Gillespie *et al.*, 2001). Similarly, anti-IL-8 activities have been described for a number of ixodid species including *Dermacentor, Amblyomma, Rhipicephalus* and *Ixodes* (Hajnicka *et al.*, 2001). IL-2 promotes differentiation of CD8 T-cells into actively cytotoxic cells (Rao and Avni, 2000) whereas IL-8 is an inflammatory chemokine normally induced by lipopolysaccharide (LPS), IL-1 and TNF-α which is involved in neutrophil chemotaxis and the attraction of effector and memory lymphocytes (Mahalingam and Karupiah, 1999; Sallusto *et al.*, 2000).

9 Anti-tick vaccines

At the present time, suppression of tick populations has relied heavily on the large-scale use of chemical acaricides. However, development of tick resistance to these chemicals together with potential environmental contamination has prompted the search for alternate means of control such as vaccination of hosts against tick antigens, i.e. transmission-blocking vaccines.

The first successful commercial vaccine against tick infestation was based on a concealed antigen from *B. microplus* midgut called Bm86 (Willadsen *et al.*, 1989). Immunization with this 89-kDa glycoprotein induced an antibody response in cattle, which dramatically affected the reproductive efficiency of ticks. Subsequently, the same group identified two more glycoproteins from *B. microplus*: the 86-kDa Bm91 (Riding *et al.*, 1994) and the 63-kDa BmA7 (McKenna *et al.*, 1998) antigens. Although these molecules were not as effective as Bm86, cocktail preparations of either one with Bm86 enhanced the anti-tick capacity of the commercial vaccine (McKenna *et al.*, 1998).

Another tick antigen that also proved to be efficient in providing protection against tick infestation was a 29-kDa protein from the salivary glands of *Haemaphysalis longicornis*, a vector of *Theileria* spp. and *Coxiella burnetii* (Mulenga *et al.*, 1999). This antigen, named p29, has a predicted structural homology to known collagen proteins and it was tentatively identified as a tick cement protein. Infestation of rabbits vaccinated with p29 resulted in decreased adult tick engorgement with little or no effect on the length of tick feeding. However, significant increases in larval and nymphal tick mortality were also observed, thus demonstrating the potential of salivary gland-derived antigens to induce anti-tick immunity.

Strategies used in developing anti-tick vaccines currently involve molecular characterization and *in vitro* expression of proteins secreted in tick saliva or associated with salivary glands (exposed antigens) or even those that do not normally interact with the host immune defences (concealed antigens) such as in tick midgut (Mulenga *et al.*, 2000; Trimnell *et al.*, 2002). In this way, defined subunit vaccines can be produced either by chromatographic purification of the antigen from saliva or salivary gland homogenates or by producing the antigen in a recombinant system (Lee and Opdebeeck, 1999; Mulenga *et al.*, 2000; Willadsen, 2001). Some of the tick antigens that have been characterized using the approach just described include an inhibitor of platelet aggregation from *O. moubata*, (Keller *et al.*, 1993), anti-haemostatic factors from *Ornithodoros savignyi* (Joubert *et al.*, 1998) and *I. scapularis* (Francischetti *et al.*, 2000, 2003; Valenzuela *et al.*, 2000), histamine- and immunoglobulin-binding proteins from *R. appendiculatus* (Paesen *et al.*, 1999; Wang and Nuttall, 1999), an immunosuppressant protein from *D. andersoni* (Alarcon-Chaidez *et al.*, 2003; Bergman *et al.*, 2000), and a number of housekeeping genes from *B. microplus* (Ferreira *et al.*, 2002a, b; Rosa de Lima *et al.*, 2002).

As mentioned above, research efforts have been directed toward development of recombinant vaccines and different types of tick antigens have been selected as potential candidates. Unfortunately, most attempts to induce protective immunity against tick infestation have met with limited success (Mulenga *et al.*, 2000). Recently, the use of DNA or gene-based immunization strategies is being considered as a more efficient approach to vaccine development and immunotherapy (Thalhamer *et al.*, 2001). DNA vaccines consist of plasmid DNA carrying genetic material encoding an antigen whose expression within the cells is under control of eucaryotic promoters (Srivastava and Liu, 2003). Important advantages of DNA vaccines over classic vaccines include their ability to induce both humoral and cellular immune responses in a range of hosts and the potential to enhance their immunogenicity by co-expression of other immunomodulatory proteins (Sharma and Khuller, 2001; Thalhamer *et al.*, 2001). Although successful induction of protective responses using genetic immunization has been demonstrated for a number of viral, protozoan and bacterial pathogens (Srivastava and Liu, 2003), there are only a handful of reports describing the use of DNA vaccination to induce immune responses against haematophagous vectors of disease (De Rose *et al.*, 1999; Foy *et al.*, 2003). Nevertheless, DNA vaccines represent an alternative strategy with enormous potential as a research tool, which may translate into faster and more efficient screening of vector antigen targets.

10 Prospects for the use of functional genomics for tick control

Recent advances in genome sequencing have resulted in a wealth of information derived from a growing number of tick species offering exciting new opportunities in the search for genes that may encode vaccine candidates. However, the immense amount of information now accumulated calls for more efficient, high-throughput methods to meet the demands of the challenging task of characterizing the expression patterns of all genes and the biological function of the proteins encoded by them. Thus, combining data mining of available sequences with expression profiles from microarray and proteomic technologies can be expected to increase the probability of identifying genes involved in host–pathogen interactions.

Currently, the use of expressed sequence tags (EST) is one of the high-throughput methods of choice to obtain information from cDNA libraries suitable for use in DNA vaccination trials. ESTs are short-tagged DNA sequences which are generated from mRNA expressed during one or more stages of the life cycle of an organism (Bashiardes and Lovett, 2001). In the study of tick–host pathogen interactions, the ultimate goal of generating sequence data from tick EST databases is to identify genes whose products will help in deducing important processes that take place during host evasion and pathogen transmission. Although information on the genomes of ticks of major veterinary and medical importance is still limited, a number of EST projects are now at a stage where enough data exist to provide new insights into host–tick–pathogen relationships.

A study on the genome of the cattle tick *B. microplus* using an EST approach (Crampton *et al.*, 1998) was aimed at finding potential genes that could aid in the control of this important ectoparasite. Differences in two developmental stages of the Lone Star tick, *A. americanum*, were studied by analysing sequence information from over 1900 ESTs collected from mRNA of adult and larval stages (Hill and Gutierrez, 2000). Similarly, sequences from a cDNA library from the salivary glands of *A. variegatum* contributed to the identification of proteins with potential anti-haemostatic and anti-inflammatory properties which might shed light into the ability of this ectoparasite to transmit disease-causing organisms (Nene *et al.*, 2002). Other recent sequencing projects involved mainly ixodid ticks and included *A. americanum* and *D. andersoni* (Bior *et al.*, 2002), *I. ricinus* (Leboulle *et al.*, 2002b), and *I. scapularis* (Valenzuela *et al.*, 2002).

In addition to identifying and targeting individual cDNA sequences for cloning and expression, the availability of entire cDNA libraries from different feeding or developmental stages of a given tick species will enable comparison of expression profiles using array methods such as cDNA or oligonucleotide microarrays. The use of microarray technologies to study tick–host interactions is still in the planning stages; however, the expression profiles of both mRNA and protein obtained from this approach can greatly contribute to understanding of the events leading to tick immunosuppression and possibly help in identifying vaccine candidates.

As stated in the previous section, genetic vaccination represents a promising approach for vaccine candidate identification; however, the need for previous knowledge about the encoded antigen restricts the use of this technique to only well-characterized antigens, which may result in the omission of less abundantly expressed and/or stage-specific antigens that may be important in eliciting protective immune responses. One of the genome-based alternatives often used to overcome these limitations is a technique called expression library immunization (ELI) (Barry *et al.*, 1995). ELI makes use of genomic libraries consisting of randomly generated DNA fragments from the genome of a given pathogen, which are cloned into appropriate immunization vectors. Gradual fractionation of protective libraries reduces the complexity of the genomic libraries and eventually results in the identification of pools of protective clones from hundreds or even thousands of candidate molecules (Johnston and Barry, 1997; Moore *et al.*, 2001; Smooker *et al.*, 2000). This strategy has led to the detection and identification of many protective antigens from many bacteria, viruses and protozoans. Using a modification of this technique, Almazan *et al.* (2003) identified several antigens from the ixodid tick *I. scapularis* that were able to induce protective immunity against tick infestation in mice. In this study, cDNA molecules were used instead of fragments from genomic DNA, further emphasizing the potential of tick EST databases to accelerate the pace of anti-tick vaccine discovery. Furthermore, complementation of functional genomics using state of the art proteomic approaches combined with high-throughput screening creates an unlimited number of possibilities for the rapid identification of new targets for therapeutics.

Proteomics has been defined as the study of the function of all proteins expressed by the genome. To date, profiling of protein expression in tick vectors involves resolution of soluble protein extracts from different tick tissues by either one-dimensional or two-dimensional polyacrylamide gel electrophoresis, followed by N-terminal sequencing, or high-pressure liquid chromatography (HPLC) coupled to mass spectrometry. A growing number of potential tick immunogens have been identified using combinations of these techniques and they continue to be the methods of choice. However, the implementation of recent developments in the science of chromatographic separation such as high-throughput multi-dimensional chromatographic systems and sophisticated array technologies can be expected to have a major impact in the study of tick immunobiology.

11 Conclusions

Knowledge of the mechanisms by which ticks elicit both innate and acquired immune responses and how these events influence transmission of tick-borne pathogens has increased dramatically over the past years. Nonetheless, progress toward characterization of antigens responsible for naturally acquired immunity in tick–host relationships is still in its early stages.

The vast amount of information obtained from the primary DNA sequences through high-throughput analyses has had a dramatic impact on the approaches taken to decipher the mechanisms of host evasion and disease transmission. However, information from mRNA and cDNA alone is often insufficient when it comes to functional characterization of a given protein antigen in terms of temporal expression, regulation, activity and post-translational modifications.

The plethora of bioactive compounds identified in tick saliva that have been implicated in both host immunity and disease transmission are a clear indication that the immunological relationships among ticks, their hosts, and the disease-causing agents they transmit are very complex. Therefore, more work is needed in order to characterize these antigens at the molecular level and the combined use of microarray and proteomic technologies represents an exciting new way to examine these processes. DNA-based vaccines offer a way to address the problems of antigen delivery, and the induction of an appropriate cell-mediated immune response needed for protection against tick-borne pathogens. Furthermore, the implementation of multi-dimensional chromatographic fractionation techniques for the analysis of the tick proteome may help in overcoming the limitations of the methods of separation currently used. Evidence accumulated over the years indicates that many molecules in tick saliva involved in successful feeding tend to be poorly immunogenic and data from vaccination trials suggest that a good strategy for developing an effective vaccine may consist of the use of a combination of antigens (multivalent) known to elicit a neutralizing immune response without compromising host immune functions.

Acknowledgements

The work of S.K.W. was supported in part by Grant AI46676 from the National Institute of Allergy and Infectious Diseases, National Institutes of Health, and Cooperative Agreement number U50/CCU119575 from the Center for Disease Control and Prevention, United States Public Health Service.

References

Alarcon-Chaidez, F.J., Muller-Doblies, U.U. and Wikel, S. (2003) Characterization of a recombinant immunomodulatory protein from the salivary glands of *Dermacentor andersoni*. *Parasite Immunol.* **25:** 69–77.

Allen, J.R., Khalil, H.M. and Graham, J.E. (1979a) The location of tick salivary antigens, complement and immunoglobulin in the skin of guinea-pigs infested with *Dermacentor andersoni* larvae. *Immunology* **38:** 467–472.

Allen, J.R., Khalil, H.M. and Wikel, S.K. (1979b) Langerhans cells trap tick salivary gland antigens in tick-resistant guinea pigs. *J. Immunol.* **122:** 563–565.

Almazan, C., Kocan, K.M., Bergman, D.K., Garcia-Garcia, J.C., Blouin, E.F. and de la Fuente, J. (2003) Identification of protective antigens for the control of *Ixodes scapularis* infestations using cDNA expression library immunization. *Vaccine* **21:** 1492–1501.

Anderson, J.F. (2002) The natural history of ticks. *Med. Clin. North Am.* **86:** 205–218.

Anguita, J., Ramamoorthi, N., Hovius, J.W., *et al.* (2002) Salp15, an *Ixodes scapularis* salivary protein, inhibits CD4(+) T cell activation. *Immunity* **16:** 849–859.

Barry, M.A., Lai, W.C. and Johnston, S.A. (1995) Protection against *Mycoplasma* infection using expression-library immunization. *Nature* **377:** 632–635.

Bashiardes, S. and Lovett, M. (2001) cDNA detection and analysis. *Curr. Opin. Chem. Biol.* **5:** 15–20.

Bastiani, M., Hillebrand, S., Horn, F., Kist, T.B., Almeida Guimarães, J. and Termignoni, C. (2002) Cattle tick *Boophilus microplus* salivary gland contains a thiol-activated metalloendopeptidase displaying kininase activity. *Insect Biochem. Mol. Biol.* **32:** 1439–1446.

Bell, J.F., Stewart, S.J. and Wikel, S.K. (1979) Resistance to tick-borne *Francisella tularensis* by tick-sensitized rabbits: allergic klendusity. *Am. J. Trop. Med. Hyg.* **28:** 876–880.

Belongia, E.A. (2002) Epidemiology and impact of coinfections acquired from *Ixodes* ticks. *Vector Borne Zoonotic Dis.* **2:** 265–273.

Bergman, D.K., Palmer, M.J., Caimano, M.J., Radolf, J.D. and Wikel, S.K. (2000) Isolation and molecular cloning of a secreted immunosuppressant protein from *Dermacentor andersoni* salivary gland. *J. Parasitol.* **86:** 516–525.

Bior, A.D., Essenberg, R.C. and Sauer, J.R. (2002) Comparison of differentially expressed genes in the salivary glands of male ticks, *Amblyomma americanum* and *Dermacentor andersoni*. *Insect Biochem. Mol. Biol.* **32:** 645–655.

Biron, C.A., Nguyen, K.B., Pien, G.C., Cousens, L.P. and Salazar-Mather, T.P. (1999) Natural killer cells in antiviral defense: Function and regulation by innate cytokines. *Ann. Rev. Immunol.* **17:** 189–220.

Black, W.C. 4th., Klompen, J.S. and Keirans, J.E. (1997) Phylogenetic relationships among tick subfamilies (Ixodida: Ixodidae: Argasidae) based on the 18S nuclear rDNA gene. *Mol. Phylogenet. Evol.* **7:** 129–144.

Bowman, A.S., Dillwith, J.W. and Sauer, J.R. (1996) Tick salivary prostaglandins: presence, origin and significance. *Trends Parasitol.* **12:** 388–396.

Brossard, M. (1982) Rabbits infested with adult *Ixodes ricinus* L.: effects of mepyramine on acquired resistance. *Experientia* **38:** 702–704.

Brossard, M. and Wikel, S.K. (1997) Immunology of interactions between ticks and hosts. *Med. Vet. Entomol.* **11:** 270–276.

Broze, G.J., Jr., Girard, T.J. and Novotny, W.F. (1990) Regulation of coagulation by a multivalent Kunitz-type inhibitor. *Biochemistry* **29:** 7539–7546.

Center for Disease Control (2002) Lyme disease – United States, 2000. *Morb. Mort. Week. Rep.* **51:** 29–31.

Champagne, D.E. and Valenzuela, J.G. (1996) Pharmacology of haematophagous arthropod saliva. In: Wikel, S.K. (ed.) *The Immunology of Host–Ectoparasitic Arthropod Relationships*, pp. 85–106. CAB International, Wallingford, UK.

Childs, J.E. and Paddock, C.D. (2003) The ascendancy of *Amblyomma americanum* as a vector of pathogens affecting humans in the United States. *Ann. Rev. Entomol.* **48:** 307–337.

Choi, E. (2002) Tularemia and Q fever. *Med. Clin. North Am.* **86:** 393–416.

Christe, M., Rutti, B. and Brossard, M. (1999) Influence of the genetic background and parasite load of mice on the immune response developed against nymphs of *Ixodes ricinus*. *Parasitol. Res.* **85:** 557–561.

Christe, M., Rutti, B. and Brossard, M. (2000) Cytokines (IL-4 and IFN-gamma) and antibodies (IgE and IgG2a) produced in mice infected with *Borrelia burgdorferi* sensu stricto via nymphs of *Ixodes ricinus* ticks or syringe inoculations. *Parasitol. Res.* **86:** 491–496.

Crampton, A.L., Miller, C., Baxter, G.D. and Barker, S.C. (1998) Expressed sequence tags and new genes from the cattle tick, *Boophilus microplus*. *Exper. Appl. Acarol.* **22:** 177–186.

Davie, E.W., Fujikawa, K. and Kisiel, W. (1991) The coagulation cascade: initiation, maintenance, and regulation. *Biochemistry* **30:** 10363–10370.

De Rose, R., McKenna, R.V., Cobon, G., Tennent, J., Zakrzewski, H., Gale, K., Wood, P. R., Scheerlinck, J.P. and Willadsen, P. (1999) Bm86 antigen induces a protective immune response against *Boophilus microplus* following DNA and protein vaccination in sheep. *Vet. Immunol. Immunopathol.* **71:** 151–160.

Dumler, J.S. and Bakken, J.S. (1998) Human ehrlichioses: newly recognized infections transmitted by ticks. *Annu. Rev. Med.* **49:** 201–213.

Dusbabek, F., Borsky, I., Jelinek, F. and Uhlir, J. (1995) Immunosuppression and feeding success of *Ixodes ricinus* nymphs on BALB/c mice. *Med. Vet. Entomol.* **9:** 133–140.

Dworkin, M.S., Schwan, T.G. and Anderson, D.E., Jr. (2002a) Tick-borne relapsing fever in North America. *Med. Clin. North Am.* **86:** 417–433, viii–ix.

Dworkin, M.S., Shoemaker, P.C., Fritz, C.L., Dowell, M.E. and Anderson, D.E., Jr. (2002b) The epidemiology of tick-borne relapsing fever in the United States. *Am. J. Trop. Med. Hyg.* **66:** 753–758.

Ferreira, B.R. and Silva, J.S. (1998) Saliva of *Rhipicephalus sanguineus* ticks impairs T cell proliferation and IFN-γ -induced macrophage microbicidal activity. *Vet. Immunol. Immunopathol.* **64:** 279–293.

Ferreira, B.R. and Silva, J.S. (1999) Successive tick infestations selectively promote a T-helper 2 cytokine profile in mice. *Immunology* **96:** 434–439.

Ferreira, C.A., Barbosa, M.C., Silveira, T.C., Valenzuela, J.G., Vaz Ida, S., Jr. and Masuda, A. (2002a) cDNA cloning, expression and characterization of a *Boophilus microplus* paramyosin. *Parasitology* **125:** 265–274.

Ferreira, C.A., Da Silva Vaz, I., da Silva, S.S., Haag, K.L., Valenzuela, J.G. and Masuda, A. (2002b) Cloning and partial characterization of a *Boophilus microplus* (Acari: Ixodidae) calreticulin. *Exp. Parasitol.* **101:** 25–34.

Finkelman, F.D. and Urban, J.F., Jr. (2001) The other side of the coin: the protective role of the TH2 cytokines. *J. Allergy Clin. Immunol.* **107:** 772–780.

Foy, B.D., Magalhaes, T., Injera, W.E., Sutherland, I., Devenport, M., Thanawastien, A., Ripley, D., Cardenas-Freytag, L. and Beier, J.C. (2003) Induction of mosquitocidal activity in mice immunized with *Anopheles gambiae* midgut cDNA. *Infect. Immun.* **71:** 2032–2040.

Francischetti, I.M.B., Ribeiro, J.M.C., Champagne, D. and Andersen, J. (2000) Purification, cloning, expression, and mechanism of action of a novel platelet aggregation inhibitor from the salivary gland of the blood-sucking bug, *Rhodnius prolixus*. *J. Biol. Chem.* **275:** 12639–12650.

Francischetti, I.M., Valenzuela, J.G., Andersen, J.F., Mather, T.N. and Ribeiro, J.M. (2002) Ixolaris, a novel recombinant tissue factor pathway inhibitor (TFPI) from the salivary gland of the tick, *Ixodes scapularis*: identification of factor X and factor Xa as scaffolds for the inhibition of factor VIIa/tissue factor complex. *Blood* **99:** 3602–3612.

Francischetti, I.M., Mather, T.N. and Ribeiro, J.M. (2003) Cloning of a salivary gland metalloprotease and characterization of gelatinase and fibrin(ogen)lytic activities in the saliva of the Lyme disease tick vector *Ixodes scapularis*. *Biochem. Biophys. Res. Commun.* **305:** 869–875.

Gayle, A. and Ringdahl, E. (2001) Tick-borne diseases. *Am. Fam. Physician* **64:** 461–466.

Gillespie, R.D., Dolan, M.C., Piesman, J. and Titus, R.G. (2001) Identification of an IL-2 binding protein in the saliva of the Lyme disease vector tick, *Ixodes scapularis*. *J. Immunol.* **166:** 4319–4326.

Gratz, N.G. (1999) Emerging and resurging vector-borne diseases. *Ann. Rev. Entomol.* **44:** 51–75.

Groux, H., O'Garra, A., Bigler, M., Rouleau, M., Antonenko, S., de Vries, J.E. and Roncarolo, M.G. (1997) A CD4+ T-cell subset inhibits antigen-specific T-cell responses and prevents colitis. *Nature* **389:** 737–742.

Gubler, D.J. (2000) Insects in disease transmission, In: Strickland, G.T. (ed.) *Hunter's Tropical Medicine and Emerging Infectious Diseases*, pp. 1005–1019, 8th edn. W.B. Saunders, Philadelphia, PA.

Hajnicka, V., Kocakova, P., Slavikova, M., Slovak, M., Gasperik, J., Fuchsberger, N. and Nuttall, P.A. (2001) Anti-interleukin-8 activity of tick salivary gland extracts. *Parasite Immunol.* **23:** 483–489.

Hajnicka, V., Kocakova, P., Slovak, M., Labuda, M., Fuchsberger, N. and Nuttall, P. A. (2000) Inhibition of the antiviral action of interferon by tick salivary gland extract. *Parasite Immunol.* **22:** 201–206.

Hannier, S., Liversidge, J., Sternberg, J.M. and Bowman, A.S. (2003) *Ixodes ricinus* tick salivary gland extract inhibits IL-10 secretion and CD69 expression by mitogen-stimulated murine splenocytes and induces hyporesponsiveness in B lymphocytes. *Parasite Immunol.* **25:** 27–37.

Hill, C.A. and Gutierrez, J.A. (2000) Analysis of the expressed genome of the Lone Star tick, *Amblyomma americanum* (Acari: Ixodidae) using an expressed sequence tag approach. *Microb. Comp. Genomics* **5:** 89–101.

Hoogstraal, H. (1979) The epidemiology of tick-borne Crimean-Congo hemorrhagic fever in Asia, Europe, and Africa. *J. Med. Entomol.* **15:** 307–417.

Horn, F., dos Santos, P.C. and Termignoni, C. (2000) *Boophilus microplus* anticoagulant protein: an antithrombin inhibitor isolated from the cattle tick saliva. *Arch. Biochem. Biophys.* **384:** 68–73.

Inokuma, H., Kerlin, R.L., Kemp, D.H. and Willadsen, P. (1993) Effects of cattle tick (*Boophilus microplus*) infestation on the bovine immune system. *Vet. Parasitol.* **47:** 107–118.

Jankovic, D., Sher, A. and Yap, G. (2001) Th1/Th2 effector choice in parasitic infection: decision making by committee. *Curr. Opin. Immunol.* **13:** 403–409.

Jaworski, D.C., Jasinskas, A., Metz, C.N., Bucala, R. and Barbour, A.G. (2001) Identification and characterization of a homologue of the pro-inflammatory cytokine macrophage migration inhibitory factor in the tick, *Amblyomma americanum*. *Insect Mol. Biol.* **10:** 323–331.

Johnston, S.A. and Barry, M.A. (1997) Genetic to genomic vaccination. *Vaccine* **15:** 808–809.

Jones, L.D., Hodgson, E., Williams, T., Higgs, S. and Nuttall, P.A. (1992a) Saliva activated transmission (SAT) of Thogoto virus: relationship with vector potential of different haematophagous arthropods. *Med. Vet. Entomol.* **6:** 261–265.

Jones, L.D., Matthewson, M. and Nuttall, P.A. (1992b) Saliva-activated transmission (SAT) of Thogoto virus: dynamics of SAT factor activity in the salivary glands of *Rhipicephalus appendiculatus*, *Amblyomma* variegatum, and *Boophilus microplus* ticks. *Exp. Appl. Acarol.* **13:** 241–248.

Joubert, A.M., Louw, A.I., Joubert, F. and Neitz, A.W. (1998) Cloning, nucleotide sequence and expression of the gene encoding factor Xa inhibitor from the salivary glands of the tick, Ornithodoros savignyi. *Exp. Appl. Acarol.* **22:** 603–619.

Keller, P.M., Waxman, L., Arnold, B.A., Schultz, L.D., Condra, C. and Connolly, T.M. (1993) Cloning of the cDNA and expression of moubatin, an inhibitor of platelet aggregation. *J. Biol. Chem.* **268:** 5450–5456.

Kjemtrup, A.M. and Conrad, P.A. (2000) Human babesiosis: an emerging tick-borne disease. *Int. J. Parasitol.* **30:** 1323–1337.

Klasco, R. (2002) Colorado tick fever. *Med. Clin. North. Am.* **86:** 435–440, ix.

Klompen, J.S., Black, W.C., 4th, Keirans, J.E. and Oliver, J.H., Jr. (1996) Evolution of ticks. *Ann. Rev. Entomol.* **41:** 141–161.

Kohl, J. (2001) Anaphylatoxins and infectious and non-infectious inflammatory diseases. *Mol. Immunol.* **38:** 175–187.

Kopecky, J. and Kuthejlova, M. (1998) Suppressive effect of *Ixodes ricinus* salivary gland extract on mechanisms of natural immunity in vitro. *Parasite Immunol.* **20:** 169–174.

Kopecky, J., Kuthejlova, M. and Pechova, J. (1999) Salivary gland extract from *Ixodes ricinus* ticks inhibits production of interferon-gamma by the upregulation of interleukin-10. *Parasite Immunol.* **21:** 351–356.

Krause, P.J. (2002) Babesiosis. *Med. Clin. North Am.* **86:** 361–373.

Krause, P.J., Telford, S.R., 3rd, Spielman, A., *et al.* (1996) Concurrent Lyme disease and babesiosis. Evidence for increased severity and duration of illness. *JAMA* **275:** 1657–1660.

Kubes, M., Fuchsberger, N., Labuda, M., Zuffova, E. and Nuttall, P.A. (1994) Salivary gland extracts of partially fed *Dermacentor reticulatus* ticks decrease natural killer cell activity in vitro. *Immunology* **82:** 113–116.

Kubes, M., Kocakova, P., Slovak, M., Slavikova, M., Fuchsberger, N. and Nuttall, P.A. (2002) Heterogeneity in the effect of different ixodid tick species on human natural killer cell activity. *Parasite Immunol.* **24:** 23–28.

Labuda, M., Jones, L.D., Williams, T. and Nuttall, P.A. (1993) Enhancement of tick-borne encephalitis virus transmission by tick salivary gland extracts. *Med. Vet. Entomol.* **7:** 193–196.

Lawrie, C.H. and Nuttall, P.A. (2001) Antigenic profile of *Ixodes ricinus*: effect of developmental stage, feeding time and the response of different host species. *Parasite Immunol.* **23:** 549–556.

Leboulle, G., Crippa, M., Decrem, Y., Mejri, N., Brossard, M., Bollen, A. and Godfroid, E. (2002a) Characterization of a novel salivary immunosuppressive protein from *Ixodes ricinus* ticks. *J. Biol. Chem.* **277:** 10083–10089.

Leboulle, G., Rochez, C., Louahed, J., Rutti, B., Brossard, M., Bollen, A. and Godfroid, E. (2002b) Isolation of *Ixodes ricinus* salivary gland mRNA encoding factors induced during blood feeding. *Am. J. Trop. Med. Hyg.* **66:** 225–233.

Lee, R. and Opdebeeck, J.P. (1999) Arthropod vaccines. *Infect. Dis. Clin. North Am.* **13:** 209–226.

Levin, M.L. and Fish, D. (2000) Acquisition of coinfection and simultaneous transmission of *Borrelia burgdorferi* and *Ehrlichia* phagocytophila by *Ixodes scapularis* ticks. *Infect. Immun.* **68:** 2183–2186.

Liew, F.Y. (2002) T(H)1 and T(H)2 cells: a historical perspective. *Nat. Rev. Immunol.* **2:** 55–60.

Liu, Y.J., Kanzler, H., Soumelis, V. and Gilliet, M. (2001) Dendritic cell lineage, plasticity and cross-regulation. *Nat. Immunol.* **2:** 585–589.

Mahalingam, S. and Karupiah, G. (1999) Chemokines and chemokine receptors in infectious diseases. *Immunol. Cell. Biol.* **77:** 469–475.

Mbow, M.L., Rutti, B. and Brossard, M. (1994a) IFN-gamma IL-2, and IL-4 mRNA expression in the skin and draining lymph nodes of BALB/c mice repeatedly infested with nymphal *Ixodes ricinus* ticks. *Cell. Immunol.* **156:** 254–261.

Mbow, M.L., Christe, M., Rutti, B. and Brossard, M. (1994b) Absence of acquired resistance to nymphal *Ixodes ricinus* ticks in BALB/c mice developing cutaneous reactions. *J. Parasitol.* **80:** 81–87.

Mbow, M., Zeidner, N., Panella, N., Titus, R. and Piesman, J. (1997) *Borrelia burgdorferi*-pulsed dendritic cells induce a protective immune response against tick-transmitted spirochetes. *Infect. Immun.* **65:** 3386–3390.

McKenna, R.V., Riding, G.A., Jarmey, J.M., Pearson, R.D. and Willadsen, P. (1998) Vaccination of cattle against the *Boophilus microplus* using a mucin-like membrane glycoprotein. *Parasite Immunol.* **20:** 325–336.

Mejri, N., Franscini, N., Rutti, B. and Brossard, M. (2001) Th2 polarization of the immune response of BALB/c mice to *Ixodes ricinus* instars, importance of several antigens in activation of specific Th2 subpopulations. *Parasite Immunol.* **23:** 61–69.

Mejri, N., Rutti, B. and Brossard, M. (2002) Immunosuppressive effects of *Ixodes ricinus* tick saliva or salivary gland extracts on innate and acquired immune response of BALB/c mice. *Parasitol. Res.* **88:** 192–197.

Mitchell, M. (1996) Acaricide resistance – back to basics. *Trop. Anim. Health Prod.* **28:** 53S–58S.

Moore, R.J., Lenghaus, C., Sheedy, S.A. and Doran, T.J. (2001) Improved vectors for expression library immunization – application to *Mycoplasma hyopneumoniae* infection in pigs. *Vaccine* **20:** 115–120.

Mulenga, A., Sugimoto, C., Sako, Y., Ohashi, K., Musoke, A., Shubash, M. and Onuma, M. (1999) Molecular characterization of a *Haemaphysalis longicornis* tick salivary gland-associated 29-kilodalton protein and its effect as a vaccine against tick infestation in rabbits. *Infect. Immun.* **67:** 1652–1658.

Mulenga, A., Sugimoto, C. and Onuma, M. (2000) Issues in tick vaccine development: identification and characterization of potential candidate vaccine antigens. *Microbes Infect.* **2:** 1353–1361.

Mulenga, A., Sugino, M., Nakajim, M., Sugimoto, C. and Onuma, M. (2001) Tick-encoded serine proteinase inhibitors (serpins); potential target antigens for tick vaccine development. *J. Vet. Med. Sci.* **63:** 1063–1069.

Narasimhan, S., Koski, R.A., Beaulieu, B., Anderson, J.F., Ramamoorthi, N., Kantor, F., Cappello, M. and Fikrig, E. (2002) A novel family of anticoagulants from the saliva of *Ixodes scapularis*. *Insect Mol. Biol.* **11:** 641–650.

Nene, V., Lee, D., Quackenbush, J., Skilton, R., Mwaura, S., Gardner, M. J. and Bishop, R. (2002) AvGI, an index of genes transcribed in the salivary glands of the ixodid tick *Amblyomma variegatum*. *Int. J. Parasitol.* **32:** 1447.

Nielsen, C.H. and Leslie, R.G. (2002) Complement's participation in acquired immunity. *J. Leukoc. Biol.* **72:** 249–261.

Nielsen, C.H., Fischer, E.M. and Leslie, R.G. (2000) The role of complement in the acquired immune response. *Immunology* **100:** 4–12.

Nithiuthai, S. and Allen, J.R. (1984a) Significant changes in epidermal Langerhans cells of guinea-pigs infested with ticks (*Dermacentor andersoni*). *Immunology* **51:** 133–141.

Nithiuthai, S. and Allen, J.R. (1984b) Effects of ultraviolet irradiation on the acquisition and expression of tick resistance in guinea-pigs. *Immunology* **51:** 153–159.

Nithiuthai, S. and Allen, J.R. (1985) Langerhans cells present tick antigens to lymph node cells from tick-sensitized guinea-pigs. *Immunology* **55:** 157–163.

Paesen, G.C., Adams, P.L., Harlos, K., Nuttall, P.A. and Stuart, D.I. (1999) Tick histamine-binding proteins: isolation, cloning, and three-dimensional structure. *Mol. Cell* **3:** 661–671.

Paesen, G.C., Adams, P.L., Nuttall, P.A. and Stuart, D.L. (2000) Tick histamine-binding proteins: lipocalins with a second binding cavity. *Biochim. Biophys. Acta* **1482:** 92–101.

Paine, S.H., Kemp, D.H. and Allen, J.R. (1983) In vitro feeding of *Dermacentor andersoni* (Stiles): effects of histamine and other mediators. *Parasitology* **86 (Pt 3):** 419–428.

Parola, P. and Raoult, D. (2001) Ticks and tickborne bacterial diseases in humans: an emerging infectious threat. *Clin. Infect. Dis.* **32:** 897–928.

Polanowski, A. and Wilusz, T. (1996) Serine proteinase inhibitors from insect hemolymph. *Acta Biochim. Pol.* **43:** 445–453.

Ramachandra, R.N. and Wikel, S.K. (1992) Modulation of host-immune responses by ticks (Acari: Ixodidae): effect of salivary gland extracts on host macrophages and lymphocyte cytokine production. *J. Med. Entomol.* **29:** 818–826.

Rao, A. and Avni, O. (2000) Molecular aspects of T-cell differentiation. *Br. Med. Bull.* **56:** 969–984.

Rathinavelu, S., Broadwater, A. and de Silva, A.M. (2003) Does host complement kill *Borrelia burgdorferi* within ticks? *Infect. Immun.* **71:** 822–829.

Reis e Sousa, C. (2001) Dendritic cells as sensors of infection. *Immunity* **14:** 495–498.

Ribeiro, J.M. (1987) *Ixodes dammini*: salivary anti-complement activity. *Exp. Parasitol.* **64:** 347–353.

Ribeiro, J.M. (1995) Blood-feeding arthropods: live syringes or invertebrate pharmacologists? *Infect. Agents Dis.* **4:** 143–152.

Ribeiro, J.M. and Francischetti, I.M. (2003) Role of arthropod saliva in blood feeding: sialome and post-sialome perspectives. *Ann. Rev. Entomol.* **48:** 73–88.

Ribeiro, J.M. and Mather, T.N. (1998) *Ixodes scapularis*: salivary kininase activity is a metallo dipeptidyl carboxypeptidase. *Exp. Parasitol.* **89:** 213–221.

Ribeiro, J.M., Makoul, G.T., Levine, J., Robinson, D.R. and Spielman, A. (1985) Antihemostatic, antiinflammatory, and immunosuppressive properties of the saliva of a tick, *Ixodes dammini*. *J. Exp. Med.* **161:** 332–344.

Riding, G.A., Jarmey, J., McKenna, R.V., Pearson, R., Cobon, G.S. and Willadsen, P. (1994) A protective "concealed" antigen from *Boophilus microplus*. Purification, localization, and possible function. *J. Immunol.* **153:** 5158–5166.

Rosa de Lima, M.F., Sanchez Ferreira, C.A., Joaquim de Freitas, D.R., Valenzuela, J.G. and Masuda, A. (2002) Cloning and partial characterization of a *Boophilus microplus* (Acari: Ixodidae) glutathione S-transferase. *Insect Biochem. Mol. Biol.* **32:** 747–754.

Rubin, H. (1996) Serine protease inhibitors (SERPINS): where mechanism meets medicine. *Nat. Med.* **2:** 632–633.

Sallusto, F., Mackay, C.R. and Lanzavecchia, A. (2000) The role of chemokine receptors in primary, effector, and memory immune responses. *Ann. Rev. Immunol.* **18:** 593–620.

Sangamnatdej, S., Paesen, G.C., Slovak, M. and Nuttall, P.A. (2002) A high affinity serotonin- and histamine-binding lipocalin from tick saliva. *Insect Mol. Biol.* **11:** 79–86.

Sauer, J.R., Essenberg, R.C. and Bowman, A.S. (2000) Salivary glands in ixodid ticks: control and mechanism of secretion. *J. Insect Physiol.* **46:** 1069–1078.

Schoeler, G.B. and Wikel, S.K. (2001) Modulation of host immunity by haematophagous arthropods. *Ann. Trop. Med. Parasitol.* **95:** 755–771.

Schoeler, G.B., Manweiler, S.A. and Wikel, S.K. (1999) *Ixodes scapularis*: effects of repeated infestations with pathogen-free nymphs on macrophage and T lymphocyte cytokine responses of BALB/c and C3H/HeN mice. *Exp. Parasitol.* **92:** 239–248.

Schoeler, G.B., Manweiler, S.A. and Wikel, S.K. (2000) Cytokine responses of C3H/HeN mice infested with *Ixodes scapularis* or *Ixodes pacificus* nymphs. *Parasite Immunol.* **22:** 31–40.

Schwan, T.G. and Piesman, J. (2002) Vector interactions and molecular adaptations of Lyme disease and relapsing fever spirochetes associated with transmission by ticks. *Emerg. Infect. Dis.* **8:** 115–121.

Sharma, A.K. and Khuller, G.K. (2001) DNA vaccines: future strategies and relevance to intracellular pathogens. *Immunol. Cell. Biol.* **79:** 537–546.

Shaw, M.K., Tilney, L.G. and McKeever, D.J. (1993) Tick salivary gland extract and interleukin-2 stimulation enhance susceptibility of lymphocytes to infection by Theileria parva sporozoites. *Infect. Immun.* **61:** 1486–1495.

Skotarczak, B., Rymaszewska, A., Wodecka, B. and Sawczuk, M. (2003) Molecular evidence of coinfection of *Borrelia burgdorferi* sensu lato, human granulocytic ehrlichiosis agent, and *Babesia microti* in ticks from northwestern Poland. *J. Parasitol.* **89:** 194–196.

Smith, T. and Kilbourne, F.L. (1893) Investigations into the nature, causation, and prevention of Texas or southern cattle fever. *Bull. Bur. Anim. Ind., US Dept. Agric.* **1:** 301.

Smooker, P.M., Setiady, Y.Y., Rainczuk, A. and Spithill, T.W. (2000) Expression library immunization protects mice against a challenge with virulent rodent malaria. *Vaccine* **18:** 2533–2540.

Sonenshine, D.E. (1991) *Biology of Ticks*, vol. 1. Oxford University Press, New York.

Spach, D.H., Liles, W.C., Campbell, G.L., Quick, R.E., Anderson, D.E., Jr. and Fritsche, T.R. (1993) Tick-borne diseases in the United States. *N. Engl. J. Med.* **329:** 936–947.

Srivastava, I.K. and Liu, M.A. (2003) Gene vaccines. *Ann. Intern. Med.* **138:** 550–559.

Stafford, K.C., 3rd, Massung, R.F., Magnarelli, L.A., Ijdo, J.W. and Anderson, J.F. (1999) Infection with agents of human granulocytic ehrlichiosis, Lyme disease, and babesiosis in wild white-footed mice (*Peromyscus leucopus*) in Connecticut. *J. Clin. Microbiol.* **37:** 2887–2892.

Steere, A.C. (2001) Lyme disease. *N. Engl. J. Med.* **345:** 115–125.

Suss, J. (2003) Epidemiology and ecology of TBE relevant to the production of effective vaccines. *Vaccine* **21 (Suppl. 1):** S19–35.

Szabo, S.J., Sullivan, B.M., Peng, S.L. and Glimcher, L.H. (2003) Molecular mechanisms regulating Th1 immune responses. *Ann. Rev. Immunol.* **21:** 713–758.

Telford, S.R., 3rd, Dawson, J.E., Katavolos, P., Warner, C.K., Kolbert, C.P. and Persing, D.H. (1996) Perpetuation of the agent of human granulocytic ehrlichiosis in a deer tick-rodent cycle. *Proc. Natl Acad. Sci. USA* **93:** 6209–6214.

Thalhamer, J., Leitner, W., Hammerl, P. and Brtko, J. (2001) Designing immune responses with genetic immunization and immunostimulatory DNA sequences. *Endocr. Regul.* **35:** 143–166.

Trimnell, A.R., Hails, R.S. and Nuttall, P.A. (2002) Dual action ectoparasite vaccine targeting 'exposed' and 'concealed' antigens. *Vaccine* **20:** 3560–3568.

Valenzuela, J.G., Charlab, R., Mather, T.N. and Ribeiro, J.M. (2000) Purification, cloning, and expression of a novel salivary anticomplement protein from the tick, *Ixodes scapularis*. *J. Biol. Chem.* **275:** 18717–18723.

Valenzuela, J.G., Francischetti, I.M.B., Pham, V.M., Garfield, M.K., Mather, T.N. and Ribeiro, J.M.C. (2002) Exploring the sialome of the tick *Ixodes scapularis*. *J. Exp. Biol.* **205:** 2843–2864.

Varde, S., Beckley, J. and Schwartz, I. (1998) Prevalence of tick-borne pathogens in *Ixodes scapularis* in a rural New Jersey County. *Emerg. Infect. Dis.* **4:** 97–99.

Walker, D.H. (1998) Tick-transmitted infectious diseases in the United States. *Ann. Rev. Public Health* **19:** 237–269.

Wang, H. and Nuttall, P.A. (1999) Immunoglobulin-binding proteins in ticks: new target for vaccine development against a blood-feeding parasite. *Cell. Mol. Life Sci.* **56:** 286–295.

Wang, X., Coons, L.B., Taylor, D.B., Stevens, S.E., Jr. and Gartner, T.K. (1996) Variabilin, a novel RGD-containing antagonist of glycoprotein Iib–IIIa and platelet aggregation inhibitor from the hard tick *Dermacentor variabilis*. *J. Biol. Chem.* **271:** 17785–17790.

Waxman, L., Smith, D.E., Arcuri, K.E. and Vlasuk, G.P. (1990) Tick anticoagulant peptide (TAP) is a novel inhibitor of blood coagulation factor Xa. *Science* **248:** 593–596.

Weiner, H.L. (2001) Oral tolerance: immune mechanisms and the generation of Th3-type TGF-beta-secreting regulatory cells. *Microbes Infect.* **3:** 947–954.

Whelen, A.C. and Wikel, S.K. (1993) Acquired resistance of guinea pigs to *Dermacentor andersoni* mediated by humoral factors. *J. Parasitol.* **79:** 908–912.

Wikel, S.K. (1982a) Histamine content of tick attachment sites and the effects of H1 and H2 histamine antagonists on the expression of resistance. *Ann. Trop. Med. Parasitol.* **76:** 179–185.

Wikel, S.K. (1982b) Influence of *Dermacentor andersoni* infestation on lymphocyte responsiveness to mitogens. *Ann. Trop. Med. Parasitol.* **76:** 627–632.

Wikel, S.K. (1985) Effects of tick infestation on the plaque-forming cell response to a thymic dependent antigen. *Ann. Trop. Med. Parasitol.* **79:** 195–198.

Wikel, S.K. (1996a) Immunology of the tick–host inteface. In: Wikel, S.K. (ed.) *The Immunology of Host-Ectoparasitic Arthropod Relationships*, pp. 204–231. CAB International, Wallingford, UK.

Wikel, S.K. (1996b) Tick modulation of host cytokines. *Exp. Parasitol.* **84:** 304–309.

Wikel, S.K. (1999) Tick modulation of host immunity: an important factor in pathogen transmission. *Int. J. Parasitol.* **29:** 851–859.

Wikel, S.K. and Alarcon-Chaidez, F.J. (2001) Progress toward molecular characterization of ectoparasite modulation of host immunity. *Vet. Parasitol.* **101:** 275–287.

Wikel, S.K. and Bergman, D. (1997) Tick–host immunology: significant advances and challenging opportunities. *Parasitol. Today* **13:** 383–388.

Wikel, S.K., Ramachandra, R.N., Bergman, D.K., Burkot, T.R. and Piesman, J. (1997) Infestation with pathogen-free nymphs of the tick *Ixodes scapularis* induces host resistance to transmission of *Borrelia burgdorferi* by ticks. *Infect. Immun.* **65:** 335–338.

Willadsen, P. (2001) The molecular revolution in the development of vaccines against ectoparasites. *Vet. Parasitol.* **101:** 353–368.

Willadsen, P. and Jongejan, F. (1999) Immunology of the tick–host interaction and the control of ticks and tick-borne diseases. *Parasitol. Today* **15:** 258–262.

Willadsen, P., Riding, G.A., McKenna, R.V., Kemp, D.H., Tellam, R.L., Nielsen, J.N., Lahnstein, J., Cobon, G.S. and Gough, J.M. (1989) Immunologic control of a parasitic arthropod. Identification of a protective antigen from *Boophilus microplus*. *J. Immunol.* **143:** 1346–1351.

Yanagihara, Y. and Masuzawa, T. (1997) Lyme disease (Lyme borreliosis). *FEMS Immunol. Med. Microbiol.* **18:** 249–261.

Yazdanbakhsh, M., van den Biggelaar, A. and Maizels, R.M. (2001) Th2 responses without atopy: immunoregulation in chronic helminth infections and reduced allergic disease. *Trends Immunol.* **22:** 372–377.

Zeidner, N., Dreitz, M., Belasco, D. and Fish, D. (1996) Suppression of acute *Ixodes scapularis*-induced *Borrelia burgdorferi* infection using tumor necrosis factor-alpha, interleukin-2, and interferon-gamma. *J. Infect. Dis.* **173:** 187–195.

Zeidner, N., Mbow, M.L., Dolan, M., Massung, R., Baca, E. and Piesman, J. (1997) Effects of *Ixodes scapularis* and *Borrelia burgdorferi* on modulation of the host immune response: induction of a TH2 cytokine response in Lyme disease-susceptible (C3H/HeJ) mice but not in disease-resistant (BALB/c) mice. *Infect. Immun.* **65:** 3100–3016.

Zhu, K., Bowman, A.S., Brigham, D.L., Essenberg, R.C., Dillwith, J.W. and Sauer, J.R. (1997) Isolation and characterization of americanin, a specific inhibitor of thrombin, from the salivary glands of the lone star tick *Amblyomma americanum* (L.). *Exp. Parasitol.* **87:** 30–38.

10

Avian coccidiosis: a disturbed host–parasite relationship to be restored

Arno N. Vermeulen

1 Introduction

The central theme of this book is host–parasite relationships with emphasis on comparative issues between different animal species. Where other contributors have highlighted similar systems, such as in fish, for a description of the interaction of avian species and their parasites there is a choice from a wide variety of parasite classes, families and species divided over all representatives of the avian kingdom. Part of the contents of this chapter was presented at the APS meeting on Comparative Physiology, San Diego, CA, August 2002 (Vermeulen *et al.*, 2002).

If only we concentrate on the domestic chicken (*Gallus gallus*), we have to conclude that it suffers from infestations by ectoparasites, such as mites (Birrenkott *et al.*, 2000), flies, lice and fleas (Ruff, 1999), helminth infections due to *Heterakis* and among others *Ascaridia galli* (Permin *et al.*, 1998, 1999; Poulsen *et al.*, 2000) and several protozoans such as *Leucocytozoon*, *Plasmodium*, *Cryptosporidium* (Ruff, 1999) and several species of the coccidial genus of *Eimeria*, which I will use as the central topic of this review. The reason for choosing this combination is mainly because of the huge economic impact this parasite has had during the 20th century on the poultry industry. This has resulted in a relatively great attention from the industry itself in combination with research from the scientific community to find ways to circumvent or resolve the problems that were experienced.

When speaking about host–parasite relationships, we refer to interactions that have evolved throughout a period of millions of years when the first representatives of each species have encountered and undergone a continuous process of co-evolution towards a balance of 'mutual understanding', whereby the parasite is proliferating to the cost of the host and the transmission of the parasite to other hosts is controlled by the virulence of the parasite. In this respect looking on population level and stating that such a relationship may have been disturbed somewhere in a time period of about 100 years is of course an example of scientific arrogance and is by no means justified by

Host–Parasite Interactions, edited by G. F. Wiegertjes & G. Flik.

in-depth epidemiological studies, whatsoever. However, when we focus in on the effect parasites may have on the behaviour and overall health of the individual host and conclude that the micro-environment this host–parasite relationship has evolved in, has considerably changed, then it is justified to study the different consequences of such changes on the virulence of the parasite and the well-being and performance of its host. Adjusting the models to explain the 'disequilibrium', should result in solutions to neutralize the effects of these changes and restore the balance (Conway and Roper, 2000). The emergence of diseases is explained by the changes in micro-environment between host and parasites and molecular evolutionary studies of the pathogens can yield data which discriminate between possible causes.

2 Evolution of the biology of the parasite and its host

The history of the domestic fowl goes back around 89 million years when *Anseriformes* (ducks) and *Galliformes* (game fowl) diverged (van Tuinen and Hedges, 2001) followed by the split between quail and domestic fowl about 50 million years later. The question is whether the parasites have co-evoluted together with their host or have crossed species barriers later in evolution. In fact, it turns out that both appear true. Most coccidial parasites harbouring each of these avian genera have segregated and developed as independent species. Especially the *Eimeria* parasites have shown a high degree of host specificity.

The genus *Eimeria* belongs to the Phylum of the *Apicomplexa*, subphylum *Sporozoa*, as shown in *Figure 1* (according to Levine, 1982). The structure of the tree is one of the many, that have been drawn up for the coccidians, although differences are minor and mainly due to interpretation and emphasis (reviewed by Tenter *et al.*, 2002). The *Eimeria* are strongly related to other coccidial protozoa such as *Toxoplasma gondii*, *Neospora caninum* and *Sarcocystis* species. Representatives of the *Piroplasmatidae* are blood parasites such as *Babesia* and *Theileria*. *Plasmodium*, the causative agent of malaria, is a representative of the *Haemospororina* suborder. The relationship with *Cryptosporidium* has recently come to debate after further study of its genome and is considered more closely related to other tissue-dwelling parasites such as *Gregarines* (Barta, 2001; Tenter *et al.*, 2002). Eimerian and cryptosporidial parasites have in common that their life cycle comprises development outside and inside the host, without the need for an intermediate host for transmission. Both harbour the gastro-intestinal tract of the host, proliferate inside epithelial cells (*Eimeria* within a parasitophorous vacuole), develop into sexual stages and are shed as oocysts with the faeces (see *Figure 2*).

The parasitic life cycle stages are all haploid apart from the early zygote which undergoes meiotic division soon after formation resulting in a haploidic oocyst (Fayer, 1980). The process outside the host is known as sporogony or sporulation. This process needs oxidative conditions and induces the formation of sporozoites inside sporocysts. Infection occurs through ingestion of the oocysts from the ground. These oocysts are crushed inside the gizzard and the sporozoites are released under the influence of digestive enzymes from the upper intestine. Sporozoites enter the enterocytes and undergo several mitotic divisions.

In chickens at least seven different species can be discerned (Shirley, 1985). *Eimeria acervulina*, *E. tenella*, *E. maxima*, *E. necatrix*, *E. brunetti*, *E. mitis* and *E. praecox* (Barta

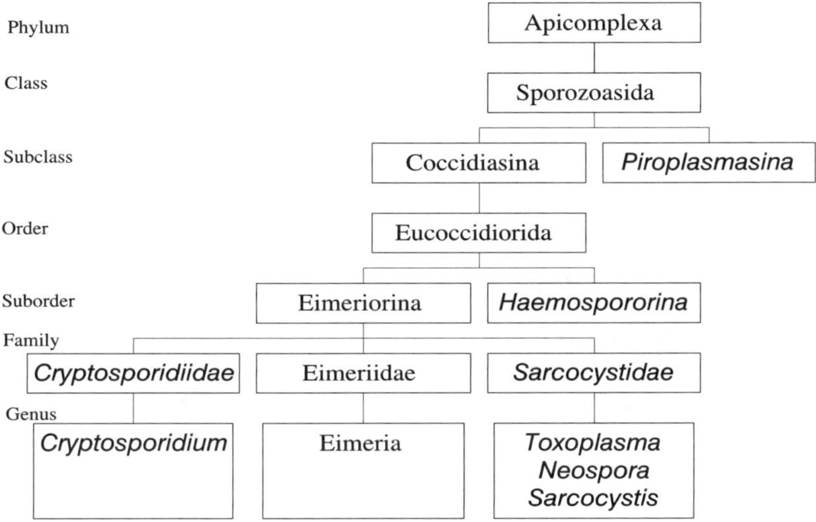

Figure 1 *Phylogenetic tree of the genus Eimeria in relation to their coccidial family members (in italics) after N.D. Levine (1982).*

et al., 1997; Fitz-Coy, 1983; Gore and Long, 1982). In the USA two more species have been described but their existence is still under debate, *E. mivati* (Edgar and Siebold, 1964) and *E. hagani*. *E. mivati* is considered a *nomina dubia* and should be named as *E. mitis* (Shirley *et al.*, 1983).

The effects these infections evoke in the chicken host are strongly associated with the site of infection and proliferation (Allen and Fetterer, 2002). All these species harbour a specific niche in the chicken intestinal tract. *E. tenella* is bound to the caeca, whereas the others parasitize the proximal (*E. acervulina, E. mitis* and *E. praecox*), the middle (*E. maxima, E. necatrix*), or the distal part of the intestine (*E. brunetti*).

From taxonomic studies and biological equivalence data, it was shown that the site specificity of each of these species had been developed before the evolutionary split between members of the Galliformes. Augustine and Danforth (1995) found higher homology between caecal parasites of turkey and chicken (*E. adenoeides* and *E. tenella*, respectively) than within the chicken species. The same was true for the duodenal species *E. meleagrimitis* and *E. acervulina* (Augustine and Danforth 1990, 1995; Barta *et al.*, 1991). Although *E. tenella* could not complete its cycle in turkey, it was able to invade turkey caecal epithelial cells and induce some part of intracellular development, which was not possible in duodenal cells of either chicken or turkey (Augustine and Danforth, 1990). According to Barta *et al.* (1997) this could indicate that *E. tenella* (and possibly also *E. necatrix*) have crossed the host species barrier from the North American turkey (*Meleagris gallopavo*) to the chicken.

Site specificity is even so strict that intravenous, intramuscular or intraperitoneal injection of sporozoites always results in development at the site where they would end up after oral inoculation (Davies and Joyner, 1962). It is hypothesized that site specificity was caused by the intrinsic characteristic of time needed for a sporozoite to be released from its sporocyst shell, while passing down the intestine (Farr and Doran,

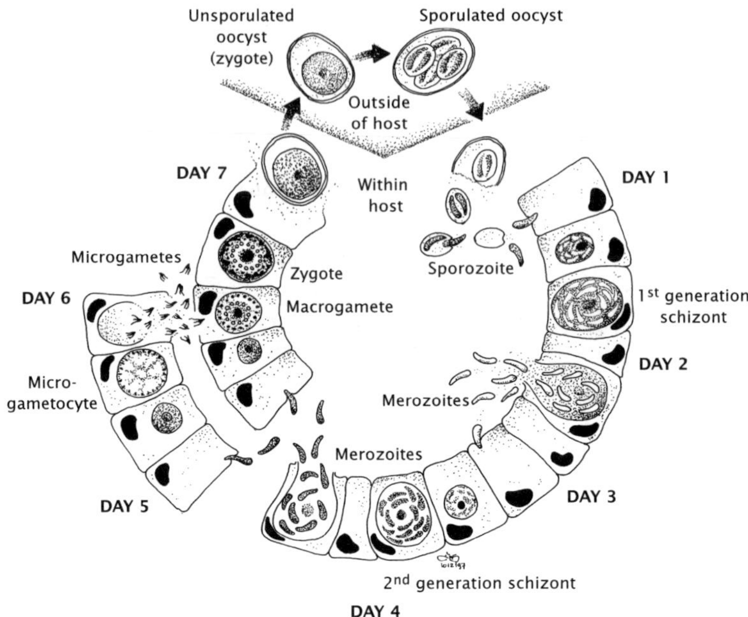

Figure 2 *Life cycle* Eimeria tenella (Courtesy of Jan Dorrestein).

1962). This, however, seems an oversimplification, since this would suggest that *Eimeria tenella* would enter anywhere in the lower intestine, however it only replicates in the specialized tissue of the caecum, meaning that more importantly the relative distribution of host cells' surface molecules is determining the successful invasion and subsequent development of the parasites is strongly influenced by the micro-environment of the host cell. The discovery of common molecular structures on both host caecal cells and *E. tenella* sporozoites was unprecedented so far. The intriguing observation that only on cells of the caecal epithelial lining of the chicken and the turkey these common structures are found suggests that sporozoites might utilize these structures as a chemo-attractant into the caeca or as a way to interact with the host cells for invasion. Anyway, it underlines the co-evolution of parasite and host (Vervelde *et al.*, 1993).

The fact that these parasites do not need an intermediate host to transfer from one host to another has had a great impact on the development of the host–parasite relationship. The potential re-infection of the same host with even more parasites than at a first occasion must make a selective pressure on the virulence of the parasite and the resistance of the host.

Although the time needed to complete its life cycle is different for all species, it only varies very marginally from 4–5 days for *E. acervulina* to 7–9 days for *E. maxima*. The sporulation process takes about 48–72 h under elevated temperatures of 26–29°C and relatively high humidity and oxygen conditions. This means that after about 7–10 days the oocysts shed from a primary infection are again infectious for susceptible hosts. On average these species have a reproduction capacity of 10^4–10^5 times per cycle (Johnston *et al.*, 2001; Williams, 2001). The time needed for the host to regenerate epithelial cells is also about 5–7 days, which appears to run in parallel with the generation of parasites.

3 Evolution of parasite virulence

The general dogma, that parasitism is always striving to moderate virulence of the parasite together with a modulated defence of the host to allow parasite replication, may be true, but is an empirically designed theory trying to explain the phenomena observed in many, but not all, host–parasite interactions as can be concluded from the numbers of papers still being published on this subject (see below).

During evolution of parasite–host interactions, parasites have a drive to produce more and more progeny, so-called exploitation of the host, which will result in higher virulence (Anderson and May, 1982; Poulin and Combes, 1999; Williams and Day, 2001). Virulence thereby is a measure of parasite fitness. The pathology induced in the host might often contribute to the successful spread of the progeny, such as the induction of diarrhoea for the shedding of intestinal parasites. The often-used term 'virulence genes' as a trait of the parasite involved in harming the host is often misused. As such no genes exist within parasites that encode damaging of the host (Poulin and Combes, 1999). Manipulation of the host towards facilitating transmission is a better description of the process. Virulence can thus be defined as the resultant of manipulation genes in some cases or the side effect of exploitation genes in other cases.

However, another theory states that well-evolved parasite–host interactions always end in a sort of symbiosis as explained theoretically by the equation for basic reproduction ratio R_0 being the result of effective transmission divided by the host's mortality and morbidity (Bowers, 2001; Charles *et al.*, 2002). Simple extrapolation of this equation results in maximum transmission at minimum host pathology and mortality (Weiss, 2002). So the trade-off for both characteristics of parasitism is an intermediate level of virulence at the maximum level of fitness for the pathogen (Gandon and Michalakis, 2000; Gandon *et al.*, 2001a, b).

This is still a general assumption and not yet taking into account the presence of other pathogens and the influence of immunity. For *Eimeria* parasites the situation is quite complicated. The fact that after these millions of years still seven species of *Eimeria* co-exist and are able to concurrently parasitize the chicken implicates from evolutionary biological assumptions, that there is no complete cross-species immunity developed between these species (Pugliese, 2002). This has also been confirmed for these different species based on cross-protection studies (Rose, 1978). Only the species with the highest reproductive rate R_0 can persist. Pugliese (2002) has elaborated in a mathematical way on the model of co-infection and superinfection and concludes that apart from factors such as cross-immunity also host demography (genetic variability of the host in relation to susceptibility) should be included. The ideas are supported from studies by Pfennig (2001).

Aspects of resistance of the host are very important factors in co-evolution. Although the host evolves at a much slower rate than the parasite is capable of, the host has developed ways to reduce susceptibility to infection. Genetic traits such as products from the major histocompatibility complex are influencing both the innate and adaptive immune response and are in most animal species highly polymorphic. If the population contains many individuals with high susceptibility for certain and high refractoriness for other pathogens, it is plausible that such populations allow the co-existence of different species of pathogens alongside. Each of these species might evolve separately and might increase in virulence. However, if multiple species are able to infect the same host concurrently, then exploitation of such a host might reduce the resources for each of the parasite species and competition occurs for these resources. Brown *et al.* (2002) have described these models in detail. They state that although general theories support the 'survival of the fittest' parasite model, whereby the most selfish species exploits more resources than the other species in the community, multiple infection can result in a differentiation of the species towards lower relatedness such that the degree of infection might stay constant. This is called a 'collective action', a sort of compromise between individuals of a group for the benefit of the community.

This does explain the co-existence of different species of *Eimeria* in the same host and might also explain a situation where new species rise and develop alongside. The lack of cross-protectivity of immune responses directed against any of these is confirmed by many studies. Lower relatedness within the same species has been shown for some species only. *E. maxima* is the most well known species having developed antigenic diversity to the extent that not all strains of this species do crossprotect (Barta *et al.*, 1998; Martin *et al.*, 1997; Smith *et al.*, 2002). Also, for some strains of *E. acervulina* some antigenic drift has been shown serologically, although not yet related to lack of cross-protection (Joyner, 1969). Along the same lines, the site specificity for each of the parasites in the intestinal tract of the chicken could be the result of the collective action. This may allow them to interfere as little as possible at maximum transmission of the parasite and full exploitation of the host, albeit by dividing 'the cake'.

A second compromise of co-infection can be read from the infection kinetics of the *Eimeria* species involved. Although in laboratory experiments multiple infections can easily be induced, in practice it is observed that infections of different species often follow a specific time-course. Species parasitizing a similar or the same region in the intestine are mostly separated in time. *E. acervulina* is often the first *Eimeria* species a chick encounters in its life on the floor in a chicken house. *E. mitis* is observed after 4–6 weeks when immunity to *E. acervulina* has reduced infection with this species. The same holds for *E. maxima* and *E. brunetti* and *E. tenella* and *E. necatrix* (Voeten *et al.*, 1988).

Finally, the availability of host cells is a major factor moderating the virulence of the parasite. At low-dose inocula a linear reproduction is observed, whereas at higher doses the reproduction of the parasite becomes impaired (the so-called 'crowding effect'), mainly due to the damage of cells or lower availability of nutrients (Williams, 2001). So evolution biology can very well explain the host–parasite interactions that have resulted from a very complex and intriguing history of the habitats of each of the players.

4 Changing the scene

Although the scenery in which this all took place has changed in time, the model holds against a background of relative equilibrium of the host and its environment. Sudden changes as have occurred in the 20th century, when the poultry industry developed very quickly by rationalizing production protocols, have disturbed the long-established balance between chicken and its *Eimerias* (see *Table 1*). Change of nutrients, in-house breeding, high-stocking densities and genetic selection for faster-growing birds can be considered as turmoils in the evolution of the parasite. Add to that the increasing consumption of chicken meat, the use of anticoccidial drugs, and money ruling all ingredients of the industry and a complete 'new' set of players has entered to 'parasitize' the chicken in direct competition with the *Eimeria* species. From 1934 onwards the production of broilers in the USA increased by a factor of 200! Consumption has grown by a factor of 4 over the last 40 years. The genetic background of the chicken has been changed considerably. In the year 2000 the average slaughterweight had increased by 50%, at the background of a shorter grow-out time. The nearly 100% increase in feeding costs per liveweight is only partly compensated by inflation and partly due to the increasing expenses to control diseases associated with high-density stocking amongst which coccidiosis is a major factor.

Nowadays, the world-wide production of broiler chickens has reached a figure of about 40 billion chickens per annum (Williams, 1999). Williams has calculated the costs of the damage due to coccidiosis in the UK as 4.5% of the gross 1995 UK sales revenue. It should be emphasized that 80% of these losses are attributable to sub-clinical coccidiosis and only 18% is costs for treatment and prophylaxis. The consequences of this revolution thus have resulted in the development of severe disease and 'overexploitation' of the chicken by the parasites. Graat *et al.* (1998) have identified all risk factors involved in the occurrence of coccidiosis in modern poultry production flocks. Bad hygienic status, dirty feeding and drinking systems, and presence of other diseases on the farm increased the risk for coccidiosis. Numerous reports describe the morbidity of the disease throughout the world. From Argentina to Canada, from Sweden and Jordan to New Zealand and Japan (Al Natour *et al.*, 2002; Jorgensen *et al.*, 1997; Karim and Trees, 1990; Lunden *et al.*, 2000; Mattielo *et al.*, 2000; McDougald *et al.*, 1997, Reza and Ali, 2000; Rodriguez, 1997; Shirley, 1997; Tsuji *et al.*, 1997).

It is hard to predict what consequences these dramatic changes in the environment of the parasites may have had on the virulence of each of the parasite species. Almost weekly passages through practically naive animals might have increased the virulence of the parasites, although the effect of coccidiostats on this parasite trait is difficult to estimate in this respect. Theoretically, a quantitative reduction of transmission either through the host or through the long-term action of external agents will increase the virulence of parasites (Gandon and Michalakis, 2000). Fifteen years of repeated laboratory passage of a pure *E. tenella* Weybridge strain appeared to have increased the pathological effects of this strain, but this might have been an intrinsic feature of this strain and not *per se* generally applicable (A. Vermeulen, unpublished observations).

Table 1 *Changes in production, end weight and production costs for broilers and their consumption statistics in the USA during the 20th century*

Year	Broilers produced	Average weight of end product (pounds, lbs)	Feed costs per liveweight (cents per pound)	Chicken meat eaten per capita (lbs)
1934	34 million	No info	No info	No info
1960	1.8 billion	3.36	8.4	23.5
2000	8.3 billion	5.00	14.6	89.6

Source: Economic Research Service USDA.

5 Pathological phenomena associated with overexploitation

The macroscopic appearance of gut lesions is the most diagnostic pathological effect of coccidiosis. These lesions, though transient, are the result of local inflammation in combination with parasite replication and associated cell death. A negative effect of these pathological reactions on the performance of the bird seems obvious, but does not always fully explain the growth rate and feed conversion observed (Conway *et al.*, 1993).

Malabsorption is generally considered as the main cause for weight gain depression of coccidian-infected chickens (Kouwenhoven, 1972). *E. acervulina* is best studied in this respect and being a duodenal parasite most involved in interference of the absorption of nutrients. Carotenoids have been used as markers for the uptake of nutrients from the small intestine. A strong correlation is found between the inoculation dose of *E. acervulina* oocysts and the reduction of plasma carotenoid levels (Yvore and Mainguy, 1972). Allen and Fetterer (2000) have shown that also the levels of L-arginine in plasma reduced in a similar manner to carotenoids and concluded that malabsorption was indeed affecting different groups of nutrients. In the 1970s similar studies were performed by Kouwenhoven and van der Horst (1972), who found lowered plasma levels for vitamin A during *E. acervulina* infection. They discovered that the uptake of vitamin A was greatly reduced due to proteinaceous sediments on the surface of the intestinal epithelial cells, which interfered with the transport of vitamin A. This protein denaturation might be due to local pH decrease caused by damage to epithelial cells (Kouwenhoven and van der Horst, 1969).

Apart from strictly local impediments on uptake of nutrients, systemic effects have also been described that could account for weight gain depression and increased feed conversion. Anorexia, changes in metabolism, febrile responses and increase in plasma nitric oxide are possibly caused by immunological responses to the infection by different species of *Eimeria*. Plasma $NO_2^- + NO_3^-$ are significantly elevated at about 6 days post infection during primary infections with *E. acervulina, E. tenella* and *E. maxima*, but not during challenge infections in well-immunized chickens (Allen 1997a, b; Allen and Lillehoj, 1998). Pro-inflammatory cytokines such as IL-1 and TNF-α could be responsible for these effects. Treatment of chickens with these cytokines induced similar phenomena (Klasing *et al.*, 1987). Allen (2000) investigated the possible role of prostaglandins in these processes, but did not find an effect of indomethacin or nimesulide on weight gain or feed conversion after *E. acervulina* infection suggesting a limited involvement for prostaglandins.

Apart from the direct effects of coccidial infection on the health of the chicken, interaction of coccidiosis with other diseases might complicate the overall disease phenomena. Ruff (1989) reviewed most interactions and discerned four types of interaction based on the graded effect each of the conditions had on the total outcome of the disease. Most important effects that *Eimeria* infections have on bacterial colonization are described for *Salmonella* in the caeca and *Clostridium perfringens* in the small intestine. In these cases the coccidia damage the mucosal barrier and allow easy access for the bacteria to the epithelial cells (Al-Sheikhly and Al-Saieg, 1980), but also metabolic parameters such as the concentrations of volatile fatty acids in the ceca could be of influence to the higher colonization of *Salmonella* (Baba *et al.*, 1985; Bradley and Radhakrishnan, 1973). The role of viruses such as IBD, reovirus in increased mortality in association with *E. tenella* or *E. mitis* infection is mostly due to their immunosuppressive characteristics. Mycotoxins also had an additive effect on mortality due to *E. tenella* or weight gain depression after *E. acervulina* infection.

In order to improve the uptake of feed constituents in the presence of coccidial infection different additions have been described that could enhance digestibility of feed, reduce pathogenic effects of the infection and compensate for weight gain depression and lowered feed conversion. Enzymes such as pentosanases were applied to reduce the viscosity of the feed bowl in the intestine, thereby increasing digestibility and decreasing the chance of pathogenic bacteria to colonize. The coccidial infection itself, however, decreases also the viscosity of the digesta, thus the net effect of these additions might be limited (Waldenstedt *et al.*, 2000).

The effect of dietary betaine is rather controversial. Betaine was applied as replacement for methionine in the feed and was described to enhance the anti-coccidial effect of ionophoric drugs, but no consistent proof was found for any of these claims (Matthews and Southern, 2000). This may be due to the small differences to be expected and the limited size of the experimental design. Larger-scale trials demonstrated a 5% beneficial effect on weight gain but not lesion scores or oocyst numbers (Waldenstedt *et al.*, 1999). Other investigators found an effect of betaine as osmotic protectant in the small intestine (Kettunen *et al.*, 2001), which might explain a beneficial effect on the outcome of the disease with respect to weight gain depression. Betaine also decreased osmolarity of the duodenum, especially in infected birds. Coccidial infection increased the thickness of and number of leucocytes in the duodenal lamina propria especially at high betaine levels and increased the activity of monocytes and production of nitric oxide (Klasing *et al.*, 2002).

Addition of omega n-3 fatty acids (linolenic acid) to the feed reduced the lesions due to *E. tenella* infection, but this needed concentrations of more than 10% flaxseed (stabilized flaxmeal), which probably induced a high intestinal oxidative stress that damaged the parasites effectively (Allen *et al.*, 1996). In conclusion, feed adjustment is still a possible way to improve the intestinal metabolism and provide a better resistance against all pathogenic micro-organisms of the gut. In this industry however, costs are a great factor and should be considered in the perspective of other, often more effective, measures. Natural resistance as evoked by constituents of feed may also involve components of the innate and adaptive immune system. The characteristics of the chicken immune system and their relation to mammals are discussed below.

6 The avian immune system

The genome sequencing projects have had a great impact on the identification of comparable physiological processes in different animal species. Presently, the functional genomic knowledge about the human genome expands almost exponentially and provides the basis and reference for characterization of the function of genes that are being identified in the chicken. The University of Michigan (http://poultry.mph.msu.edu) is compiling all data around the chicken genome; the University of Delaware has an EST project running more specifically on immune-response genes and their products (http://www.chickest.udel.edu/). From the data published it is clear that chicken genomes comprise most genes involved in immune responses that are also active in mammalian species. The main difference lies in the genomic organization of these genes. Chickens have a much smaller genome (1.2×10^9

bp, only about 20 times *Eimeria*) compared to mammals, mainly due to less non-coding stretches and thus a more condensed organization of genes.

This also is valid for the immune response genes organized in the major histocompatibility complex (MHC), the reason for Kaufman to call it the 'minimal essential' MHC (Kaufman *et al.*, 1999a). The immediate consequence of such a limited space in non-coding regions is less possibilities for recombination of chromosomes or parts thereof. Chicken T- and B-cell receptors comprise only two families of Vα and Vβ genes, which allow for a reduced recognition repertoire compared to mammals (Goebel, 2000). It has been shown that chicken MHC class I and class II genes only segregate into a few families compared to the complex superfamilies that have evolved in the mammal MHC. The predominant expression of only a single MHC class I molecule, that determines the immune response to several pathogens, shows that the chickens have a reduced capacity to react to non-self structures (reviewed in Kaufman *et al.*, 1999b). It has been shown that some chicken lines differ in their capacity to resist some microbial agents, such as viruses (Rous Sarcoma virus, Marek's disease virus) but also *Eimeria* infection of different species (Emara *et al.*, 2002). Broiler lines seem to differ in these respects from White Leghorn lines, which reflects their difference in genetic background (Li *et al.*, 1999).

The fact that chickens, however, do develop solid immunity against most occurring diseases might indicate that other genes apart from the MHC are involved in genetic resistance against the agents causing these problems (reviewed by Vainio and Imhof, 1995). The presence of other genes, not present in mammals, involved in NK cell activity might suggest that these cells play a prominent role in the first line of innate defence against different pathogens (Goebel, 2000). The gamma-delta T-cell repertoire seems to be more diverse than the mammalian counterpart. The involvement of unique MHC-associated loci, such as the Rfp-Y and the B-G MHC class IV type genes is of interest in this respect, though far from clear. B-G-derived proteins are present on the (nucleated) erythrocytes of the chickens and play a role in antigen presentation and recognition of self structures on these cells. The Rfp-Y region contains both class I and class II genes, of which at least the class I loci are functional and produce biologically active molecules present on lymphocytes, erythrocytes and other cell types (Miller *et al.*, 2000).

Although orthologues between mammalian and chicken immune response genes can be identified, the relative low sequence homology between the counterparts has hampered the discovery of such genes in the chicken. Especially cytokine genes have shown about 25% sequence homology with mammalian counterparts, which is generally too low to detect these by hybridization (Kaiser, 2000). Mostly biological assays have been utilized to unravel the role of cytokines in the chicken. Nowadays, several chemokines and interleukins are cloned (reviewed by Staeheli *et al.*, 2001) such as IL-1β, IL-2, IL-6, IL-8, IL-15, IL-16 and IL-18 and interferons α, β and IFNγ.

Very interesting is that no typical type II cytokines, such as IL-4, IL-10 or IL-5 have been cloned or identified in the chicken genome so far. With the genomic absence of the ε-heavy chain gene for cytophilic antibody IgE this suggests that chickens are highly impaired in developing a Th2-dependent immune response and that other mechanisms exist in acquired immunity against typical type II organisms such as parasitic helminths. Chickens are able to mount high antibody titres, although differences occur between different lines of chickens in this respect. The absence of IgE could be compensated by higher IgG or IgA antibody levels.

Another hypothesis is that the low expression level of MHC class I molecules on CD4+ lymphocytes of several chicken lines is determining the 'Th2-like' response, whereas a high expression level would skew the response through a strong cell proliferation towards a more Th1-like response (Ewald, 2000). In a similar fashion the relatively low expression of class I molecules on lymphocytes of the Marek's disease resistant B-21/B-21 line would allow for a better killing of MD-infected cells by appropriate NK cells (Kaufman *et al.*, 1999b). Using antigen delivery through scavenger receptors on avian macrophages, it was shown that the Th1/Th2 type responses could be manipulated in chickens (Vandeveer *et al.*, 2001), suggesting a role for class II molecules and co-stimulatory factors. Broiler-type chickens seem to display a blunted inflammatory response compared to layers, which might reflect their faster growth potential by suffering less from decreased food intake at an acute phase response of inflammation (Leshchinsky and Klasing, 2001).

Most cell differentiation (CD) markers found in mammalian species have been described for chickens as well. B- and T-cell receptors are organized in a comparable way, CD4 and CD8 molecules in association with CD3 complex are present on mature T-cells and on immature thymocytes. Breed *et al.* (1996) described CD4-CD8 (dim) positive T-cells in the circulation of White Leghorn chickens and hypothesized that these function as memory cells being a CD4 cell with some CD8αα remaining. The organization of the chicken lymphoid system is also not very different from that in mammals, but a few functional characteristics are worth mentioning. The famous segregation of B-cell and T-cell ontogeny and maturation in bursa of Fabricius and the thymus respectively made the chicken initially a very attractive model for the early immunologists. The nomenclature of these cells refers to this.

Chickens lack however, distinct lymph nodes although the lymph system exists. The mucosal-associated lymphoid tissue is organized in a similar fashion, although the number of organized lymphoid centres is lower in the chicken especially along the intestine, where Peyer's patches are identifiable but at low density. The caecal tonsils are amongst the few larger foci of lymphoid cells and are thought to be important for the response to non-self antigens passing the intestine, through a back-flow of material toward the caeca. Germinal centres present in the lamina propria mainly contain antibody-producing cells in association with antigen-presenting cells. The intestinal lymphoid cells are concentrated in two geographically separated areas, the intra-epithelial lymphocytes (IEL) and the lamina propria (LP) lymphocytes separated by the basement membrane. The IEL function as a first line of defence and comprise NK-cells, γδT-cells (TCR1) and a lower percentage, αβT-cells (TCR2) mostly of the CD8-positive subset. The LP contain high concentrations of neutrophils, macrophages and B- and T-lymphocytes derived from the peripheral blood leucocyte pool. Young chickens are able to respond to non-self antigens but are immunologically still immature until the age of around 3 weeks post-hatch. Vervelde and Jeurissen (1993) reported that the numbers of IEL in chickens increase over a period of about 8 weeks, which indicates that in this period the immune system comes into contact with increasing amounts of antigen both from food and micro-organisms and the responsiveness against these molecules matures. In practice chickens are vaccinated already at day of hatch (see under 'vaccination') or even still *in ovo*, the latter mostly with replicating viruses that will stimulate the immune system over a longer period of time (Sharma *et al.*, 2002).

7 Immunity against coccidial infections

7.1 *Induction of protection*

Infection with *Eimeria* parasites induces a solid and relatively long-lasting immunity in chickens. The sensitivity for infection and the induction of resistance are strongly genetically determined. Cloned lines of chickens show different resistance patterns for different *Eimeria* species (Brake *et al.*, 1997; Bumstead and Millard, 1992; Bumstead *et al.*, 1995b). Thus, the induction of immunity against all *Eimeria* species does not seem to be linked to a single locus, precluding successful selection of coccidiosis-resistant lines. The establishment of protection is dependent on the infectious dose and the number of repeated infections, which differs for the individual species. *E. maxima* is regarded as one of the most immunogenic species, since inoculation of only a few oocysts already induced protection. This might also refer to the stage that is most involved in the induction of protective immunity against this species. If induction of immunity is associated with a minimal biomass of antigen available for stimulation, it might be concluded that the later stages of this species are most important for the induction of protection. Protection is often defined as the status of shedding a minimal numbers of oocysts following challenge infection. Blocking of oocyst formation at the gametocyte stage might be the first step in inducing immunity against this species. For other species such as *E. tenella* more rounds of infection apparently are needed to induce a status of solid immunity (Rose and Long, 1980). Immunity after infection is species-specific.

The stages that are responsible for the induction of the immunity can be different from the targets of the effector branch of the immune system. From studies with irradiated oocysts or sporozoites, that could invade host cells but not proliferate, it was shown that early development is determining the induction of the appropriate immune response possibly due to the correct presentation of antigens in the context of class I MHC (Jenkins, 1991a, b). These results were corroborated by investigations using precocious strains, that have a shortened life cycle due to absence of one or more mitotic stages, that showed that first-generation schizonts were more immunogenic when compared on biomass than later developmental stages (McDonald *et al.*, 1988). From our own work using Toltrazuril (Baycox) as a drug to kill intracellular stages (Greif, 2000) it was shown that when giving the drug already 24 h prior to infection a protective immunity could be induced at rather low dosages of inoculation of *E. tenella* or *E. acervulina* oocysts (see *Table 2*). At a low dose of 100 oocysts *E. acervulina* a reduction of protection (only 50%) seems to have been induced when the first generation was allowed to complete its proliferation. This suggests that these first stages are immunosuppressive to some extent. In all, these data show that indeed the sporozoite and/or the first intracellular stage is decisive for good induction of immunity.

Table 2 Early developmental stages are important for induction of protection

Treatment	Priming dose	Challenge	Protection
Baycox® T-24 hr	10^2 *E. acervulina*	10^5 *E. acervulina*	85%
	10^5 *E. acervulina*		99%
	$5*10^3$ *E. tenella*	10^4 *E. tenella*	95%
Baycox® T+24 hr	10^2 *E. acervulina*	10^5 E. *acervulina*	50%
	10^5 *E. acervulina*		95%
	$5*10^3$ *E. tenella*	10^4 *E. tenella*	95%
		10^5 *E. acervulina*	0%
None	None	10^4 *E. tenella*	0%

Groups of 5 birds were kept in wired floor cages. They were treated with 10x curative dose of toltrazuril (Baycox®) for 48 hr, starting at T–24 or T+24 relative to priming infection (T = 0). Four weeks later all birds were challenged with a high dose of the homologous species. None of the chickens excreted any oocysts prior to challenge. Protection is set as percent reduction of numbers of oocysts shed relative to the untreated challenged birds.

7.2 *The effector phase of immunity*

Sporozoites are most often seen as the target stage for the effector branch of the immune system. After entry of sporozoites of *E. tenella* into the region of the villus tip, they have to pass the lamina propria to reach the crypts where the first intracellular stage develops. In naive and immune chickens invasion of villus epithelium occurs with similar efficiency. In immune chickens, however, few sporozoites reach the crypt epithelium, which means that most sporozoites are neutralized by LP leucocytes (Jeurissen *et al.*, 1996). Similar findings were reported for other species. Beattie *et al.* (2001) found that, at homologous challenge, *E. maxima* sporozoites in immune chickens do not reach the crypts, whilst most sporozoites of a non cross-protective heterologous strain of *E. maxima* were allowed to passage to the crypts and develop. This again indicates that sporozoites are the target stages for the protective immune response, although it is not known whether the immune response cannot act against later stages as well.

Antigenic diversity has become a major issue when studying immune responses and protection against heterologous strains of different species. For the most studied species strain differences have been detected both in virulence and in cross-protective characteristics (Fitz-Coy, 1992). For *E. maxima* this feature appears best documented. Long and Millard (1979) described immunological differences in *E. maxima* strains and applied mixed inocula to protect against both strains. Smith *et al.* (2002) reported that the lack of cross-protection against *E. maxima* strains was not only influenced by the antigenic make-up of the parasite but also the genetic background of the host determined the capacity of chickens to mount a strong and protective immune response against the heterologous strain. Line 15i chickens were most peculiar in this aspect since primary infection with the Houghton strain of *E. maxima* induced 100% protection against a challenge with the Weybridge strain of this species whereas the inverse order of infection did not induce any cross-protection. Chickens with identical B-locus (B2, B4, B12 or B15) background differed in their capacity to mount a good cross-protective response suggesting that the chicken MHC system is not a major factor in this. Clare *et al.* (1985, 1989) demonstrated this also for *E. tenella*.

7.3 *Cells involved in the immunity against* Eimeria

The immunity against the intracellular stages is strongly dependent on T-cell-mediated mechanisms (Lillehoj and Trout, 1996). Although antibodies are produced during and after infection their role in reducing the damage of a challenge infection is limited (Davis *et al.*, 1978; Rose and Hesketh, 1983; Rose *et al.*, 1984; Zigterman *et al.*, 1993). Antibodies against gametocyte stages of *E. maxima* could however reduce the production of oocysts after transfer to naive animals (Rose, 1972).

Studies investigating the role of different cell populations in immunity against *Eimeria* describe the process in the intestine or the systemic events. In the intestine the responses of IEL and LPL differ in cell types involved and functional aspects associated to these cell types (reviewed by Yun *et al.*, 2000a). The IEL-derived γδT-cells increase sharply in numbers early after infection with *E. acervulina*. Most IEL are CD8+ cells and also this cell type is preferentially seen in the direct environment of sporozoites of *E. acervulina*, but also CD4 cells are increased. The role of IEL is however not clear. Lillehoj and Bacon (1991) found a correlation between increasing numbers of CD8+-IEL and resistance against secondary infection in different MHC-inbred lines. A putative role for IEL as a transport vehicle for sporozoites through the lamina propria is very doubtful (Rose *et al.*, 1984). These cells undergo apoptosis as part of the selective exclusion and are not moving up and down the LP. Sporozoites of *E. tenella* were not seen in association with IEL or LPL but migrated independently through the LP to the crypts (Vervelde *et al.*, 1995). γδT cells in the IEL fraction are potentially involved in the repair of damaged tissue by influencing epithelial cell growth and secrete cytokines that regulate inflammation (Boismenu *et al.*, 1996). IEL contain a subpopulation of NK cells that could play a major role in the innate response to *Eimeria* infection by cytolysis of infected cells or as a source of IFNγ (Lillehoj, 1989). NK cell activity was correlated with genetic background of susceptible or resistant chicken lines.

Following primary infection with *E. tenella* granulocytes (neutrophils), macrophages and lymphocytes massively infiltrated the LP. Sporozoites that had not reached the crypts by 48 h were often detected inside or at close proximity to macrophages, suggesting that these cells are involved in the non-specific defence and reduce the numbers of developing parasites in the crypts (Vervelde *et al.*, 1996). CD4 cells are entering the LP in large numbers and outnumber the CD8 cells. Although the CD4 cells are numerous they are hardly detected in association with sporozoites. Later in development the CD4 cells were found associated with first-stage schizonts. The time needed to infiltrate the LP was shorter in immune chickens compared to naive chickens and the numbers were less. After 48 h when the infection was greatly neutralized most LP were normal again. In immune chickens CD8 cells appear to be involved in inhibiting the passage of sporozoites to the crypts. More CD8 than CD4+ cells were present in the LP of immune chickens. Sporozoites were often detected inside or in direct proximity to CD8+ cells (Jeurissen *et al.*, 1996). The sequence of events seen in the intestine is strongly correlated to the development of the intestinal stages as might be understood. Strong influxes of cells immediately after infection, peak events around days 5–7, the most pathogenic phase of a primary infection, and reducing numbers after day 7 of infection when the parasites are leaving the body.

All the phenomena seen in the chickens are strongly genetically determined. Chickens selected for divergent *Eimeria* pathology do show differences in numbers of CD4 versus CD8 cells present shortly after primary infection and also the cytokine profiles of the cell populations appear to differ (Yun *et al.*, 2000b). This again could relate to the limited MHC class I and class II repertoire.

When the process of infection and cellular immune response is studied in the pool of peripheral blood lymphocytes again the profile of the intestinal processes can be observed. During the first hours of infection a leucopenic response is detected in the peripheral blood indicating an immediate acute inflammatory response upon the release of sporozoite/oocyst material in the intestinal lumen, probably attracting neutrophils to the site of infection (Rose *et al.*, 1979). Upon restoration of that response probably by recruitment of cells from spleen, bursa and thymus, cells in the peripheral blood seem unresponsive to mitogen and *Eimeria* antigen up to day 6. At day 8 a peak of CD8 cells is observed in some models (Breed *et al.*, 1996), in combination with a strong proliferative responsiveness of CD4 cells. Others report a drop in spleen CD4 cells on day 8 and drop of PBL CD8 cells around day 6 (Bessay *et al.*, 1996). But both studies show that the events in the intestine are opposing those in the peripheral blood, which indicates the strong relationship between the lymphocytes homing in the different lymphoid organs of the chicken.

7.4 *The role of cytokines*

With the better availability of avian cytokine genetic information and bio-assays (Staeheli *et al.*, 2001), the role of the different cytokines in the immune response against *Eimeria* infections became heavily studied (Myamoto *et al.*, 2002; Ovington *et al.*, 1995). In fact, all cytokines that have been studied and are known to be functional in chickens have played a role in the onset and the effector phase of the immune response against *Eimeria*. The inflammatory responses associated with activation of NK cells, macrophages involve IL-1, IL-6, TNF-α, TGF-β and a series of chemokines such as cMGF (Byrnes *et al.*, 1993; and reviewed by Qureshi *et al.*, 2000). These cytokines stimulate T-cell responses (IL-2 production), wound repair, haematopoietic maturation of monocytes and could be involved in pathological phenomena observed (IL-6). Major effects however are described for IFNγ.[1] Its functional role in neutralization of intracellularly developing *Eimeria* has been widely documented (Heriveau *et al.*, 2000; Kogut and Lange, 1989; Laurent *et al.*, 2001), but also *in vivo* injection or presentation of IFNγ did appear to influence the *Eimeria* infection such that improved performance could be noticed from treated chickens (Lillehoj, 1998; Lowenthal *et al.*, 1997). Both CD4 and CD8 cells (under the influence of CD4 produced factors) have been shown to produce IFNγ, especially by cells at the site of infection and the PBL appearing in the blood shortly after clearance of the intestinal infection (days 8–10) (Bessay *et al.*, 1996; Breed *et al.*, 1996, 1997; Yun *et al.*, 2000a, b).

---

[1] The recent discovery of chicken IL-12 has great relevance for studying Eimeria-induced responses in chickens. It is generally believed that this pro-inflammatory cytokines drives the Interferon-gamma response, which has been shown to be of utmost importance in acquisition of resistance to the parasite. Degen W.G.J. *et al.* (2004) Identification and molecular cloning of functional chicken IL-12. J. Immunol. 172,: 4371-4380.

The role IFNγ has in the neutralization of *Eimeria* parasites is, however, not yet clear. Intracellular starvation of tryptophan levels is a well-known effect, as described for *Toxoplasma*-infected cells (Halonen and Weiss, 2000). It is known that activation of macrophages by IFNγ increases inducible nitric oxide synthase (iNOS) activity and produces high concentrations of nitric oxide. In *Eimeria*-infected chickens plasma NO levels sharply increase around days 4–8 after infection for all species. Laurent *et al.* (2001) have utilized the quantitative RT-PCR to study the up-regulation of cytokines and chemokines during infection with *E. tenella* and *E. maxima*. Their findings could corroborate the immunohistological data that IFNγ and iNOS were highly (>200 fold) up-regulated in the caeca during a primary *E. tenella* infection and peaked by day 7. *E. maxima* induced the up-regulation of intrajejunally expressed K203, and IFNγ again peaking around day 7. The functional role of these cytokines in reduction of the effect of the primary infection has been suggested from the role of macrophages in controlling this first infection. Additionally, the cytokines produced maturate the, adaptive immune response to act in a secondary infection. IL-1β, IFNγ and other chemokines acting on the maturation and proliferation of T-cells are involved in the induction phase of the memory response.

The mechanism by which the parasites are neutralized is however not defined. Macrophages were shown in close proximity of intra-intestinal sporozoites and could act through the action of nitric oxides (Jeurissen *et al.*, 1996). CD8 cells might act through the production of IFNγ on newly invaded sporozoites, but also on intracellularly developing parasites through the classical MHC class I restricted cytotoxicity. The role of more recently discovered cytokines such as IL-18 and IL-15 in the protection against *Eimeria* infection has not been studied conclusively.

8 Restoring the balance between host and parasite

Due to the excessive pressure of parasite numbers on the immune system of the host, the responsiveness of the host has become very inefficient, resulting in induction of pathology, disease and lower performance. This situation has long been controlled by the use and application of different drugs often given prophylactically in the feed or drinking water. During the last 60 years more anti-coccidials were developed and often with a shortened life span. Cross-resistance with older drugs quickly induced refractoriness to the action of the newly launched compounds resulting in a rat race between parasite and drug companies with the ever-winning parasite having the best odds (reviewed in Chapman, 1984; Chapman, 1998; Sangster, 2001). Immunological control seems nowadays the best option to restore the balance between host and parasite under the extreme conditions installed to obtain maximal profit for the farmer and high-quality food for the consumer.

8.1 *Vaccination*

By inducing a memory pool in young chickens comprising T-cells that have experienced previous contact with major antigens of the parasites of the different species, the chickens might better respond on high-dose field challenges. The process of vaccination as a way to prevent losses has been applied since the early 1950s, although at a very limited scale. Coccivac™ (see *Table 3*) was developed in the USA as one of the first officially registered vaccines. The vaccine comprised live oocysts of all relevant species for a specific poultry management. The broilers were most sensitive for the action of the proliferative parasite around weeks 3–4 of age, since in that period the growth was exponential and the amount of feed consumed was relatively low (Voeten *et al.*, 1988). Infection in this period induces a so-called sub-clinical coccidiosis often due to *E. acervulina* and *E. maxima* which results in less efficient feed conversion and growth depression. Vaccination in the first week of age could induce a status of immunity to better control infections later in life.

Table 3 *Live chicken coccidiosis vaccines (updated from Vermeulen et al., 2001)*

Tradename	Attenuated	Ionophore tolerance	Species[$]	Application[#]	Manufacturer (year of introduction)
Coccivac D	no	no	A,T,M,N,B, P,H,Miv	DW	Schering Plough (1952)
Coccivac B	no	no	A,T,M,Miv	DW/SB	Schering Plough (1952)
Immucox C1	no	no	A,T,M,N,	DW/G	Vetech Labs (1985)
Immucox C2	no	no	A,T,M,N, B,Mi,P	DW	Vetech Labs (1985)
Paracox	yes	no	A,T,M2,N, B,Mi,P	DW	Schering Plough (1989)
Paracox 5	yes	no	A,T,M2,Mi	SF, SB	Schering Plough (2001)
Livacox D	yes	no	A,T,	DW	Biopharm (1992)
Livacox T	yes	no	A,T,M,	DW	Biopharm (1992)
Livacox Q	yes	no	A,T,M,B	DW, ocular	Biopharm (1992)
Viracox	yes	no	A,T,M,P	DW	Stallen AH (2001)
Nobilis COX ATM	no	yes	A,T,M2,	DW/SB	Intervet. (2001)

[$.] Species abbreviated: *E. acervulina* (A), tenella (T), maxima (M), necatrix (N), brunetti (B), mitis (Mi), mivati (Miv), hagani (H), praecox (P).
M2 means 2 antigenically different strains of *E. maxima*.
[#] Application: drinking water (DW), Spray on birds (SB), Spray on feed, (SF), Oral gel (G).

The development of other similar types of vaccines has occurred rapidly in the last 20 years due to the increasing resistance of parasites against the most commonly applied drugs (reviewed by Chapman *et al.*, 2002; Vermeulen *et al.*, 2001). Williams has recently summarized, in a comprehensive review, most of the studies and experience obtained by using live *Eimeria* vaccines (Williams, 2002). Live vaccines can be divided into fully virulent and low virulent strains (see *Table 3*). Fully virulent strains are usually hot field strains that can be immunogenic after careful administration to young chicks, whereby the dose given to an individual chick is very critical. Firstly, the number of oocysts has to be sufficient to induce an infection and evoke the induction of an adaptive immune response, secondly the adverse effects of this infection should be minimal to prevent to much damage of the intestine, that might result in lesions, and loss of performance. The window for this kind of vaccine is usually rather limited, as these vaccines have long been ignored and only applied on a small scale under controlled conditions.

Strains of lower virulence could be selected from natural populations by harvesting oocysts 1–2 days prior to the peak oocyst shedding. These strains were called precocious and related to a genetic trait of early gametogenesis, by omitting one of the mitotic replication stages. The reduction of the proliferative capacity made these strains suitable as vaccines with a better safety margin. Especially for use in older birds such as the breeders these vaccines have been widely accepted and applied (Williams, 2002). Nobilis COX ATM™ is somewhat unique in its concept. The strains have been selected from field strains for a naturally acquired lower proliferative capacity comparable with the precocious strains. These strains however were also selected for a relative tolerance to widely used ionophores. The ionophores are not only utilized as anti-coccidials but also demonstrate some bacteriostatic activity. The combination of these aspects made it possible to apply ionophores in conjunction with a live vaccine. The widely existing resistance of *Eimeria* field strains for the common drugs might justify the use of this vaccine under specific conditions (Vermeulen *et al.*, 2001).

As discussed above most live vaccines contain two different *E. maxima* strains to cover the antigenic drift observed in the field. Nobilis COX ATM contains the ACM and ACVM strains, both isolated in The Netherlands during the early 1990s. The capacity of these strains to protect against heterologous challenge is illustrated in *Table 4*. These data suggest that both strains even show an additive effect or synergy towards their protection to heterologous strains (T. Schetters, personal communication).

Table 4 *Synergy between antigenically diverse strains of E.maxima derived from Nobilis COX ATM!™ (T. Schetters, unpublished observations)*

Vaccine strain and dose	Challenge strain	% reduction to control
ACM 100 ooc	*E. maxima* Weybridge	42%
ACVM 100 ooc.	*E. maxima* Weybridge	73%
ACVM+ACM 50+50 ooc.	*E. maxima* Weybridge	87%

Groups of 20 chickens were kept in wired floor isolators. They were orally inoculated twice with 1 week interval with 100 sporulated oocysts of *E. maxima* strain (Nobilis COX ATM™) and challenged 2 weeks later with a heterologous strain from UK.

In general, all live vaccines show excellent immunogenicity, but their success in the field is hampered by the problem of unequal administration and obligatory recirculation of the parasites to stimulate sustainable protection. Spray application at 1 day of age seems to be the most successful and reproducible application, although other routes are still used (Williams, 2002). Last but not least should be mentioned the limited production capacity, for these large numbers of vaccine oocysts have to be generated by chickens themselves kept under GMP conditions and processing and quality control need to be made economically feasible.

Subunit vaccines have been studied since the early 1980s. The use of parasite-derived antigens has been considered as not economically of interest for the broiler-type chicken market. Their shear numbers and low cost margins do not allow individual vaccination with this kind of vaccine. Wallach *et al.* (1995) have however recently released a subunit vaccine for application as a maternal vaccine CoxAbic™ (Abic, Israel), inducing specific antibody levels in the egg yolk of their broiler offspring. These antibodies should protect the young chick against too-high challenges and allow immunity to develop against the species in the house. The antigens are derived from gametocytes of *E. maxima* and appear to induce antibodies against conserved epitopes on gametogonic stages of other species. Recently the gene has been described corresponding to the 56-kDa antigen, which is a component of this vaccine (Belli *et al.*, 2002). The 56-kDa protein appears to be localized in the oocyst wall-forming bodies and has all features of a structural protein.

Other vaccination strategies have focused on the delivery of distinct antigens of the most important species of *Eimeria* by using recombinant vectors. Viral vectors such as fowlpox virus, herpes viruses (HVT), have been used to express *Eimeria* genes of interest (Cronenberg *et al.*, 1999). Similarly, reports of bacterially produced antigens mention their application as vaccine candidates (Vermeulen *et al.*, 2001). The search for the appropriate genes involved in the induction of a protective response against the individual species is still on-going, but several candidates have been described, selected from different strategies (reviewed by Jenkins, 1998; Vermeulen, 1998). Most promising seems to be the selection of antigens based on utilization of the relevant immune mechanisms evoked by infections in chickens.

Antibodies have been shown to play only a minor role in protection and thus are not a good tool in discovering the right candidates. T-cell stimulation has been used as a read-out by different investigators and several interesting candidates could be selected from sporozoites of *E. acervulina* (Jenkins, 1998), *E. maxima* (Bumstead *et al.*, 1995b) and *E. tenella* (Vermeulen, 1998). Breed *et al.* (1999) have shown that by using lymphoproliferation in combination with production of IFNγ-protective antigens were selected from sporozoites of *E. tenella*. The cells involved in this response were obtained from the pool appearing in the blood shortly after clearance of the infection and consisted of both CD4- and CD8-positive T-cells. These proteins were recently identified as belonging to the anti-oxidant pool of the parasite, expressed in different stages of the development (Kuiper *et al.*, 2001). Protection could be shown when presented as live recombinant *Salmonella* by reducing the caecal lesions considerably after virulent *E. tenella* challenge.

Apart from these anti-oxidant enzymes a lactate-dehydrogenase (LDH) has been shown partially protective when presented as recombinant *Salmonella* or as live herpes virus turkey (HVT) vector (Schaap *et al.*, 1999).[2] The *E. tenella* LDH was able to reduce caecal lesions in heavily challenged birds (*Figure 3*). In this experiment it was compared with a well-known candidate EtSO7 (Crane *et al.*, 1991) and showed superior protection. The *E. acervulina* (EaLDH) was tested for weight gain of the challenged chickens, kept under semi-field conditions (*Table 5*). After 19 days of challenge infection both HVT-EaLDH groups demonstrated a moderate, but significantly better weight than the birds receiving the control vector. This might indicate that especially cytosolic proteins are inducing T-cell-mediated responses, probably due to

[2] The data on the molecular characterization of Eimeria tenella LDH will shortly be published by Schaap D. *et al.* (2004) An Eimeria vaccine candidate appears to be Lactate Dehydrogenase; characterization and comparative analysis. Parasitology vol. 128 (6), pp 603-616.

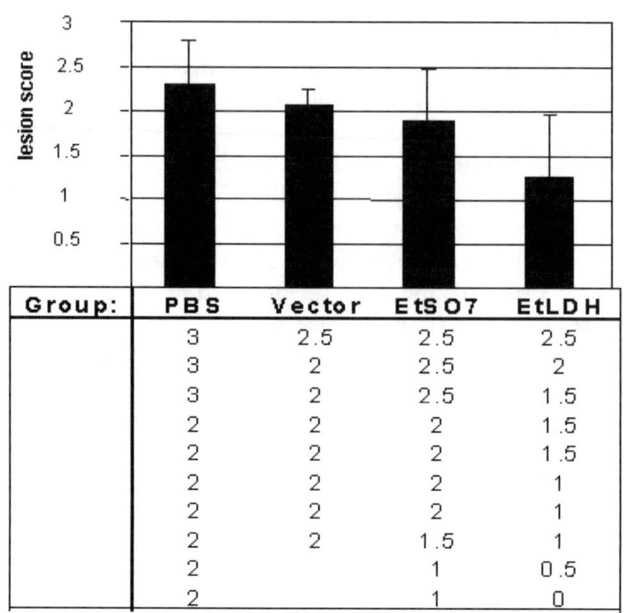

Group:	PBS	Vector	EtSO7	EtLDH
	3	2.5	2.5	2.5
	3	2	2.5	2
	3	2	2.5	1.5
	2	2	2	1.5
	2	2	2	1.5
	2	2	2	1
	2	2	2	1
	2	2	1.5	1
	2		1	0.5
	2		1	0

Figure 3 *Vaccination with EtLDH reduces* E. tenella *induced caecal lesions. Chickens were vaccinated with live* S. typhimurium *vaccination vectors, containing empty plasmid, EtSO7 or EtLDH expression plasmids (and an additional PBS control). Groups of 8–10 chickens were subsequently challenged with* E. tenella, *and induced caecal lesions were analysed and semi-quantified. Shown are the individual lesion scores, and the means with standard deviations are plotted on top. A Kruskal–Wallis test showed that the lesion score in the EtLDH group was significant reduced compared with the PBS group* (P = *0.0033) and the empty vector group* (P = *0.011).*

Table 5 *Improved weight gain upon EaLDH vaccination.*

			day 0	day 6	day 13	day 19
Pen-1	HVT/EaLDH	Avg.	1133	1371	2010	2596
	(n=51)	Std.	91	110	146	170
	HVT	Avg.	1123	1344	1944	2514
	(n=46)	Std.	93	110	175	246
Pen-2	HVT/EaLDH	Avg.	1150	1371	1991	2601
	(n=45)	Std.	93	134	176	208
	HVT	Avg.	1174	1398	2010	2570
	(n=45)	Std.	77	112	183	288

Day old chickens were vaccinated once with recombinant HVT expressing EaLDH, or with HVT parent strain as control. Twenty-four days later all chickens were challenged with a mixed dose containing both *E. acervulina* and *E. maxima* oocysts. Weight gain of all animals was individually followed for 19 days and average weights per group are indicated (gram per chicken, including standard deviations). Total number of animals per group varied between 45 and 51 chickens as indicated. Average weight gain in HVT/EaLDH vaccinated animals (compared with HVT control) 19 days after challenge was 62 gram for both pens ($P < 0.04$).

release of these components at schizogony or presentation in the context of class I on infected cells. It is intriguing that a normally conserved enzyme such as LDH is so diverged in different *Eimeria* species with only 66–70% identity in amino-acid composition (*Table 6*). This might suggest that this protein has undergone immunological pressure, especially if it is taken into account that most other proteins that were vaccine candidates demonstrated homologies around 90% and higher (*Table 7*). Protection studies were reported using an *E. acervulina* 3-1E antigen of merozoites that induced also IFNγ in stimulated immune T-cells (Lillehoj *et al.*, 2000).

Recently also DNA vaccine strategies have been applied in chickens using genes of *E. tenella* and *E. acervulina*. The *E. tenella* SO7 gene was applied in a pcDNA plasmid and administered to chickens in different doses but with little or no success (Kopko *et*

Table 6 *Amino acid conservation of LDH between species. Shown is the percentage amino acid identity between species*

	E. acervulina	E. maxima	E. tenella	T. gondii LDH1	T. gondii LDH2	P. falciparum	Human	Chicken	Dogfish
E. acervulina	x								
E. maxima	80	x							
E. tenella	70	66	x						
T. gondii LDH1	59	58	54	x					
T. gondii LDH2	60	55	56	71	x				
P. falciparum	50	50	47	46	48	x			
Human	27	26	27	26	27	27	x		
Chicken	25	23	26	26	28	27	84	x	
Dogfish	27	24	25	26	25	25	76	74	x

Table 7 *Homology between* Eimeria *genes from different species.*

Gene		E. acervulina	E. tenella	E. maxima	Length	Identity
SO7 (S05575)	E. acervulina	x	AF073460		217	69%
Hsp70 (S37165)	E. acervulina	x		S51682	520	92%
Transhydrogenase (L08392)	E. acervulina	x	T81520		418	91%
Asp. Protease (CAA80843)	E. acervulina	x	CAC20152		468	97%
Em100 (A48569)	E. maxima		A44638	x	727	69%
Calmodulin kinase (CAA96438)	E. maxima		Z71757	x	413	91%

Shown are full length *Eimeria* genes, with known homologues in other *Eimeria* species. Gene names and genbank accession numbers are indicated. Amino acids lengths are given, and the percentage identity is shown.

al., 2000). The *E. acervulina* 3-1E gene was applied in a pMP13 plasmid and induced different cellular immune responses resulting in partial protection after homologous challenge. The addition of cytokine genes was reported to enhance the effects obtained (Min *et al.*, 2001).

Although much work has been focused on proteins associated with the invasion process, such as microneme proteins (Bumstead and Tomley, 2000; Ryan *et al.*, 2000; Tomley and Soldati, 2001), these might not appear the best candidates for a vaccine due to the vast redundancies and overlapping functions of these proteins. However, it should be said that some investigators have reported some partial protection after using some of these antigens (Danforth *et al.*, 1989). With the vast amount of new data becoming available on the genomic organization of *Eimeria* and related protozoa, a new challenge begins for the bio-informatics to process the information and for the biologists to utilize it to come forward with better solutions for the problems described (Ng *et al.*, 2002).

9 Summary

The co-evolution of *Eimeria* and its host the domestic chicken has resulted in a delicate balance of mutual understanding and respect. This balance has been broken by the complete change of the environment in which the parasite was able to reproduce to such an extent that the host, stressed and weakened by heat, crowding and concurrent infections could not combat the shear numbers of organisms. The use of drugs to control the situation has been shown to only temporarily create relief. Resistance widely developed by the flexible genome of the parasite returned new drugs at a greater speed than they had been developed. Improved hygienic measures, better facility management and good understanding of epidemiology of the parasites spreading and proliferation seem the first and most promising set of tools to control the balance.

Reduction of stock density may only provide any relief if this is done at a factor of 10 or higher and this is not a realistic measure in relation to the profit. Free-range chickens are an alternative if only animal welfare is at stake. However, in terms of prevalence of parasitic infections, such as coccidia, helminthes or ectoparasites, chickens do not seem to be better off (Permin *et al.*, 2002).

Immunological surveillance and the development of safe, effective and economical vaccines are further refinements that can be used to restore the relationship between parasite and host. Several live vaccines are effective and applied, but certainly have drawbacks in safety and production. New technology such as recombinant vectors together with a better understanding of the cell biology of the parasite from biological and genomic information should provide improved vaccines for the future.

The strong genetically determined characteristics involved in the induction and maintenance of a sustainable protective immune response might turn out to be of decisive importance for the success of these strategies. The consequences for the physiology of the parasite remain to be understood.

Acknowledgements

I am greatly indebted to my colleagues in the Department of Parasitology R&D, Intervet for their technical skills and scientific support in the experimental work that

forms the basis for my humble attempts to unravel and understand host–parasite relationships. Dick Schaap I thank for his critical review of the manuscript and useful comments. Theo Schetters is acknowledged for his inspiring ideas on the positioning of vaccines (e.g. Nobilis Cox ATM) in the relationship with the host. Virgil Schijns (Dept of Vaccine Technology and Immunology, Intervet) has been a great help in the discussions on the avian immune system.
Gert Flik and Declan Nolan (University of Nijmegen) are thanked for inviting me to the San Diego conference and to write this review. My family Moon, Niels, Menno, Eline en Nicole must have missed me for many weekends during the completion of this manuscript, I hope to compensate for that shortly.

References

Al Natour, M.Q., Suleiman, M.M. and Abo-Shehada, M.N. (2002) Flock-level prevalence of Eimeria species among broiler chicks in northern Jordan. *Prev. Vet. Med.* **53:** 305–310.

Al Sheikhly, F. and Al Saieg, A. (1980) Role of Coccidia in the occurrence of necrotic enteritis of chickens. *Avian Dis.* **24:** 324–333.

Allen, P.C. (1997a) Nitric oxide production during *Eimeria tenella* infections in chickens. *Poult. Sci.* **76:** 810–813.

Allen, P.C. (1997b) Production of free radical species during *Eimeria maxima* infections in chickens. *Poult. Sci.* **76:** 814–821.

Allen, P.C. (2000) Effects of treatments with cyclooxygenase inhibitors on chickens infected with *Eimeria acervulina*. *Poult. Sci.* **79:** 1251–1258.

Allen, P.C. and Fetterer, R.H. (2000) Effect of *Eimeria acervulina* infections on plasma L-arginine. *Poult. Sci.* **79:** 1414–1417.

Allen, P.C. and Fetterer, R.H. (2002) Recent advances in biology and immunobiology of *Eimeria* species and in diagnosis and control of infection with these coccidian parasites of poultry. *Clin. Microbiol. Rev.* **15:** 58–65.

Allen, P.C. and Lillehoj, H.S. (1998) Genetic influence on nitric oxide production during *Eimeria tenella* infections in chickens. *Avian Dis.* **42:** 397–403.

Allen, P.C., Danforth, H.D., Morris, V.C. and Levander, O.A. (1996) Association of lowered plasma carotenoids with protection against cecal coccidiosis by diets high in n-3 fatty acids. *Poultry Sci.* **75:** 966–972.

Anderson, R.M. and May, R.M. (1982) Coevolution of hosts and parasites. *Parasitology* **85:** 411–426.

Augustine, P.C. and Danforth, H.D. (1990) Avian eimeria: invasion in foreign host birds and generation of partial immunity against coccidiosis. *Avian Dis.* **34:** 196–202.

Augustine, P.C. and Danforth, H.D. (1995) *Eimeria tenella* and *E. acervulina* – differences in ability to elicit cross-species protection as compared with the turkey coccidium, e-adenoeides. *Avian Dis.* **39:** 709–717.

Baba, E., Fukata, T. and Arakawa, A. (1985) Factors influencing enhanced *Salmonella typhimurium* infection in *Eimeria tenella*-infected chickens. *Am. J. Vet. Res.* **46:** 1593–1596.

Barta, J.R., Jenkins, M.C. and Danforth H.D. (1991) Evolutionary relationships of avian *Eimeria* species among other apicomplexan protozoa – Monophyly of the Apicomplexa is supported. *Mol. Biol. Evol.* **8:** 345–355.

Barta, J. R., Martin, D.S., Liberator, P.A., Dashkevicz, M., Anderson, J.W., Feighner, S.D., Elbrecht, A., Perkinsbarrow, A., Jenkins, M.C. and Danforth, H.D. (1997) Phylogenetic relationships among eight *Eimeria* species infecting domestic fowl inferred using complete small subunit ribosomal dna sequences. *J. Parasitol.* **83:** 262–271.

Barta, J.R., Coles, B.A., Schito, M.L., Fernando, M.A., Martin, A. and Danforth, H.D. (1998) Analysis of infraspecific variation among five strains of *Eimeria maxima* from North America. Intern. J. Parasitol. **28:** 485–492.

Barta, J.R., Martin, D.S., Carreno, R.A., Siddall, M.E., Profous-Juchelkat, H., Hozza, M. Powles, M.A. and Sundermann, C. (2001) Molecular phylogeny of the other tissue coccidia: Lankesterella and Caryospora. *J. Parasitol.* **87:** 121–127.

Barta, J.R. (2001) Molecular approaches for inferring evolutionary relationships among protistan parasites. *Vet. Parasitol.* **101:** 175–186.

Beattie, S.E., Fernando, M.A. and Barta, J.R. (2001) A comparison of sporozoite transport after homologous and heterologous challenge in chickens immunized with the Guelph strain or the Florida strain of *Eimeria maxima*. *Parasitol. Res.* **87:** 116–121.

Belli, S.I., Lee, M., Thebo, P., Wallach, M.G., Schwartsburd, B. and Smith, N.C. (2002) Biochemical characterisation of the 56 and 82 kDa immunodominant gametocyte antigens from *Eimeria maxima*. *Int. J. Parasitol.* **32:** 805–816.

Bessay, M., Le Vern, Y., Kerboeuf, D., Yvore, P. and Quere, P. (1996) Changes in intestinal intra-epithelial and systemic T-cell subpopulations after an *Eimeria* infection in chickens: comparative study between *E. acervulina* and *E. tenella*. *Vet. Res.* **27:** 503–514.

Birrenkott, G.P., Brockenfelt, G.E., Greer, J.A. and Owens, M.D. (2000) Topical application of garlic reduces northern fowl mite infestation in laying hens. *Poult. Sci.* **79:** 1575–1577.

Boismenu, R., Feng, L., Xia, Y.Y., Chang, J.C. and Havran, W.L. (1996) Chemokine expression by intraepithelial gamma delta T cells. Implications for the recruitment of inflammatory cells to damaged epithelia. *J. Immunol.* **157:** 985–992.

Bowers, R.G. (2001) The basic depression ratio of the host: the evolution of host resistance to microparasites. *Proc. Roy. Soc. London: Series B Biol. Sci.* **268:** 243–250.

Bradley S. and Radhakrishnan C.V. (1973) Coccidiosis in chickens: obligate relationship between *Eimeria tenella* and certain species of cecal microflora in the pathogenesis of the disease. *Avian Dis.* **17:** 461–475.

Brake, D.A., Fedor, C.H., Werner, B.W., Miller, T.J., Taylor, R.L. and Clare, R.A. (1997) Characterization of immune response to *Eimeria tenella* antigens in a natural immunity model with hosts which differ serologically at the b locus of the major histocompatibility complex. *Infect. Immun.* **65:** 1204–1210.

Breed, D.G.J., Dorrestein, J. and Vermeulen, A.N. (1996) Immunity to *Eimeria tenella* in chickens – phenotypical and functional changes in peripheral blood t-cell subsets. *Avian Dis.* **40:** 37–48.

Breed, D.G.J., Dorrestein, J., Schetters, T.P.M., Waart, L.V.D., Rijke, E. and Vermeulen, A.N. (1997) Peripheral blood lymphocytes from *Eimeria tenella* infected chickens produce gamma-interferon after stimulation in vitro. *Parasite Immunol.* **19:** 127–135.

Breed, D.G., Schetters, T.P., Verhoeven, N.A., Boot-Groenink, A., Dorrestein, J. and Vermeulen, A.N. (1999) Vaccination against *Eimeria tenella* infection using a fraction of *E. tenella* sporozoites selected by the capacity to activate T cells. *Intern. J. Parasitol.* **29:** 1231–1240.

Brown, S.P., Hochberg, M.E. and Grenfell, B.T. (2002) Does multiple infection select for raised virulence? *Trends Microbiol.* **10:** 401–405.

Bumstead, N. and Millard, B.J., (1992) Variation in susceptibility of inbred lines of chickens to 7 species of *Eimeria*. *Parasitology* **104:** 407–413.

Bumstead, J. and Tomley, F. (2000) Induction of secretion and surface capping of microneme proteins in *Eimeria tenella*. *Mol. Biochem. Parasitol.* **110:** 311–321.

Bumstead, J.M., Bumstead, N., Rothwell, L. and Tomley, F.M. (1995) Comparison of immune responses in inbred lines of chickens to *Eimeria maxima* and *Eimeria tenella*. *Parasitology* **111:** 143–151.

Bumstead, J.M., Dunn, P.P. and Tomley, F.M. (1995) Nitrocellulose immunoblotting for identification and molecular gene cloning of *Eimeria maxima* antigens that stimulate lymphocyte proliferation. *Clin. Diagnost. Lab. Immunol.* **2:** 524–530.

Byrnes, S., Eaton, R. and Kogut, M. (1993) In vitro interleukin-1 and tumor necrosis factor-alpha production by macrophages from chickens infected with either *Eimeria maxima* or Eimeria tenella. *Intern. J. Parasitol.* **23:** 639–645.

Chapman, H.D. (1984) Drug resistance in avian coccidia (a review). *Vet. Parasitol.* **15:** 11–27.

Chapman, H.D. (1998) Evaluation of the efficacy of anticoccidial drugs against *Eimeria* species in the fowl. *Intern. J. Parasitol.* **28:** 1141–1144.

Chapman, H.D., Cherry, T.E., Danforth, H.D., Richards, G., Shirley, M.W. and Williams, R.B. (2002) Sustainable coccidiosis control in poultry production: the role of live vaccines. *Intern. J. Parasitol.* **32:** 617–629.

Charles, S., Morand, S., Chasse, J. and Auger, P. (2002) Host patch selection induced by parasitism: basic reproduction ratio r(0) and optimal virulence. *Theor. Pop. Biol.* **62:** 97.

Clare, R.A., Strout, R.G., Taylor, R.L., Jr., Collins, W.M. and Briles, W.E. (1985) Major histocompatibility (B) complex effects on acquired immunity to cecal coccidiosis. *Immunogenetics* **22(6):** 593–599.

Clare, R.A., Taylor, R.L., Jr., Briles, W.E. and Strout, R.G. (1989) Characterization of resistance and immunity to *Eimeria tenella* among major histocompatibility complex B-F/B-G recombinant hosts. *Poult. Sci.* **68:** 639–645.

Conway, D.P., Sasai, K., Gaafar, S.M. and Smothers C.D. (1993) Effects of different levels of oocyst inocula of *Eimeria acervulina*, *E. tenella*, and *E. maxima* on plasma constituents, packed cell volume, lesion scores, and performance in chickens. *Avian Dis.* **37:** 118–123.

Conway, D.J. and Roper, C. (2000) Micro-evolution and emergence of pathogens. *Intern. J. Parasitol.* **30:** 1423–1430.

Crane, M.S., Goggin, B., Pellegrino, R.M., Ravino, O.J., Lange, C., Karkhanis, Y.D., Kirk, K.E. and Chakraborty, P.R. (1991) Cross-protection against four species of chicken coccidia with a single recombinant antigen. *Infect. Immun.* **59:** 1271–1277.

Cronenberg, A.M., van Geffen, C.E., Dorrestein, J., Vermeulen, A.N. and Sondermeijer, P.J. (1999) Vaccination of broilers with HVT expressing an *Eimeria acervulina* antigen improves performance after challenge with *Eimeria*. *Acta Virol.* **43:** 192–197.

Danforth, H.D., Augustine, P.C., Ruff, M.D., McCandliss, R., Strausberg, R.L. and Likel, M. (1989) Genetically engineered antigen confers partial protection against avian coccidial parasites. *Poult. Sci.* **68:** 1643–1652.

Davies, S.F.M. and Joyner, L.P. (1962) Infection of the fowl by the parenteral inoculation of oocysts by *Eimeria*. *Nature* **194:** 996–997.

Davis, P.J., Parry, S.H. and Porter, P. (1978) The role of secretory IgA in anti-coccidial immunity in the chicken. *Immunology* **34:** 879–888.

Edgar, S.A. and Seibold, C.T. (1964) A new coccidium of chickens *Eimeria mivati* sp. n. (Protozoa: Eimeriidae) with details of its life history. *J. Parasitol.* **50:** 193–204.

Emara, M.G., Lapierre, R.R., Greene, G.M., Knieriem, M., Rosenberger, J.K., Pollock, D.L., Sadjadi, M., Kim, C.D. and Lillehoj, H.S. (2002) Phenotypic variation among three broiler pure lines for Marek's disease, coccidiosis, and antibody response to sheep red blood cells. *Poult. Sci.* **81:** 642–648.

Ewald, S.J. (2000) MHC-determined polymorphism in the amount of class I molecules expressed on lymphocytes of broiler breeder chicken haplotypes. Current progress on avian immunology research. In: Schat, K.A. (ed.) *Proceedings of the 6th Avian Immunology Research Group Meeting* pp. 197–200. Cornell University, Ithaca, N.Y.

Farr, M. and Doran, D. (1962) Comparative excystation of four species of poultry coccidia. *J. Protozool.* **9:** 403–407.

Fayer, R. (1980) Epidemiology of protozoan infections: The coccidia. *Vet. Parasitol.* **6:** 75–103.

Fitz-Coy, S.H. (1983) The life history, pathogenicity, pathology, immunogenic properties, ultrastructure and control of two isolates of *Eimeria mitis*, Tyzzer 1929. *Dissert. Abst. Intern. B* **43:** 3411–3425.

Fitz-Coy, S.H. (1992) Antigenic variation among strains of *Eimeria maxima* and *E. tenella* of the chicken. *Avian Dis.* **36:** 40–43.

Gandon, S. and Michalakis, Y. (2000) Evolution of parasite virulence against qualitative or quantitative host resistance. *Proc. Roy. Soc. London Ser. B Biol. Sci.* **267:** 985–990.

Gandon, S., Mackinnon, M.J., Nee, S. and Read, A.F. (2001a) Imperfect vaccines and the evolution of pathogen virulence. *Nature* **414:** 751–756.

Gandon, S., Jansen, V.A. and van Baalen, M. (2001b) Host life history and the evolution of parasite virulence. *Evolution: Intern. J. Organ. Evol.* **55:** 1056–1062.

Goebel, T. (2000) Antigen-specific receptors on chicken T- and NK-cells. Current progress on avian immunology research. In: Schat, K.A. (ed.) *Proceedings of the 6th Avian Immunology Research Group Meeting* pp. 104–109. Cornell University, Ithaca, NY.

Gore, T.C. and Long, P.L. (1982) The biology and pathogenicity of a recent field isolate of *Eimeria praecox* Johnson, 1930. *J. Protozool.* **29:** 82–85.

Graat, E.A.M., van der Kooij, E., Frankena, K., Henken, A.M., Smeets, J.F.M. and Hekerman, M.T.J. (1998) Quantifying risk factors of coccidiosis in broilers using on-farm data based on a veterinary practice. *Prevent. Vet. Med.* **33:** 297–308.

Greif, G. (2000) Immunity to coccidiosis after treatment with toltrazuril. *Parasitol. Res.* **86:** 787–790.

Halonen, S.K. and Weiss, L.M. (2000) Investigation into the mechanism of gamma interferon-mediated inhibition of *Toxoplasma gondii* in murine astrocytes. *Infect. Immun.* **68:** 3426–3430.

Heriveau, C., Dimier-Poisson, I., Lowenthal, J., Naciri, M. and Quere, P. (2000) Inhibition of *Eimeria tenella* replication after recombinant IFN-gamma activation in chicken macrophages, fibroblasts and epithelial cells. *Vet. Parasitol.* **92:** 37–49.

Jenkins, M.C. (1998) Progress on developing a recombinant coccidiosis vaccine. *Intern. J. Parasitol.* **28:** 1111–1119.

Jenkins, M.C., Augustine, P.C., Danforth, H.D. and Barta, J.R. (1991a) X-irradiation of *Eimeria tenella* oocysts provides direct evidence that sporozoite invasion and early schizont development induce a protective immune response(s). *Infect. Immun.* **59:** 4042–4048.

Jenkins, M.C., Augustine, P.C., Barta, J.R., Castle, M.D. and Danforth, H.D. (1991b) Development of resistance to coccidiosis in the absence of merogonic development using X-irradiated *Eimeria acervulina* oocysts. *Exp. Parasitol.* **72:** 285–293.

Jeurissen, S.H.M., Janse, E.M., Vermeulen, A.N. and Vervelde, L. (1996) *Eimeria tenella* infections in chickens: aspects of host–parasite interaction. *Vet. Immunol. Immunopathol.* **54:** 231–238.

Johnston, W.T., Shirley, M.W., Smith, A.L. and Gravenor, M.B. (2001) Modelling host cell availability and the crowding effect in *Eimeria* infections. *Intern. J. Parasitol.* **31:** 1070–1081.

Jorgensen, W. K., N. P. Stewart, P. J. Jeston, J. B. Molloy, G. W. Blight, and R. J. Dalgliesh. 1997. Isolation and pathogenicity of Australian strains of Eimeria praecox and Eimeria mitis. Australian Veterinary Journal. **75**:592-595.

Joyner, L.P. (1969) Immunological variation between two strains of *Eimeria acervulina*. *Parasitology* **59:** 725–732.

Joyner, L.P. (1982) Host and site specificity. In: Long, P.L. (ed.) *The Biology of the Coccidia,* pp. 35–62. Edward Arnold (Publ.) London, UK.

Kaiser, P. (2000) Cytokine workshop. Current progress on avian immunology research. In: Schat, K.A. (ed.) *Proceedings of the 6th Avian Immunology Research Group Meeting*, pp. 357–361. Cornell University, Ithaca, NY.

Karim, M.J. and Trees, A.J. (1990) Isolation of five species of *Eimeria* from chickens in Bangladesh. *Trop. Anim. Health Prod.* **22:**153–159.

Kaufman, J., Milne, S., Gobel, T.W., Walker, B.A., Jacob, J.P., Auffray, C., Zoorob, R. and Beck, S. (1999a) The chicken B locus is a minimal essential major histocompatibility complex. *Nature* **401:** 923–925.

Kaufman, J., Jacob, J., Shaw, I., Walker, B., Milne, S., Beck, S. and Salomonsen, J. (1999b) Gene organisation determines evolution of function in the chicken MHC. *Immunol. Rev.* **167:** 101–117.

Kettunen, H., Tiihonen, K., Peuranen, S., Saarinen, M.T. and Remus, J.C. (2001) Dietary betaine accumulates in the liver and intestinal tissue and stabilizes the intestinal epithelial structure in healthy and coccidia-infected broiler chicks. *Comp. Biochem. Physiol. A: Mol. Integ. Physiol.* **130:** 759–769.

Klasing, K.C., Laurin, D.E., Peng, R.K. and Fry, D.M. (1987) Immunologically mediated growth depression in chicks: influence of feed intake, corticosterone and interleukin-1. *J. Nutr.* **117:** 1629–1637.

Klasing, K.C., Adler, K.L., Remus, J.C. and Calvert, C.C. (2002) Dietary betaine increases intraepithelial lymphocytes in the duodenum of coccidia-infected chicks and increases functional properties of phagocytes. *J. Nutr.* **132:** 2274–2282.

Kogut, M.H. and Lange, C. (1989) Recombinant interferon-gamma inhibits cell invasion by *Eimeria tenella*. *J. Interferon Res.* **9:** 67–77.

Kopko, S.H., Martin, D.S. and Barta, J.R. (2000) Responses of chickens to a recombinant refractile body antigen of *Eimeria tenella* administered using various immunizing strategies. *Poult. Sci.* **79:** 336–342.

Kouwenhoven, B. (1972) The significance of infection dose, body weight and age of the host on the pathogenic effect of *Eimeria acervulina* in the fowl. *Zeitsch. Parasiten.* **39:** 255–268.

Kouwenhoven, B. and van der Horst, C.J. (1969) Strongly acid intestinal content and lowered protein, carotene and vitamin A blood levels in *Eimeria acervulina* infected chickens. *Zeitsch. Parasiten.* **32:** 347–353.

Kouwenhoven, B. and van der Horst, C.J. (1972) Disturbed intestinal absorption of vitamin A and carotenes and the effect of a low pH during *Eimeria acervulina* infection in the domestic fowl (*Gallus domesticus*). *Zeitsch. Parasiten.* **38:**152–161.

Kuiper, C.M., Roosmalen-Vos, S.V., Beek-Verhoeven, N., Schaap, T.C. and Vermeulen, A.N. (2001) *Eimeria tenella* anti-oxidant proteins: differentially expressed enzymes with immunogenic properties. In: Ellis, J.T., Johnson, A.M., Morrison, D.A., Smith, N.C. (eds) *Proceedings of the VIIIth International Coccidiosis Conference*, pp. 102–103. Palm Cove, Australia.

Laurent, F., Mancassola, R., Lacroix, S., Menezes, R. and Naciri, M. (2001) Analysis of chicken mucosal immune response to *Eimeria tenella* and *Eimeria maxima* infection by quantitative reverse transcription-PCR. *Infect. Immun.* **69:** 2527–2534.

Leshchinsky, T.V. and Klasing, K.C. (2001) Divergence of the inflammatory response in two types of chickens. *Develop. Comp. Immunol.* **25:** 629–638.

Levine, N.D. (1982) Taxonomy and life cycles of Coccidia. In: Long, P.L. (ed.) *The Biology of the Coccidia*, pp. 1–33. Edward Arnold (Publ) London, UK.

Li, L., Johnson, L.W., Livant, E.J. and Ewald, S.J. (1999) The MHC of a broiler chicken line: serology, B-G genotypes, and B-F/B-LB sequences. *Immunogenetics* **49:** 215–224.

Lillehoj, H.S. (1989) Intestinal intraepithelial and splenic natural killer cell responses to eimerian infections in inbred chickens. *Infect. Immun.* **57:** 1879–1884.

Lillehoj, H.S. (1998) Role of T lymphocytes and cytokines in coccidiosis. *Int. J. Parasitol.* **28:** 1071–1081.

Lillehoj, H.S. and Bacon, L.D. (1991) Increase of intestinal intraepithelial lymphocytes expressing CD8 antigen following challenge infection with *Eimeria acervulina*. *Avian Dis.* **35:** 294–301.

Lillehoj, H.S. and Trout, J.M. (1996) Avian gut-associated lymphoid tissues and intestinal immune responses to eimeria parasites [Review]. *Clin. Microbiol. Rev.* **9:** 349.

Lillehoj, H.S., Choi, K.D., Jenkins, M.C., Vakharia, V.N., Song, K.D., Han, J.Y. and Lillehoj, E.P. (2000) A recombinant *Eimeria* protein inducing interferon-gamma production: comparison of different gene expression systems and immunization strategies for vaccination against coccidiosis. *Avian Dis.* **44:** 379–389.

Long, P.L. and Millard, B.J. (1979) Immunological differences in *Eimeria maxima*: effect of a mixed immunizing inoculum on heterologous challenge. *Parasitology* **79:** 451–457.

Lowenthal, J.W., York, J.J., O'Neil, T.E., Rhodes, S., Prowse, S.J., Strom, D.G. and Digby, M.R. (1997) In vivo effects of chicken interferon-gamma during infection with *Eimeria*. *J. Interferon Cytokine Res.* **17:** 551–558.

Lunden, A., Thebo, P., Gunnarsson, S., Hooshmian-Rad, P., Tauson, R. and Uggla, A. (2000) *Eimeria* infections in litter-based, high stocking density systems for loose-housed laying hens in Sweden. *Br. Poult. Sci.* **41:** 440–447.

Martin, A.G., Danforth, H.D., Barta, J.R. and Fernando, M.A. (1997) Analysis of immunological cross-protection and sensitivities to anticoccidial drugs among five geographical and temporal strains of *Eimeria maxima*. *Intern. J. Parasitol.* **27:** 527–533.

Matthews, J.O. and Southern, L.L. (2000) The effect of dietary betaine in *Eimeria acervulina*-infected chicks. *Poult. Sci.* **79:** 60–65.

Mattiello, R., Boviez, J.D. and McDougald, L.R. (2000) *Eimeria brunetti* and *Eimeria necatrix* in chickens of Argentina and confirmation of seven species of *Eimeria*. *Avian Dis.* **44:** 711–714.

McDonald, V., Wisher, M.H., Rose, M.E. and Jeffers, T.K. (1988) *Eimeria tenella*: immunological diversity between asexual generations. *Parasite Immunol.* **10:** 649–660.

McDougald, L.R., Fuller, L. and Mattiello, R. (1997) A survey of Coccidia on 43 poultry farms in Argentina. *Avian Dis.* **41:** 923–929.

Miller, M.M., Goto, R.M., Bacon, L.D. and Hunt, H.D. (2000) Serologic evidence for *RFP-Y* class I expression. Current progress on avian immunology research. In: Schat, K.A. (ed.) *Proceedings of the 6th Avian Immunology Research Group Meeting* pp. 171–176. Cornell University, Ithaca, NY.

Min, W., Lillehoj, H.S., Burnside, J., Weining, K.C., Staeheli, P. and Zhu, J.J. (2001) Adjuvant effects of IL-1beta, IL-2, IL-8, IL-15, IFN-alpha, IFN-gamma TGF-beta4 and lymphotactin on DNA vaccination against *Eimeria acervulina*. *Vaccine* **20:** 267–274.

Miyamoto, T., Min, W. and Lillehoj, H.S. (2002) Kinetics of interleukin-2 production in chickens infected with *Eimeria tenella*. *Comp. Immunol. Microbiol. Infect. Dis.* **25:**149–158.

Ng, S.T., Sanusi Jangi, M., Shirley, M.W., Tomley, F.M. and Wan, K.L. (2002) Comparative EST analyses provide insights into gene expression in two asexual developmental stages of *Eimeria tenella*. *Exper. Parasitol.* **101:** 168–173.

Ovington, K.S., Alleva, L.M. and Kerr, E.A. (1995) Cytokines and immunological control of *Eimeria* spp. *Intern. J. Parasitol.* **25:** 1331–1351.

Permin, A., Nansen, P., Bisgaard, M. and Frandsen, F. (1998) *Ascaridia galli* infections in free-range layers fed on diets with different protein contents. *Br. Poult. Sci.* **39:** 441–445.

Permin, A., Bisgaard, M., Frandsen, F., Pearman, M., Kold, J. and Nansen, P. (1999) Prevalence of gastrointestinal helminths in different poultry production systems. *Br. Poult. Sci.* **40:** 439–443.

Permin, A., Esmann, J.B., Hoj, C.H., Hove, T. and Mukaratirwa, S. (2002) Ecto-, endo- and haemoparasites in free-range chickens in the Goromonzi District in Zimbabwe. *Prevent. Vet. Med.* **54:** 213–224.

Pfennig, K.S. (2001) Evolution of pathogen virulence: the role of variation in host phenotype. *Proc. Royal Soc. London Ser. B: Biol. Sci.* **268:** 755–760.

Poulin, R. and Combes, C. (1999) The concept of virulence: interpretations and implications. *Parasitol. Today* **15:** 474–475.

Poulsen, J., Permin, A., Hindsbo, O., Yelifari, L., Nansen, P. and Bloch, P. (2000) Prevalence and distribution of gastro-intestinal helminths and haemoparasites in young scavenging chickens in upper eastern region of Ghana, West Africa. *Prevent. Vet. Med.* **45:** 237–245.

Pugliese, A. (2002) On the evolutionary coexistence of parasite strains. *Math. Biosci.* **177–178:** 355–375.

Qureshi, M.A., Heggen, C.L., and Hussain, I. (2000) Avian macrophage: effector functions in health and disease. *Dev. Comp. Immunol.* **24:** 103–119.

Reza Razmi, G. and Ali Kalideri, G. (2000) Prevalence of subclinical coccidiosis in broiler-chicken farms in the municipality of Mashhad, Khorasan, Iran. *Prevent. Vet. Med.* **44:** 247–253.

Rodriguez, J.C., Segura, J.C., Alzina, A. and Gutierrez, M.A. (1997) Factors affecting mortality of crossbred and exotic chickens kept under backyard systems in Yucatan, Mexico. *Trop. Anim. Health Prod.* **29:** 151–157.

Rose, M.E. (1972) Immunity to coccidiosis: maternal transfer in *Eimeria maxima* infections. *Parasitology* **65:** 273–282.

Rose, M.E. (1978) Immune responses of chickens to coccidia and coccidiosis. In: Long, P.L., Boorman, K.N. and Freeman, B.M. (eds) *Avian Coccidiosis*, pp. 297–338. British Poultry Science Ltd., Edinburgh, UK.

Rose, M.E. and Hesketh, P. (1983) Infections with *Eimeria* species: the role of bile. *J. Parasitol.* **69:** 439–440.

Rose, M.E and Long, P.L. (1980) Vaccination against coccidiosis in chickens. In: Taylor, A.E.R. and Muller, R. (eds) *Vaccination against Parasites*, pp. 57–74. Blackwell Scientific Publications, Oxford, UK.

Rose, M.E., Hesketh, P. and Ogilvie, B.M. (1979) Peripheral blood leucocyte response to coccidial infection: a comparison of the response in rats and chickens and its correlation with resistance to reinfection. *Immunology* **36:** 71–79.

Rose, M.E., Lawn, A.M. and Millard, B.J. (1984) The effect of immunity on the early events in the life-cycle of *Eimeria tenella* in the caecal mucosa of the chicken. *Parasitology* **88(Pt 2):** 199–210.

Ruff, M.D. (1989) Interaction of avian coccidiosis with other diseases: a review. Coccidia and intestinal coccidiomorphs, Vth Int. Cocc. Conference, Tours (France). INRA Publ, Pp. 173–181.

Ruff, M.D. (1999) Important parasites in poultry production systems. *Vet. Parasitol.* **84:** 337–347.

Ryan, R., Shirley, M. and Tomley, F (2000) Mapping and expression of microneme genes in *Eimeria tenella*. *Intern. J. Parasitol.* **30:**1493–1499.

Sangster, N.C. (2001) Managing parasiticide resistance. *Vet. Parasitol.* **98:** 89–109.

Schaap, D., Arts, G., Verhoeven, N., Kroeze, H., Spreeuwenberg, K., Kuiper, K. and Vermeulen, A. (1999) Adaptation of *Eimeria lactate* dehydrogenase to its parasitic life style. *Proc. Mol. Parasitol. Meet. Woods Hole* **239C.**

Sharma, J.M., Zhang, Y., Jensen, D., Rautenschlein, S. and Yeh, H.Y. (2002) Field trial in commercial broilers with a multivalent in ovo vaccine comprising a mixture of live viral vaccines against Marek's disease, infectious bursal disease, Newcastle disease, and fowl pox. *Avian Dis.* **46:** 613–622.

Shirley, M.W. (1985) New methods for the identification of species and strains of *Eimeria*. Research in avian coccidiosis. In: McDougald, L.R. (ed.) *Proc. Georgia Coccidiosis Conf.*, 18-20 November 1985, pp. 1–35.

Shirley, M.W. (1997) *Eimeria* spp. from the chicken: occurrence, identification and genetics. *Acta Vet. Hung.* **45:** 331–347.

Shirley, M.W., Jeffers, T.K. and Long, P.L. (1983) Studies to determine the taxonomic status of *Eimeria mitis*, Tyzzer 1929 and *E. mivati*, Edgar and Seibold 1964. *Parasitology* **87:** 185–198.

Smith, A.L., Hesketh, P., Archer, A. and Shirley, M.W. (2002) Antigenic diversity in *Eimeria maxima* and the influence of host genetics and immunization schedule on cross-protective immunity. *Infect. Immun.* **70:** 2472–2479.

Staeheli, P., Puehler, F., Schneider, K., Gobel, T.W. and Kaspers, B. (2001) Cytokines of birds: conserved functions – a largely different look. *J. Interferon Cytokine Res.* **21:** 993–1010.

Tenter, A.M., Barta, J.R., Beveridge, I., Duszynski, D.W., Mehlhorn, H., Morrison, D.A., Thompson, R.A. and Conrad, P.A. (2002) The conceptual basis for a new classification of the coccidia. *Intern. J. Parasitol.* **32:** 595–616.

Tomley, F.M. and Soldati, D.S. (2001) Mix and match modules: structure and function of microneme proteins in apicomplexan parasites. *Trends Parasitol.* **17:** 81–88.

Tsuji, N., Kawazu, S., Ohta, M., Kamio, T., Isobe, T., Shimura, K. and Fujisaki, K. (1997) Discrimination of eight chicken *Eimeria* species using the two-step polymerase chain reaction. *J. Parasitol.* **83:** 966–970.

Vainio, O. and Imhof, B.A. (1995) The immunology and developmental biology of the chicken. *Immunol. Today* **16:** 365–370.

van Tuinen, M. and Hedges, S.B. (2001) Calibration of avian molecular clocks. *Mol. Biol. Evol.* **18:** 206–213.

Vandaveer, S.S., Erf, G.F. and Durdik, J.M. (2001) Avian T helper one/two immune response balance can be shifted toward inflammation by antigen delivery to scavenger receptors. *Poult. Sci.* **80:** 172–181.

Vermeulen, A.N. (1998) Progress in recombinant vaccine development against coccidiosis A review and prospects into the next millennium. *Intern. J. Parasitol.* **28:** 1121–1130.

Vermeulen, A.N., Schaap, D.C. and Schetters, T.P. (2001) Control of coccidiosis in chickens by vaccination. *Vet. Parasitol.* **100:** 13–20.

Vermeulen, A.N., Schetters, T.P.M. and Schaap, T.C. (2002) Avian coccidiosis; a host parasite relationship to be restored. *The Physiologist* **45(4):** 342.

Vervelde, L. and Jeurissen, S.H. (1993) Postnatal development of intra-epithelial leukocytes in the chicken digestive tract: phenotypical characterization in situ. *Cell Tissue Res.* **274:** 295–301.

Vervelde, L., Vermeulen, A.N. and Jeurissen, S.H. (1993) Common epitopes on *Eimeria tenella* sporozoites and cecal epithelium of chickens. *Infect. Immun.* **61:** 4504–4506.

Vervelde, L., Vermeulen, A.N. and Jeurissen, S.H.M. (1995) *Eimeria tenella*: sporozoites rarely enter leukocytes in the cecal epithelium of the chicken (*Gallus domesticus*). *Exper. Parasitol.* **81:** 29–38.

Vervelde, L., Vermeulen, A.N. and Jeurissen, S.H.M. (1996) In situ characterization of leucocyte subpopulations after infection with *Eimeria tenella* in chickens. *Parasite Immunol.* **18:** 247–256.

Voeten, A.C., Braunius, W.W., Orthel, F.W. and van Rijen, M.A. (1988) Influence of coccidiosis on growth rate and feed conversion in broilers after experimental infections with *Eimeria acervulina* and *Eimeria maxima*. *Vet. Q.* **10:** 256–264.

Waldenstedt, L., Elwinger, K., Thebo, P. and Uggla, A. (1999) Effect of betaine supplement on broiler performance during an experimental coccidial infection. *Poult. Sci.* **78:** 182–189.

Waldenstedt, L., Elwinger, K., Lunden, A., Thebo, P., Bedford, M.R. and Uggla, A. (2000) Intestinal digesta viscosity decreases during coccidial infection in broilers. *Br. Poult. Sci.* **41:** 459–464.

Wallach, M., Smith, N.C., Petracca, M., Miller, C.M., Eckert, J. and Braun, R. (1995) *Eimeria maxima* gametocyte antigens: potential use in a subunit maternal vaccine against coccidiosis in chickens. *Vaccine* **13:** 347–354.

Weiss, R.A. (2002) Virulence and pathogenesis. *Trends Microbiol.* **10:** 314–317.

Williams, P.D. and Day, T. (2001) Interactions between sources of mortality and the evolution of parasite virulence. *Proc. Royal Soc. London B Biol. Sci.* **268:** 2331–2337.

Williams, R.B. (1998) Epidemiological aspects of the use of live anticoccidial vaccines for chickens. *Intern. J. Parasitol.* **28:** 1089–1098.

Williams, R.B. (1999) A compartmentalised model for the estimation of the cost of coccidiosis to the world's chicken production industry. *Intern. J. Parasitol.* **29:** 1209–1229.

Williams, R.B. (2001) Quantification of the crowding effect during infections with the seven *Eimeria* species of the domesticated fowl: its importance for experimental designs and the production of oocyst stocks. *Intern. J. Parasitol.* **31:** 1056–1069.

Williams, R.B. (2002) Anticoccidial vaccines for broiler chickens: pathways to success. *Avian Pathol.* **31:** 317–353.

Yun, C.H., Lillehoj, H.S. and Lillehoj, E.P. (2000a) Intestinal immune responses to coccidiosis. *Develop. Comp. Immunol.* **24:** 303–324.

Yun, C.H., Lillehoj, H.S. and Choi, K.D. (2000b) *Eimeria tenella* infection induces local gamma interferon production and intestinal lymphocyte subpopulation changes. *Infect. Immun.* **68:** 1282–1288.

Yvore, P. and Mainguy, P. (1972) Influence de la coccidiose duodenale sur la tenure en carotenoides du serum chez le poulet. *Annal. Recherche Vet.* **3:** 381–387.

Yvore, P., Mancassola, R., Naciri, M. and Bessay, M. (1993) Serum coloration as a criterion of the severity of experimental coccidiosis in the chicken. *Vet. Res.* **24:** 286–290.

Zigterman, G.J., d. van, V., van Geffen, C., Loeffen, A.H., Panhuijzen, J.H., Rijke, E.O. and Vermeulen, A.N. (1993) Detection of mucosal immune responses in chickens after immunization or infection. *Vet. Immunol. Immunopathol.* **36:** 281–291.

11

Conclusions

Geert F. Wiegertjes and Gert Flik

This book brought together several research groups working on host–parasite interactions. Not only do the parasites described in this book range from uni- to multicellular and from endo- to ectoparasites, their hosts range from fish to chickens and mammals. United are these contributions by their comparative approach and the affection of the different authors for 'their' parasites. The first chapters in this book dealt with a number of fundamental processes important in host–parasite interactions, among which the major physiological events leading to cell death (Steinhagen). Here, the author discussed apoptosis or necrosis possibly being 'different executions of the same death'. Oxidative stress, for example, does not always directly determine whether apoptosis or necrosis results from the insult; more important for the final outcome can be factors such as the insult severity, cell type and metabolic status. For fish, structural changes following cell death are well documented, but they are post mortem changes. Regretfully, studies on the physiological responses preceeding cell death and on the immunological implications of pathogen-induced cell demise are vitually lacking in fish, but certainly needed for a proper understanding of pathogen–host interactions on the physiological level. The roles of apoptosis in the interactions between parasite and hosts are multifaceted and highly dependent on individual associations between the two organisms involved (Hoole and Williams). Whilst there are instances where both organisms appear to gain, in the majority of cases apoptosis appears to favour only one of the two parties. In the instances when the parasite benefits, apoptosis has been related to infectivity and virulence, an interruption of the killing mechanism of the host and liberation of the pathogen. In the instances where the host benefits, controlled cell death has been associated with limiting the pathogen population, parasite migration within the host and, in some instances, actual killing of the invading organism.

Thionine-positive cells probably evolved from a common ancient cell type: mast cells and basophilic granulocytes, both involved in host defences against parasites (Nielsen, Lindenstrøm, Sigh and Buchmann). Thionine-positive cells; metachromatically stained with the basic aniline dye thionine, can comprise several cell types, however. This hampers an unambiguous identification of mast cells and basophils in species other than mammals. Thus, although eosinophilic granular cells in teleost fish display metachromasia, suggesting they could be putative mast cells, further functional

Host–Parasite Interactions, edited by G. F. Wiegertjes & G. Flik.

characterization is needed to clarify their contribution to anti-parasitic responses. Another cell type involved in the innate immune response to parasites macrophages, was discussed by Joerink, Saeij, Stafford, Belosevic and Wiegertjes. To date, the existence of polarized macrophages that differ in terms of receptor expression, effector function and cytokine and chemokine production is acknowledged in mice. These polarized macrophages are thought to play an important role in the differentiation of T-helper lymphocytes. For cold-blooded vertebrates such as teleost fish, the information on T-lymphocytes is sparse and prototypical type I or type II cytokine profiles cannot be detected. Macrophages, however, are considered an evolutionary ancient cell type, and can easily be recognized in fish. If macrophage polarization indeed does also occur in fish it may determine the type of immune response against foreign parasites such as the haemoflagellates dealt with in this chapter. The following chapter (Woo) continued on haemoflagellate parasites with reviewing the pathophysiology caused by two groups of parasites that cause disease; comparing cryptobiosis in salmonid fish with trypanosomiasis in African livestock. It is certainly surprising how the very significant differences between these two diseases can be caused by two closely related (Order Kinetoplastida) parasites.

The next two chapters both dealt with ectoparasitic crustacean parasites as opposed to the endoparasitic kinetoplastid parasites in the previous two chapters. Walker, Wendelaar Bonga and Flik started with reviewing the current ecological knowledge for freshwater *Argulus* with a focus on the general biology. Argulids, although obligate ectoparasites of fish, are also frequently encountered swimming freely in the water column. Off-host survival, therefore, is a crucial factor in the successful location of a new host. When studying the stress response to ectoparasites it is important to distinguish between direct and indirect effects of the stressor: low-level infections do not have major direct effects but indirectly do cause stress-induced changes in the skin. Johnson and Fast reviewed the morphology and biology of seawater lice in addition to their host–parasite interactions and the process of disease: there is good evidence that sea lice, like other parasites, maintain themselves on their hosts by using a variety of strategies including immunomodulation of the host responses. In general, the future seems to lie with molecular approaches, aiming at the expression of identified and cloned genes from parasites to obtain the relatively large amounts of purified proteins required for further characterization. Good future strategies would be to prepare tissue-specific cDNA libraries from important host tissues or to use subtractive libraries from parasites collected prior to and post-attachment and feeding on the host.

Another group of ectoparasites of fish are the monogeneans; one of the largest helminth groups existing (Buchmann, Lindenstrøm and Bresciani). Recent research has elucidated clear host responses against monogeneans: humoral factors such as complement, lectins and to some extent immunoglobulins, may affect the parasites. The cytokine network seems to have a regulatory role and, with this in mind, it would be interesting to study putative macrophage polarization in infected fish. Interestingly, host substances can be recognized by monogeneans suggesting a role for chemotactic mechanisms controlled by the parasite.

The last two chapters dealt with parasites that do not occur on or in fish. Alarcon-Chaidez and Wikel (Chapter 9) described the mechanisms by which ticks elicit both innate and acquired immune responses in warm-blooded animals and how these events influence transmission of tick-borne pathogens. Again, progress toward characterization of antigens responsible for (naturally acquired) immunity is in its early stages. Although information obtained from primary DNA sequences has provided a lot of information, functional characterization of these antigens in terms of temporal expression, regulation, activity and post-translational modifications is required for further studies. The co-evolutionary balance of *Eimeria* and its host the domestic chicken (Chapter 10) has been disturbed by heat, crowding and concurrent infections in the present chicken environment (Vermeulen). This resulted in a new situation in which the parasite was able to reproduce to such an extent that the host could not combat the shear numbers of organisms. Drug resistance widely developed by the flexible genome of the parasite and returned new drugs at a greater speed than they had been developed. More sustainable approaches such as improved hygienic measures, better facility management and good understanding of epidemiology of the parasites spreading and proliferation are now the most promising set of tools to control the balance between parasite and host.

In general, parasites have evasion mechanisms that depend on factors such as their life-cycle stage, penetration route and their micro-environment inside the host. Despite the generally large amount of parasite antigen and the host immune and inflammatory response to the parasite, parasites manage to survive within the host for long periods. To be able to do so, parasites make use of multiple evasion mechanisms that have been acquired during millions of years of evolution. The parasite's successful survival depends mainly on evading the host's immune system. This can be achieved, for example, by penetrating and multiplying within host cells, varying or eliminating the surface coat or modulating the host immune response. Roughly, parasites with extracellular stages are the target of the humoral immune response, whereas those with intracellular stages are susceptible to attack by the cell-mediated immunity. The general agreement is that, for most parasitic infections, successful immunoprophylaxis requires detailed knowledge about host–parasite relationships and the complexity of effector mechanisms. The future should bring a dissection and characterization of the different immune responses in natural infections but especially after immunization. We hope that this book contributed to the understanding of host–parasite interactions and will stimulate and focus further research on the host (immune) responses to parasites.

In conclusion, owing to the ongoing large-scale sequencing projects we are presently experiencing a rapid increase in both our knowledge of genes involved in the host's innate and acquired immune responses and of parasite genes. However, true understanding of host–parasite interactions models requires more information of proteomics. Where genomics seeks to profile genes being expressed at a particular developmental stage or under a particular set of physiological conditions, proteomics seeks to profile proteins synthesized. *In vivo* host–parasite models will remain essential to this understanding. Most mouse model systems are often used simply because they have been in use for many years, their features have been well documented, because work on them has defined the current paradigm or because they are easy to work with and readily available from local suppliers. The animal models described in this book are unique in that they without exception represent natural infections rather than model systems. The outcomes therefore, we believe, can be interpreted and applied in a much more direct manner.

Index